SHEEP
MANAGEMENT
AND WOOL

SHEEP MANAGEMENT AND WOOL TECHNOLOGY

J. B. D'Arcy BEd, Dip Teach

Former Head, Division of Rural Studies
Sydney Technical College

Published by
NEW SOUTH WALES UNIVERSITY PRESS LTD
PO Box 1 Kensington NSW Australia 2033
Phone (02) 697-5452 Fax (02) 398 3408

First published 1972
Second edition 1979
Third edition 1990

Available in North America through:
International Specialized Book Services
5602 N.E. Hassalo Street
Portland Oregon 97213-3640
United States of America

ISBN 0 86840 036 X

CONTENTS

1 Brief History of the Sheep and Wool Industry in Australia   1

2 The Australian Merino   7

3 British Breeds of Sheep   23

4 Australian and New Zealand Breeds   45

5 Sheep Areas of Australia   56

6 Biology of the Skin and the Wool Fibre   69

7 Properties of Wool and Woollen Goods   79

8· Characteristics of Wool in Relation to Classing   84

9 Defects in Wool and Wool Faults   90

10 Vegetable Matter in Wool   95

11 Picking Up, Skirting and Rolling   99

12 Australian Clip Preparation Standards   108

13 Classer's Duties   117

14 Marketing   122

15 Types and Yield   131

16 Wool Testing   158

17 Carbonising and Scouring 213

18 Textile Manufacture 221

19 Sheep Classing and Merino Breeding 237

20 Fertility and Mating of Sheep 248

21 Care and Management of the Ewe, and Lambing 252

22 Lamb Marking 257

23 Weaning and Care of Young Animals 262

24 Shearing, Crutching and Wigging 265

25 Pasture Improvement and Management, Fodder Conservation and Supplementary Feeding 279

26 Mineral Deficiencies 287

27 Drought Feeding 289

28 Crossbreeding and Fat Lamb Raising 295

29 Anatomy and Physiology of Sheep 304

30 External Parasites of Sheep: Blowfly Strike, Mulesing, Jetting, Dressing 312

31 Dipping Sheep 324

32 Internal Parasites 328

33 Some Common Sheep Diseases 333

34 Poisonous Plants 340

Glossary 344

Index 350

1

BRIEF HISTORY OF THE SHEEP AND WOOL INDUSTRY IN AUSTRALIA

The beginning of the sheep and wool industry in Australia coincided with the landing of the First Fleet in 1788 when Governor Phillip arrived with seventy 'fat tail' sheep from the Cape of Good Hope. These were used for food and were later replaced by thirty sheep from Calcutta and some coarse-woolled sheep from Ireland which arrived in 1793. Captain Macarthur, who came to the colony in 1790, observed how adaptable fine-woolled sheep were and he conceived the idea of producing fine wool grown under conditions of cheap land and cheap labour, rather than the breeding of mutton types to feed the convict settlement. Although ambitious and far-sighted, Macarthur could not have visualised what an industry would develop from the humble beginnings of his day. Macarthur, as the wealthiest and most influential man in Australia, brought pressure to bear on Governor Hunter to import more fine-woolled sheep, and in 1797 Captains Waterhouse and Kent returned from the Cape with sixteen Spanish escurial merinos. Of these, Macarthur was fortunate enough to secure three rams and five ewes. The Rev. Samuel Marsden and William Cox also obtained some of this shipment and began breeding for wool.

The year 1797, then, can be taken as the beginning of Australia's wool industry, for although there is little doubt that the previous importation of sheep from Calcutta contained pure merinos, this shipment from the Cape in 1797 was the first authentic record of the arrival of pure-blooded merinos. In 1803 Macarthur sold 111 kg (245 lb) of wool in London at $1.03 (10/4d) per 454 grams (one pound) and this encouraged him to continue producing wool from Spanish strains. This same year he returned from England with negrettis from the Kew stud and he also secured more merinos from Captain Waterhouse's flock.

Macarthur's stud, now situated on a 4000 hectare (10 000 acre) land grant at Camden

Park, supplied escurial-negretti strains of merinos to numerous flocks in Victoria, Tasmania and west of the mountains in New South Wales for at least thirty years.

Expansion of Industry

Governor Macquarie encouraged exploration of the colony and in 1813 the Blue Mountains were crossed, making it possible for the sheep industry to extend into the vast expanse of the now world-famous sheep country of the hinterland. In 1810 there were 34 000 sheep in the colony but with the opening up of the back country the sheep industry leaped ahead, so that by 1820 there were 156 000 and by 1843 these numbers had increased to 12 million.

Spreading to Bathurst and Mudgee (tablelands districts) and then following the rivers, settlement began in the western areas. Settlement in western Victoria expanded more quickly than in the western region of New South Wales and the settlers in the latter area used the type of merino that was already giving satisfactory results in Victoria and the New South Wales Tablelands. It was not until many years later that a more satisfactory type evolved for the drier areas of New South Wales.

Up until 1831 many free grants of land were made, and by 1850 the best of the country was under occupation. The discovery of gold at Ophir in 1851 and the cessation of the transportation of convicts (which occurred in 1849) temporarily interfered with land settlement, but the former later resulted in an influx of cheap labour.

Mudgee Type

Captain Waterhouse started his flock with the earliest importations and kept his flock fairly pure escurial blood. Captain Cox of *Burrundella*, Mudgee, bought this flock and introduced saxony blood, and although he tried rambouillet and vermont he reverted to the saxony, later using much Tasmanian saxony blood. His rams were in demand and did much to popularise the famous Mudgee type. Many breeders followed with this stylish, short, fine, dense, soft, but rather heavy-conditioned, beady-tipped, type of wool, and for sixty years (until about 1860), the majority of importations were saxony or silesian, while Tasmanian rams were popular about 1900. Some owners of flocks founded on saxony or silesian blood declare to the present day that they have not introduced rambouillet or Peppin, and in some cases no vermont either, although they have increased length of staple and body weight.

The fine-woolled type was very suitable and profitable in the high-altitude, high-rainfall country, but it soon became apparent that larger framed, more robust sheep, carrying a longer, stronger wool, were better suited to the open plains country.

Peppin Types

In 1852, Mr Salter of Cunningham Plains began breeding with rambouillet rams and demonstrated how they added size, staple length and fleece weight. Sheep men of the day liked the stylish, square, beady-tipped, fine wool of the Mudgee type and any variation was liable to be dubbed inferior, or even 'cross-bred'. It took bold men like the Peppin brothers and Bayley to prove the worth of large-framed, strong-woolled sheep for drier areas. In 1885, although the Peppin brothers produced highly profitable sheep, the type was not accepted as a champion merino at the Sydney Show until over fifty years later. One reason why the Peppin type did not progress more rapidly from the start was the success of the fine wool types on the Tablelands. Another reason was the predominance of fine wools in Victoria where flock numbers at that time (1860) were greater than in New South Wales. Even to this day, although the Victorian type has changed, the finer types still give excellent results in parts of that state. Today the sheep distribution in Australia is entirely different — New South Wales carries over one-third of the total sheep, Western Australia carries one-fifth, and then come, in order of importance, Victoria, South Australia, Queensland, Tasmania, the Northern Territory and the Australian Capital Territory.

The Peppin brothers, at *Wanganella*, near Deniliquin, did more to produce the modern wanganella than any other person or stud. As far as can be ascertained, they started with a mixture of selected large-framed saxony ewes and some South Australian blood ewes from *Canally*. They used a rambouillet ram, 'Emperor', and a superior type of plain-bodied vermont ram, 'Grimes'. How these were used is not known, but it is certain that the Peppins were very definite in selecting a type which, in their opinion, was suitable to the district. Austin, in *The Merino, Past, Present and Probable*, draws attention to the radical change of type and is convinced that English leicester blood was used. This could account for the early variations and 'throwbacks'

Early type merino, 1830

Grimes, 1866

Wanganella, 1874

Vermont, 1896

Australian vermont, 1910

Modern-type Australian merino

objected to by many breeders. The Peppin brothers, however, kept to their type — large-framed, strong-boned animals, with less body wrinkles but good frontal development, and with a good length of staple.

Although the type was slow to catch on, it was destined to become the most popular type in Australia. It is now recognised throughout the world as the 'wanganella', and in Australia many breeders still refer to it as the Peppin strain. The big majority of medium-to-strong stud sheep are of this strain, proving very suitable for our plains country, being good travellers, hardy and with the capacity to respond both in carcass and wool growth to a high plane of nutrition. Today the wanganella sheep are open-faced, strong-boned, large-framed, square-built animals carrying heavy, deep-grown fleeces of a bold 60/64s quality number. The staple is bulky, long and bright in colour, the fleece being light-conditioned and of good density for its type.

The Vermont

In 1883 Samuel McCaughey, the 'sheep king' of that time, with a chain of properties throughout western New South Wales and Queensland, sought additional weight of fleece and size of sheep. He was disappointed that the Peppins were not throwing 'like' stock and he was impressed by the progeny of a vermont ram 'Matchless', who was 91 kg (200 lb) live weight, fairly plain-bodied and carried 12.7 kg (28 lb) of wool, although he had the heavy yellow yolk peculiar to the latter-day vermonts. As a result, McCaughey purchased the good vermont rams at the Sydney sales and started a craze for wrinkly vermonts. He made big profits by importing and then selling inferior merino rams of this strain from the United States. In 1889, at the height of the boom, McCaughey spent $100 000 (£50 000) on vermonts in the United States. The popularity of the vermonts continued until 1900–1902, when Australia fell victim to the worst drought and the heaviest loss of sheep in its history.

The poorest type of vermont imported from the United States was very wrinkly (wrinkles for density being advocated by enthusiasts), carried short-stapled, yellow, pasty, fatty wool of approximately a 28 per cent yield, and was constitutionally weak. The fleece was rendered uneven by the presence of wrinkles, but the character was good and the quality fine, ranging in bulk from 74s to 66s — a little stronger than the saxony strains.

Table 1.1   Australian Sheep Numbers as at March 31 in thousands.

Year	NSW	Vic.	Qld	SA	WA	Tas.	Aust.(a)
1989	61,043	28,487	14,580	17,340	36,143	4,946	162,639(p)
1988	54,932	26,997	14,362	17,352	33,951	4,746	152,443
1987	52,192	26,586	14,627	17,234	33,463	4,954	149,157
1986	51,498	25,716	14,208	17,506	32,907	4,822	146,776
1985	55,498	26,471	14,043	17,263	31,574	4,780	149,747
1984	50,997	24,632	13,033	16,368	29,518	4,583	139,242
1983	48,095	22,748	12,225	15,448	30,164	4,452	133,237
1982	48,700	25,341	12,344	16,709	30,268	4,513	137,976
1981	46,000	25,487	10,620	17,056	30,764	4,380	134,407
1980	48,600	24,400	12,163	16,046	30,431	4,245	135,985
1979	48,400	23,445	13,611	15,102	30,307	4,153	135,139
1978	48,000	22,021	13,438	14,073	29,823	3,969	131,445
1977	49,700	21,925	13,304	15,132	31,158	4,015	135,360
1976	53,200	25,395	13,599	17,279	34,771	4,249	148,643
1975	54,983	26,410	13,908	17,621	34,476	4,136	151,652
1974	53,296	25,787	13,119	16,431	32,451	3,964	145,173
1973	52,000	24,104	13,346	15,651	30,919	3,824	140,029
1972	62,000	29,496	14,604	17,970	34,405	4,237	162,910
1971	70,600	33,761	14,774	19,166	34,709	4,517	177,792
1970	72,284	33,156	16,446	19,747	33,634	4,560	180,079
1969	68,153	30,185	20,324	18,392	32,901	4,395	174,602
1968	67,786	27,909	19,948	16,405	30,161	4,428	166,912
1967	63,848	31,239	19,305	17,864	27,370	4,321	164,237
1966	61,396	30,968	18,384	17,993	24,427	4,127	157,563

Source: Australian Bureau of Statistics
(a) Includes Northern Territory and Australian Capital Territory; (p) subject to revision

In 1891 Australia was carrying 106 million sheep, but by 1902 the number was reduced to nearly half. Rabbits, fires, a type of sheep unsuited to withstand drought conditions and wool blindness were all factors in this drop in numbers. The use of the vermont was popular in New South Wales, Queensland and Tasmania, but it was not so popular in South Australia and Victoria. Although a slight infusion of vermont of the better type (e.g. 'Grimes' or 'Matchless') may have been a benefit to our finer woolled flocks, unfortunately much damage was caused by the craze for excessive wrinkles, which led to the use of inferior quality vermonts. Today plain-bodied sheep (e.g. wanganella and South Australian types) are popular in the western areas and a finer plain-bodied sheep of saxony–wanganella blood is preferred for the Tablelands district.

2

THE AUSTRALIAN MERINO

The Australian merino is the result of selection designed to produce a high-grade wool under comparatively harsh environmental conditions. The more angular merino is better able to stand up to these conditions than the more rounded and better fleshed larger, quicker-maturing breeds.

Although the merino is essentially a wool breed, the surplus sheep must sometimes be slaughtered for mutton and for this reason, and generally to improve constitution, the original fine-woolled type of merino has been improved by embodying some of the good features of conformation of the British longwools, giving the wanganella and South Australian types. This has also improved resistance to fly strike, increased fleece weight and evened up fibre length and diameter. Nevertheless, in comparison with mutton types, the merino is still angular in appearance, being more thinly fleshed, with longer and finer bones; the withers are higher, the ribs longer and not so well arched, the loins lighter, and the rump is inclined to droop, with flat thighs and narrow or V-shaped 'twist'. Merino breeders like a reasonable length of leg to produce good travellers, especially in the stronger-woolled types, but this is not conducive to the production of a good mutton type. Although extreme wrinkling has been eliminated the merino still carries a relatively loose and heavy pelt, with tiny wrinkles which can only be seen when the wool is clipped very close. On this account, a 'plain-bodied' fine-woolled merino has a wool-growing body surface of approximately $1.07\,\mathrm{m}^2$ ($11\frac{1}{2}$ square feet), which is the same area as that of the lincoln, despite the great difference in size. By way of interest the corriedale is usually completely plain-bodied and, even with its large frame, has less wool-growing area — approximately $1.02\,\mathrm{m}^2$ (11 square feet), whereas the polwarth may be set down as approximately $0.93\,\mathrm{m}^2$ (10 square feet).

Irrespective of the type of sheep, fine or strong, mutton or dual-purpose, it is recognised as being important that the conformation should embody certain points which are believed to indicate good constitution (i.e. ability to thrive and produce under any given environment). These points are:

1. large muzzle with large, full, open nostrils;
2. a strong, deep jaw with incisor teeth meeting the dental pad squarely;
3. strong masculine head in the male and finer features in the ewe (these are normal, but evidence of masculinity does not reflect a sheep's prepotency as is commonly believed);
4. a deep, broad chest, well-sprung ribs and straight back, giving adequate lung and heart room;
5. sound feet and legs (particularly important in the merino, which is sometimes expected to travel long distances in search of feed and water).

In spite of deficiencies in the conformation of the merino, breeders aim to produce sheep with all evidences of sound constitution, together with well-balanced appearance — good topline, good depth, reasonably well-sprung ribs and well-developed quarters.

The Merino Ram

In selecting a ram, the first consideration is the suitability of the strain for the locality and for the type of ewes with which it is intended to be mated. The second consideration is the constitution of the animal and, finally, the covering and overall appearance. Moreover, in considering the strain, attention would be paid not only to the predominant type, but also to the degree of evenness of type apparent in the flock as a whole.

We shall now briefly study each point as applied to a merino ram:

The muzzle. This should be of a full, bold formation, not pointed. The lips should be fleshy, not thin or papery. The nostrils should be reasonably dilated. There should be no obstruction to the breathing operation. The mucous membrane of the nostrils and lips should be a bright pink colour — not pallid. Whilst the presence of coloured spots on the lips may be minor in importance, a solid uniform flesh colouring is more desirable. A good bold muzzle indicates strength of constitution and generally more abundant wool production. The extremities of both upper and lower jaws should meet accurately. Avoid animals showing undershot or overshot malformations.

The face. The face of a merino ram should present a bold, masculine appearance. It should be of medium length, slightly arched or convex in contour. A concave or 'dished' formation indicates weakness and lack of vitality. The face should be open and covered with short hair. This hair should be a pale honey or cream colour, not white. White in the case of merinos is known as 'frosty face', and is an objectionable feature. Such a fault indicates deterioration in the quality of the wool. Harshness usually develops, ultimately revealing the presence of kempy fibres. The latter seriously depreciates the manufacturing qualities of the wool. The hairy covering of the face should handle with a soft, velvety texture.

The forehead. Should be nicely rounded, possessing width between the eyes. A flat or straight forehead is undesirable.

The crest. Should present a continuation of the curving formation of the forehead. There should be a space between the base of the horns.

The poll. Should be strong, full and well-rounded.

The scrag. Should be of medium length, not long, possessing a muscular formation. A thin narrow neck is undesirable. A good strong neck not only enhances the animal's

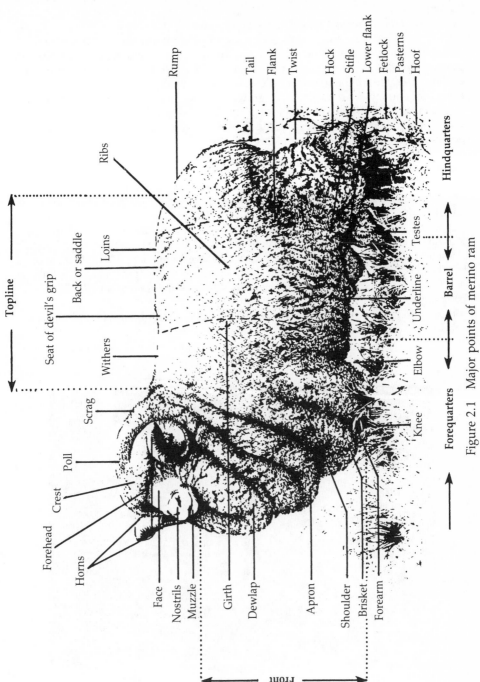

Figure 2.1 Major points of merino ram

appearance, but is an indication of masculinity. It should taper slightly towards the horns.

The horns. The formation and general appearance of the horns is of considerable importance. First, in merino rams, they must show substance without coarseness. A strong horn indicates robustness and vigour. In shape they stand out from the head in a lateral direction, with a regular spiral formation. The first turn should be a sufficient distance from the head to permit of the free growth of wool. The bases should also be placed far enough apart to allow the growth of wool to continue from the forehead to the crest. The horns should be of a uniform dull amber colour, with a fresh, well-nourished appearance, especially in young rams. Black streaks or marks are objectionable. The transverse undulations or corrugations should be pronounced and even. Plain, smooth, whitish-coloured horns indicate common origin.

The eyes. Should be of an active, alert, yet mild appearance. The eyelids should show the same creamy-coloured markings as the face, and be free from tan.

The ears. Are comparatively short. They should, however, be thick and velvety to handle.

The withers. Across the withers is generally the weak link in the animal's conformation. Merino sheep are perhaps more susceptible to weakness in this region. In judging or classing, therefore, this point requires very careful examination. The withers should be well filled, and as broad as possible. They should be neither low nor high, although of the two evils the latter is preferable. The whole of the topline should, however, be straight.

The back. Should be straight and as broad as possible throughout. A good top is the product of well-sprung ribs. Merino sheep do not, as a rule, possess a well-rounded barrel, but any improvement in this direction adds to the development of a broad

A champion strong-wool merino ram at the Albury Sheep Show

back, and in turn enhances the worth of the sheep. From a mutton point of view, a broad back is an important asset.

The loins. Should also be broad and full. The section of the back should be quite straight, and in line with those regions of the back already described.

The rump. A good length is desirable. Length of quarter not only adds to the general appearance of the build, but is responsible for a heavier hindquarter and better joints. The rump should be nicely rounded on top, falling away gradually to the tail. It should be broad and full. The hindquarters should be deep.

The twist. This should be well let down and broad or full. Extra development here also lends greater weight to the quarter. Breadth of twist means well set-apart hind legs. This adds to the beauty of the conformation.

The hocks. Should be strong, and straight when viewed from behind. Cow-hocked or turkey-legged sheep are unattractive; besides, good straight legs indicate soundness, strength of constitution and good travelling capacity.

The pastern. Should be strong. The animal should stand up well on its pasterns. Down at the fetlocks indicates weakness.

The hoofs. Should be of good shape and of a constant dull amber colour. Black markings are objectionable.

The flanks. The outer flank (i.e. the lower portion of the thigh) should be well filled and rounded, not flat. The inner flank should be deep and not slack.

The ribs. The barrel should be as well-rounded as possible. This will only happen where the ribs are well-sprung. Plenty of depth is necessary, particularly at the girth. The depth combined with thickness produces girth measurement. Good girth indicates a roomy, well-developed forequarter. This signifies robustness. The upper flank should be well filled. Slackness here detracts from the symmetry of the conformation.

The devil's grip. This is a well-known malformation often occurring in merino sheep. It is one of the most objectionable defects of the conformation. The animal affected is not only ill-shaped and unattractive, but is generally regarded as being extremely weak in constitution. This fault takes two forms: the first — the most common — shows a sharp and pronounced depression immediately behind the shoulders. In some cases, two noticeable depressions appear on either side of the spinal column. The second form gives one the impression that the sheep has a cord tied tightly around the body just behind the girth.

The underline. Should be well let down and deep, and as straight as possible, between the fore and hind legs. This straight effect is the outcome of a well-developed deep chest and flank.

The forelegs. Should be well set apart, indicating a well-developed, roomy chest. Legs should be of medium length, with strong, broad knees.

The shoulders. Should be deep, full and well rounded. Thickness through the forequarters is most desirable.

The apron. On merino rams a well-developed, bold apron not only adds to the general appearance and character of the ram but, moreover, is an indication of more abundant wool production. The apron should encircle the neck, being very bold in appearance at the throat, and generally diminishing in size as it approaches the top of the neck. It is the largest fold, and situated immediately in front of the shoulders.

One or two other folds may occur on either side of the apron. These are minor folds. Uniformity of formation on both sides of the neck is a desirable feature in the presence of these folds.

The throat. Should be deep, with well-developed apron, folds and dewlap. The latter is

a deep wrinkle or fold appended to the throat, and extending from the underjaw to the point of the brisket.

Merino Ewe

After making allowances for sex, the above points would apply to the ewe. The ewe is hornless and is of finer build throughout, but must be of good conformation and free of constitutional weakness previously mentioned. The face is of finer cut, the neck lighter and a lighter bone structure is noticeable in the head and legs. The fleece does not carry the frontal development of the ram and the wool is softer, lighter in condition and is a beautiful creamy white colour.

Having briefly described the most important points from the judging aspect, it is now necessary to study them rather more collectively, from the viewpoint of greater practical value. For this purpose the conformation may be appraised as follows:

1. general appearance and size;
2. symmetry;
3. constitution;
4. robustness or masculinity.

The value of the conformation will very largely depend upon the development of the sheep in the various respects already mentioned.

General Appearance and Size

In judging or classing sheep, these two factors play a most important part in determining the animal's worth. Either in the show ring or stud sales, the finish of the sheep's general appearance goes a long way towards influencing the opinion of the judge or buyer. A well-balanced, well set-up appearance is the result of a well-proportioned conformation. The character or style of the animal, which also contributes very considerably to its general appearance, will be indicated by a good active carriage, good action when moving, good straight, sound legs, well-developed front, a well-shaped, symmetrical head, well-carried. The face should be strong, open and soft.

In either judging or classing, the expert will be guided to a great extent by the impression created in his mind by both general appearance and style. As this is best gauged when the animal is standing naturally, most judges and classers prefer to see them when they are standing as free as possible. It is also necessary to have the sheep standing apart from each other: when crowded, it is not possible to make such a precise analysis of appearance and style. The favourable or unfavourable appearance will depend upon its general make-up, and the way it is furnished throughout. The trained eye of the expert is attracted immediately by outstanding quality. Equally, defects just as readily create a poor impression. Certainly more searching examination may reveal faults of a more or less serious nature that did not affect the general appearance, which were not so visible in this long-distance appraisal. However, as a general rule, a well-developed frame, good carriage, style and deportment will be found to have associated equally good qualities. For exhibition purposes, a good appearance is a valuable asset.

Size

This refers to the development of frame, or conformation. In merino sheep, and more especially in modern types, size of frame is expected. In determining the merits of a sheep in this respect, weight is the object. Height from the ground does not always indicate size. Often much of the supposed size can be attributed to long legs. The animal shows too much daylight underneath. In such cases the conformation may not be very heavy at all. The thickset, close-coupled type of sheep may, on the first appearance, look somewhat small beside the more leggy class, but in the majority of cases the advantage of actual size and weight would rest with the former, the barrel being thicker, deeper, better filled and lower set.

Size of frame not only adds to the value of the animal for mutton purposes, but also means greater wool-producing area. Moreover, and especially in the low-set type, it indicates sturdiness, robustness and virility. Size also enhances the value of the sheep in either the store or fat-stock markets.

Symmetry

Beauty of conformation will be governed by symmetry. This refers to the proportion of development existing between the various parts of the conformation. Proportion is the basis of beauty of form. In the development of conformation in stock, symmetry is most essential. Certainly it varies in different classes, breeds and types, but in all cases the objective of the breeder should be to approach as near as possible to perfection. It not only adds considerably to the attractiveness of the animal, but even more importantly, it enhances the utility, because a well-proportioned conformation is generally necessary for the proper and free functioning of the internal and vital organs. It thus follows that all points should be uniformly well developed.

Often, in the desire to achieve a special objective, some parts or points are exaggerated in development, resulting in a disturbance of the balance, or sense of proportion, of the conformation. Such an exaggeration immediately becomes conspicuous, and detracts from the beauty of form of the animal. In many instances, the over-developed part will appear coarse. Not only should all parts be proportionately well-developed, but the development should be consistent with quality. So long as this exists, the development cannot become excessive. Coarseness should at all times be avoided.

Constitution

This relates to the build of animal, with special reference to physical well-being, or soundness. Healthy, virile sheep are termed sound. This soundness is the outcome of a good constitution. Constitutional weaknesses and defects produce many undesirable features in the form, and, moreover, restrict the proper functioning of the organs or members of the body.

When selecting sheep, and particularly rams, for breeding purposes, too much consideration cannot be given to constitution. Hereditary defects not only impair the utility and value of the ram but, more importantly, will invariably be transmitted to the progeny. In this reproduction they often become more acute. To place the flock upon the most stable breeding basis, constitution should be the first care of the breeder. Improvements in covering the fancy points might be affected subsequently, but the breeding-out, or elimination, of constitutional defects is a very tedious and costly process. The 'drag' from such weaknesses is generally very active, and is likely to reappear even after it seemed to have been removed. Sound, vigorous stock will resist the more common diseases that sheep are prone to, whilst weak ones are more susceptible. Unsound sheep are also unprofitable. For wool production, reproduction or economical keeping, the best results can only be derived from the soundest stock, capable of converting the food consumed to the greatest advantage.

Robustness and Masculinity

These terms allude also to the development of the animal, but have special reference to that strong, vigorous growth that represents great strength. Both the conformation and covering will be examined for evidence of this outstanding development. The vigorous growth of all parts, together with the manly, attractive carriage and firm stance are all features suggesting masculinity. It will also be represented by a strong, masculine head, substance in the horns and bone, and a deep, broad and thick frame, with quarters well filled and compact. Deep girth and flanks and a good broad straight back are important. The covering, too, should show plenty of body and substance throughout.

The principal value of robustness and masculinity lies in the fact that it is indicative of great strength. For breeding purposes, this is very desirable. These features should be well developed in the sire. An effeminate ram, notwithstanding other good qualities, will rarely be reliable, and may even introduce serious troubles into the flock.

The prepotency of the ram, or the power to transmit his good points or qualities to the progeny, will, in no small measure, depend upon his superior strength. It is the result of the inherent superiority of strength over weakness. As the ram is not only responsible for a great number of offspring, but, moreover, is generally selected with the object of introducing improvement, it is vitally important that the ram should, as far as possible, be capable of transmitting to the progeny his outstanding qualities. This is, of course, not always reliable. Indeed, even masculinity does not guarantee the ram will be prepotent. Every stud-breeder is aware of the fact that some sires never reproduce their good qualities, mainly due to their not being the most potent factor in the breeding process. However, generally speaking, masculinity is the most reliable indication of prepotency.

The Covering

The covering must next be considered. In all breeds of sheep, the fleece is regarded as one of the most important products. Indeed, it is the covering that has been mainly responsible for the value and utility of the animal. When determining the quality of the animal, therefore, the fleece is most carefully examined. Full appraisal requires the most expert knowledge. While many stock breeders possess a great amount of aptitude and skill for recognising symmetry and beauty of build, the number who would be capable of gauging the merits of a fleece of wool would be more limited. Whilst a stud wool will differ from a commercial wool to some extent, a good general knowledge of wool is still essential in order to detect either merit or defect.

The value of the covering naturally becomes more accentuated in the case of those breeders that are bred mainly for wool production. The development of quality and abundance of fleece has been given much consideration during the evolution of the Australian merino. This is also the case with the British breeds of sheep. Even with the purely mutton breeds, it is the rule to pay very close attention to the style, type and quality of the fleece.

Whilst quality and quantity of wool production is the principal objective of the breeder, it also offers a most reliable guide as to the animal's breeding and general quality. The charcteristics peculiar to the various breeds and types will be found to be very definitely fixed in the covering of well-bred sheep. Indeed, they should be just as stable as the markings or general features of the conformation. As the most prominent product, and a very reliable index as to the purity of the breeding of the sheep, covering must be one of the most important factors determining the value of the animal, either for breeding or general commercial purposes.

In judging the covering of a merino sheep, it would be considered under the following headings:
1. trueness of type;
2. evenness of covering;
3. density;
4. quality.

Trueness of Type

In estimating trueness of type, the fleece forms the principal subject of examination. It is, however, not the only one. Both conformation and other general characteristics, such as markings, must also be considered in conjunction with the covering. As the term would suggest, it means that the sheep under consideration should reveal, in its general make-up — wool, conformation and characteristics — all the most desirable qualities of the breed developed to a pronounced degree. Naturally, to determine the merit of the sheep in this respect, the judge or breeder must be well acquainted with the distinguishing features; it naturally follows that trueness of type can only be promptly identified by those who have carefully studied the development of the breed or type.

Types in New South Wales

In the evolution of the Australian merino sheep, three distinct types have sprung into being. Each type embodies special features. These features have been developed, very largely, with the idea of conforming to the local requirements, especially climate. In the successful breeding of all classes of livestock, the maximum profit is, perhaps, only obtainable when the most suitable type for the locality is chosen. This is accentuated in the case of sheep because the wool or covering is most susceptible to the effects of climate, particularly heat. Thus in the evolution of the merino sheep in New South Wales, three very distinct types have been developed. In each instance the fundamental reason for the variation was the prevailing climatic conditions. As this factor influences the production of wool to such a marked extent, the breeder has selected breeding stock to conform with the local conditions, and then concentrated upon the improvement and development of the type. It will be found that, from a wool-growing aspect, the state can be aptly divided into three main divisions — the Tablelands, the Western Slopes and the Plains. Not only does the climate in these areas vary, but the character of the country and pasture also shows a great amount of variation. The Tablelands are elevated, ranging from 610 to 1220 m (2000 to over 4000 feet) above sea level. The climate is cold and bracing, the winter comparatively long, and the pasture finer and more sparse. The Western Slopes, and the eastern portion of the Central Division is mostly undulating country. The climate is warmer, and the winter shorter; consequently there is a much longer period of pasture growth. The country is generally more fertile, with a correspondingly more abundant growth of grass and herbage.

On the Plains country, the climate is warmer still and the winter is comparatively mild. The more fertile country, more luxuriant (in good seasons) growth of pasture, together with much edible scrub all tend to develop a very virile, strong type of sheep.

From this brief reference it will readily be perceived that no one type of sheep would give uniformly good results under these several conditions. This, then, has been the fundamental cause of the development of the three distinct types: the fine-woolled, medium-woolled and strong-woolled merino.

Fine-woolled type

The fine-woolled type has been extensively produced in New South Wales, Victoria and Tasmania. In the other states the conditions have generally favoured the stronger-woolled types.

Prior to the beginning of the present century, the principal objective of sheep-breeders was the production of fine wool. Up to that period the carcass was of little value, while manufacturers evinced the keenest desire for fine-fibred wool. With the introduction of cold storage facilities on ships, and freezing processes, the export of mutton increased enormously. Indeed, it was through these facilities that the export of beef and mutton became practicable. The very noticeable shrinkage in the flocks of the state from 1895 onwards was an additional factor towards inflating the value of mutton, and consequently breeders were induced to turn their attention to the development of a frame for the production of a more suitable type of mutton sheep. This had the effect of increasing the popularity of the strong-woolled types, and caused a corresponding reluctance to keep the finer-woolled sheep. However, notwithstanding the very considerable influence brought about by the expansion of the export mutton trade, it still remains necessary to very carefully recognise the prevailing demands of locality.

It will, therefore, be observed that the fine-woolled type gives the best results under the influence of the colder conditions of the Tablelands, the maximum standard of quality being obtainable in these districts. On the Tablelands, especially the Northern Tablelands, are found some of the best flocks of fine-woolled merino sheep in the state. This high standard of quality has been developed under the very suitable natural conditions, together with the breeding skill and enterprise of the flockmaster. The fine-woolled type is distinguished by a comparatively smaller frame. The bone is finer, and the general constitution of the sheep would appear more refined. The covering, however, will show the most distinctive features. The fleece is generally very dense and compact. The length is, as a rule, shorter, ranging from 51 to 90 mm (2 to 3½ inches). The staple formation is compact or rounded. The fibre is fine, ranging from 70s upwards. The crimps are very numerous and well defined from base to tip. Being more numerous, the crimps are more acute, or sharper, in formation. The wool is usually very well preserved, having only a small amount of tip, which is generally black, due to the richer condition of the wool, together with the cooler, more moist climatic conditions. The wool is very soft, silky and bright.

Medium-woolled type

The medium-woolled type has been evolved mainly on the Western Slopes of this state, where the climatic conditions and pastures are a mean between the Tablelands and the Plains: the country is more fertile, the pastures are thicker and the carrying capacity greater. These factors, combined with a much shorter winter season, have tended to produce a more vigorous growth of frame and wool. Consequently the extreme fineness of the Tablelands wool is rarely obtained.

Sheep breeding in these districts has, therefore, encouraged the evolution of the medium-woolled type. This type of merino sheep would be distinguished by a more robust conformation. The frame is larger and the bones stronger. The face and horns will also have a stronger, more vigorous appearance. The covering, too, will also show decidedly more robust growth. The length would vary from 90 to 102 mm (3½ to 4 inches). In build, the staple shows a greater body, and is more square rather than round. The character, or crimp, is also broader, but still regular. The tips will conform to the formation of the staple and, if true to type, will be square and compact. The fibre is stronger, being a 64s count. In the case of rams, the wool will be a robust or masculine type of 64s. This wool rarely carries as much condition as the fine wool, nor is it quite as soft in texture.

Strong-woolled Type
The Plains country of New South Wales and South Australia has provided the conditions necessary for the evolution and development of the strong-woolled type. The summer is hot and dry and, in exposed country, the finer-woolled types would deteriorate in quality and abundance. From a very early period in the history of the sheep-breeding industry in New South Wales, it became evident that the stronger-woolled types were the most suitable and profitable to keep. To withstand the conditions, both sheep and wool must possess substance. That these opinions were well founded is confirmed by the success achieved by such well-known Riverina studs as *Wanganella, Boonoke* and *Uardy*. These strains are noted for their soundness, size of frame and wool production, and are extensively used in various parts of Australia where the natural conditions are similar to the Western Division of New South Wales.

The strong-woolled type possesses a very large, well-developed frame for a merino. The length, breadth and depth is greater than that of the medium-woolled type. All features appear decidedly more robust and stronger. The covering also reveals strong distinguishing features. The staple formation is very large, or full-bodied. It has a blocky appearance. The tips are correspondingly large and oblong-shaped. The length is considerable for merino wool, ranging from 102 to 130 mm (4 to 5 inches) long. The fibre is strong, averaging a 60s quality. Notwithstanding the length, the wool should be compactly or densely grown. The back, particularly, should be well-covered. It is essential that the back should be compactly covered in order to minimise, as far as possible, the injurious effect of the heat and dust upon the wool. The back portion of the fleece is, of course, more exposed to damage from such causes, and the shorter stapled or more open-woolled fleeces will be more seriously affected.

The crimps of the strong-woolled type are fewer, and have a more undulating formation. It is more important, however, that the character should be well-defined and regular. Straight-fibred wools in even this type are undesirable. The texture is firmer than in the case of the medium type of wool, but should nevertheless be soft. The colour should be bright.

From these brief descriptions, it will be seen that 'trueness to type' refers to the manner in which the peculiar features of each type are displayed in the animal under examination. To be true to type, the sheep should represent, in both conformation and covering, those distinctive characteristics of the type to which it belongs. Should the general character show signs of a blending, or mixing of the outstanding features of two different types, then the distinctiveness would have vanished, and the animal would, under such circumstances, not be truely representative. From either a breeding or an exhibition point of view, this would detract from its value. Trueness to type is most important because it indicates, to a very marked extent, the breeding of the animal. By good breeding the distinctive features of the breed or type become more permanently fixed, and consequently the units may be relied upon to transmit their progeny with a greater degree of certainty. The presence of these representative features is, to the judge, the sign of reliable breeding.

Evenness of Covering
This relates to the length, density and quality of the fleece. It is applied in a comparative sense, and implies that the fleece, as regards those features quoted, should be proportionately even or uniform throughout. In all sheep the fleece will

show more or less variation in the various component parts. The more common the breeding of the stock the more pronounced such variation will be.

The maximum quality will usually be found in the wool grown within the region of the shoulders. The fleece will gradually depreciate in all characteristics in almost all directions from this part. The most gradual falling-off will be along the sides in the direction of the points. The wool grown on these parts — the points — is shorter, more irregular in quality, and less abundant. The back is another part where the fleece is likely to show some defects in length or density. To be evenly covered, all those parts which are generally recognised as being more or less faulty should be comparatively well-covered with wool of proportionately good quality and length.

Those parts of the fleece which produce the more inferior wool are usually submitted to the most critical examination by the judge or classer. This is because scientific breeding has succeeded in improving the covering of the merino sheep underneath, on the points as well as the fleece proper. Perhaps the production of wool on the points is one of the most outstanding forms of improvement; consequently, if any reversion or retrogression is taking place, such parts will reveal the first symptoms. Thus the judge looks to these for information that enables him to arrive at a more precise analysis of the animal's merit as regards evenness of covering.

The modern types of merino sheep are particularly well-covered, and 'rosellas' and 'barebellies' are now very rarely seen. The belly-wool, at the present time, is of comparatively high manufacturing quality, whilst the wool from the points is very free from objectionable fibres such as gare and kemp. The greatest consideration should also be given to the covering of the back in merino sheep. Generally speaking, this is a very common seat of weakness. Frequently the staple will be shorter, or more open or loose in growth, defects which are most undesirable: firstly because, in such cases, the manufacturing quality and value of the wool deteriorates to a great degree through exposure to the heat, dust and rain; and secondly because, in order to counter these conditions, greater effort and care in both selection and breeding must be made in order to overcome, as far as possible, the inherent tendency to develop defects on the lines mentioned. The length, body or compactness of staple and density should be well maintained all over the back portion of the fleece.

Another fault frequently encountered in the covering of merino sheep will be found across the withers. It is more common in the finest-woolled types. This part of the fleece often assumes a discoloured, most unattractive appearance and the fibres lack life and character. The wool will generally be yolk-stained. The presence of this patch is often attributed to the effects of a rainy season, but in many instances it will be found to be hereditary. Whilst it might be difficult to definitely locate the cause of this eyesore, it would appear to be very closely allied to the constitution of the animal, a contention strengthened by the fact that it is most frequent and pronounced in sheep inclined to be of a weaker constitution. Special attention should be given to this part of the covering when selecting breeding stock. The withers should be well covered, with typical wool length and density being well maintained.

The character and quality of the wool throughout the whole fleece is another most important factor when appraising the merits of the covering as regards evenness. The importance of a comparatively high standard of quality throughout rests first and mainly with the elevation of the manufacturing quality of the wool grown on the points. This must materially increase the commercial value of the fleece as a whole. Secondly, a fleece showing the character and quality well developed on the lower parts would at once indicate good breeding. It is well known that the fleeces of the more common-bred stock reveal a greater variation in quality. When determining quality in these respects, the standard of perfection will be decided on a comparative basis. As mentioned before, the wool on the forequarters, especially the shoulders, will be the best. A close examination of the fleece will show that the fibre gradually becomes stronger, or thicker, as the breech-end is approached. Nevertheless the wool on this part should show the same relatively high standard of quality and character as the recognised finer-points portions. The character particularly should be well sustained throughout these regions. It would indicate, as regards wool production, that the animal is well-finished. As already mentioned, the wool produced by common bred stock on these parts is not only lighter, but is frequently comparatively low in quality. Objectionable fibres, such as kemp and gare, first appear about the breech-end, the points and folds. The presence of such fibres very greatly depreciates the manufacturing value of the wool; consequently selection of breeding stock should aim at the production of a fleece showing a high standard of quality throughout. The attainment of this object would, as a matter of course, eliminate the objectionable fibres. In the process of retrogression, pronounced coarseness of wool fibres on the britch and points might be regarded as the forerunner of the development of still more objectionable fibres. The degeneration, as a rule, occurs in graduated stages.

For these reasons it is necessary, when judging merino sheep, to carefully note the character and quality of the covering of the points, particularly the breech, the lower part of the thigh, around the tail, and along the hind leg from the top of the rump downwards. The character of the wool on the folds and head should be carefully noted. It should show a comparatively good crimp, be soft in texture, and not be straight or hard. The face should be open and soft.

Density

As mentioned earlier, density refers to the thickness, or compactness, of the component fibres of the fleece. Obviously, the animal producing the maximum number of fibres on given area of pelt will have the densest fleece. The importance of density is that, first, it very considerably influences the weight of the fleece. The development of density is one of the most noteworthy achievements in the evolution of the modern merino sheep. It has greatly increased wool production on a per-head basis, and consequently added considerably to the value of the individual animal.

The quality of the wool is also affected, being better preserved in dense fleeces. As the wool is more compact, the fleece resists, to a greater extent, the perishing influences of the weather, with the result that a smaller proportion of wasty tip occurs. The substance and quality of the staple will be better preserved also. This adds to the manufacturing value, which is reflected in the high realisations made for the raw, or greasy, wool.

In actual judging practice, various tests are made when examining the covering for the purpose of ascertaining the degree of density existing. A dense, compact wool will, when gathered in the hand, feel springy and full. By simply applying pressure with the palm of the hands on the surface it will not only resist such pressure to a noticeable extent, but will feel more solid and firm. Again, upon opening up the wool, a dense fleece would offer more resistance and feel more springy, at the same time revealing a very small expanse of pelt. It is very necessary that the fleece should be comparatively dense throughout. The parts generally more subject to slackness are the forearms, lower parts of the body, especially on the girth line immediately behind the forearms, the lower flank, thighs, belly and back. It is most important that the back should be thickly covered, whilst the other parts referred to should be proportionately dense.

Quality

Having considered the covering from those aspects which largely determine quantity of wool production, it now becomes necessary to pay attention to those factors relative to quality. The quality of any wool is most important, but in the case of the merino it is vitally so. Modern developments in the breeding of merino sheep have not been altogether conducive to the production of the highest standard of quality. This does not imply that the quality of our merino sheep is degenerating, but it has to be admitted that the finer qualities of merino wool are annually becoming rarer. Seasonal conditions undoubtedly play an important part in framing the general quality of the clip, but nevertheless there is no denying the fact that the general trend of breeding favours the production of a more robust type of sheep, carrying, of course, a more robust type of wool. Certainly some districts still produce some exceptionally fine-fibred clips, but these are fast becoming the exception to the rule. Undoubtedly it is quite possible to produce excellent wools of the broader-fibred descriptions, but only by most careful breeding. The high-grade manufacturing qualities of merino wool tend to depreciate as the fibre becomes broader unless the greatest amount of vigilance is exercised. Naturally a broader-fibred wool will readily incline towards a firmer texture, and that highly important soft, pliable handle diminishes to such an extent that in a comparatively short period it might be quite pronounced, to the obvious detriment of the wool. It follows that in the production of such wools, quality must be given the closest attention in order to counter such undesirable tendencies.

The wool of stud sheep, and more especially rams, would not be expected to develop such high manufacturing qualities as flock wool. The stud wool will always be more robust, as quality of wool is an important asset. Moreover, the feeding, care and keeping generally encourages substance. With rams, the wool would be expected to show very pronounced robustness or masculinity. This would denote strength and soundness of constitution. A wool showing the high manufacturing qualities would, in all probability, be much too effeminate for a desirable ram's wool. However, after making due allowance for these requirements, the wool should still give evidence of good breeding and the capacity for improving the standard of quality of the flock wool. The ram is perhaps the most important unit in moulding the quality of the wool of the progeny, and the quality of the wool of the progeny will very largely determine its manufacturing and commercial value. The realisation from the sale of the clip will, in a large measure, be governed by its standard of quality.

From these deductions it is quite evident that it is essential to pay the most careful attention to the quality of the covering when selecting breeding stock, and especially in the case of rams. Strictly avoid rams showing lack of character in their wool, frosty-white faces, or kempy fibres through the fleece.

The principal properties contributing to quality in a ram's wool are:
1. character;
2. softness;
3. elasticity;
4. length of staple; and
5. colour.

As previously mentioned, substance, or body, is also essential. The length will, of course, vary according to type, but it must be comparatively good. If the sire's wool is short, good length cannot be expected in the flock wool.

The Character

The character of the wool indicates breeding as well as quality. The crimpy formation in all well-bred wools is noted for its regularity throughout the whole length of staple.

It should also be pronounced. Straight-fibred wools do not, as a rule, possess the same measure of softness, pliability or elasticity. Moreover, quite apart from the influence it has upon the development of other properties, a pronounced, regular crimp gives a more attractive appearance, which is appreciated by those familiar with wool. It is the attractiveness of something particularly good.

Softness
Softness is so important a manufacturing property in merino wool that its development always exercises a very considerable effect upon the wool's value. It is a very desirable feature to foster in the clip, and so long as it is well-maintained in the wool of the rams used, it is likely to be a prominent feature of the flock wool. The wool of a merino should have a soft, mellow and pliable handle. Such wools are of higher manufacturing quality, and besides will not readily develop undesirable fibres such as kemp. The ram's fleece should have a springy and full handle. This indicates that the wool possesses plenty of life and elasticity, which are desirable properties. Moreover, such wools are produced by sheep possessing plenty of vitality and soundness.

Colour
The colour of the wool is also most important. Nothing adds more to the appearance and commercial value of wool than a good colour. Merino wool is noted for its pure, bright, creamy-white colour, and this is the most desirable shade. It is often contended that the golden tint indicates robustness, and consequently must be evident in ram's wool. There can be no doubt, however, of the superiority of the good bright, soft, white colour from a manufacturer's point of view, and where this can be developed with the required amount of robustness, it is certainly elevating the value of the clip. Climate is indeed a very potent factor in the development of colour, but breeding can do, and has done, much towards improving the quality of wool in this direction. Yellow fleece is unattractive and undesirable.

South Australian Merino

South Australian merinos were specifically bred to thrive and provide an economic return from wool in the arid pastoral conditions found in much of South Australia. Rainfall in these arid districts is mostly in the vicinity of 250 mm or less per year, and plants such as the saltbush (*Atriplex* spp) make up a large part of the natural vegetation.

The South Australian merino is physically the biggest of the strains of merino sheep in Australia. This breed is generally longer, taller and heavier of body than the Peppin types and tends to have less loose skin, in the form of skin wrinkles, than other strains. Their wool is at the most coarse end of the range of merino wool types. It also tends to carry a higher proportion of natural grease which has been specifically sought by breeders to provide protection to the fibre under the most adverse grazing conditions.

Apart from South Australia, this strain of merino is found in significant numbers in the pastoral regions of Western Australia, Queensland and New South Wales.

Poll Merino

The poll or hornless merino is a comparatively new development. Hornless merinos have occurred as mutants in all the main strains as this is a feature of their genetic structure. Breeding over many years, with careful selection for this characteristic, has developed a hornless animal with the following advantages:

- elimination of fly strike in the poll area of the head;
- easier shearing and handling;
- less damage to the animals during yarding and transportation.

Polled merinos are gaining in popularity and are found in all the merino areas of Australia.

3

BRITISH BREEDS OF SHEEP

The United Kingdom is remarkable for the number of distinct breeds of sheep evolved in such a small area. The variations of climate and pasture and the amount of intensive farming practised were, no doubt, factors which encouraged enterprising breeders to cross and select in the hope of producing types more suited to their particular environmental conditions. Although it must not be imagined that all British breeds are extensively used, there are over thirty in existence claiming the United Kingdom as their country of origin. Since they are always bred in a cold climate, the conditions suitable for their development are limited in Australia. They thrive in the colder parts of the continent, on the slopes and highlands, although they are also found on the better class plains country.

The value of the British breeds in Australia is dependent not so much upon their usefulness in their pure state, as upon their adaptability for mating with the merino. They are what may be termed 'agricultural breeds' and in management and systems of breeding should be treated as such. Whilst rich pasture and good food supply will help to develop them, a scanty and non-nutritious food supply will tend to diminish qualities most desired.

For convenience, the English breeds will be considered under three headings, long-woolled, short-woolled and mountain breeds.

The Long-woolled Group

Breeds of this group mostly originated on the rich lowlands of England and, as the name implies, they carry wool of a long staple ranging from 178–203 mm (7 or 8 inches) in the case of romney marsh to over 305 mm (12 inches) in the lincoln. The wool is coarse and lustrous with a bold crimp. These sheep have black hooves and nostrils, and white faces and legs, although several show a bluish tint on the face, caused by the dark skin showing through the hair covering. All are hornless. They have large frames, are heavier boned than the downs breeds and develop heavy weights at maturity. The mutton is of good quality, but compared with the downs breeds is coarse-grained and not up to the standard in flavour or colour. The fat is too heavy and rather badly distributed and the joints are not as shapely.

Breeds of this group blend or 'nick' very well with the merino and this, together with their heavy carcass and heavy wool production, brings them into prominence in Australia for crossing with the merino to produce dual-purpose sheep or mothers of fat lambs. For best results they require good feed conditions, but are more adaptable to adverse conditions and drought than the downs breeds. They are heavy milkers, good mothers and are found in the cooler parts of Australia, particularly in agricultural districts. As with all British breeds, with the possible exception of the dorset horn, their breeding is confined to the autumn months.

The Short-woolled Group

Compared with the long-wools, the wool of the sheep of this group is short. It is fine, lacks style, is harsh and of a chalky-white colour, lacking lustre. The fleeces are of light weight, mushy and of a relatively low market value. The wool is particularly useful for hosiery and certain types of paper felt manufacture. In a suitable cool climate with an abundance of sweet, short, nutritious grasses, or well-drained country (carefully avoiding sour, wet conditions) they are invaluable as mutton sheep. Because of their early maturity and high fecundity, they are largely used in fat lamb raising. The mutton is firm and outstanding in its rich colour, flavour, grain, shape and light, even distribution of fat. The downs group of breeds are also noted for their high killing-out figures. The markings are from light brown to black on face and legs with a few exceptions (e.g. dorset horn and ryeland). The dorset horn, although always used as such, is not a true downs. It shows characteristics of both down and long-woolled types and, in breeding habit and horn development, the features of a merino.

The ryeland is another white-faced short-woolled breed showing some features of the long-woolled group. This breed is not so fastidious about grazing conditions as is the true downs, doing better on rank growth, under wet conditions. The ryeland is considered to be unsuitable for hot climatic conditions.

The Mountain Breeds

As would be expected of sheep most suited to the bleak heath conditions on the high-lands of the United Kingdom, the mountain breeds are noted for their extreme hardi-ness. These breeds are of little consequence to Australia, as their outstanding features do not conform to our requirements. The cheviot, however, one of the favourites for crossbreeding in England, and one of the most likely to suit our hardest mountain conditions, has been tried in Tasmania, Victoria and South Australia. Results have not been convincing and its use is still not very extensive, though at the present time a few enthusiastic breeders are still trying to popularise this breed. There is a great variation of types, both in wool and carcass, within this group. Their ancestry in many cases is hard to trace but it appears that both downs and long-wool blood have been used. Some breeds are horned, others hornless, and their markings are of wide variety, from white faces and legs, as in the Welsh mountain, to a mixture of black and brown or jet black in others. The wool is usually coarse, long and intermixed with hair and kemp. These sheep are slow maturing and of a light build. The mutton is of excellent flavour, lean and generally considered to be of good quality, but it lacks the shortness of bone and depth of fleshing (muscle) which are characteristic of the highly improved mutton types.

Cheviot wool lacks the character of other long-wools and usually contains kemp fibres.

British Breeds in Use in Australia

Breeds in use in Australia are listed in order of their popularity in New South Wales:

- *Long-wools:* border leicester, romney marsh, English leicester, lincoln.
- *Short-wools:* dorset horn, southdown, suffolk, ryeland, hampshire, dorset down, shropshire.
- *Mountain breeds:* cheviot.

The hampshire and dorset down are rarely used in New South Wales, but are more popular in Victoria. The cheviot is used in Victoria, South Australia and Tasmania, but there are only a few in New South Wales.

Long-woolled group: Border Leicester, Romney Marsh, English Leicester, Lincoln

Border Leicester

The border leicester is the most extensively used long-wool in Australia. It is very popular in New South Wales and is used for crossing with the merino for the production of fat lambs in the more marginal areas, and the half-bred ewes are frequently mated to produce second-cross lambs in the first-class fat lamb districts. The first-cross ewe lamb has the advantage that if it is not got away as a lamb, it can be carried through to make a very satisfactory breeder. The breed originated in the border country between England and Scotland, where Culley Bros. used Bakewell's improved English leicester to cross with the cheviot. A hardy breed, it grows rapidly; the ewes are prolific and are good mothers. The border leicester stands hot climates as well, but through its adaptability it competes strongly with other British breeds in the cooler districts. The width of the head in the lamb sometimes causes some difficulty at lambing.

Points

Head	The head is larger than that of the English leicester, carried high, and hornless. Free of wool and covered with white hair with an occasional black spot. Wide eyes and narrows from the eyes towards the crown of the head.
Face	Is rather long and narrow with convex or Roman nose. Strong deep jaw, wide nostrils and the lips are dark to black in colour.
Eyes	Full and prominent.
Ears	Lively, mobile, fine, soft, medium size, set high up on the head, erect, pointing backwards and not drooping, with inside and out; black spots sometimes appear with age.
Neck	Fairly long, tapering nicely from the head and set strongly on wide shoulders; there must be no slackness at the junction.
Chest	Broad and deep with full rounded appearance. The brisket is carried well forward.
Back	Is straight, very broad throughout and rather long; with high rounded rump carried well out behind the back legs.
Ribs	Are particularly well sprung giving a full well-rounded appearance.
Hindquarters	Loins well-filled and firm and the quarters of good length.

Legs	Are clean and free of wool, covered with white hair and are finer in bone than the English leicester. The belly is carried high, giving the animal a very straight underline and a rather leggy appearance. Legs are set apart and squarely.
Hoofs	Dark in colour.
Skin	Pink.
Testes	Firm and even and scrotum covered with wool.
Covering	The staple is long ranging up to 254 mm (10 inches) in length and quite distinct in form. It is rounded, broad at the base and tapering to the tip, which is not as curly as that of the English leicester. The fleece is comparatively even and the quality number is 44/46

Border Leicester

Romney Marsh

The romney or kent sheep is another of England's old long-wool breeds evolved under special conditions, namely the inhospitable, swampy lands of Kent, and it is natural that the breed carries an inbred hardiness under heavy rainfall conditions.

The original romney was considerably improved in conformation and fleece by the introduction of English leicester blood, and by continuous selection its ability to stand up to cold wet conditions has been preserved in the improved romney. The romney is now well spread in the United Kingdom and is popular in Australia, New Zealand, North and South America and South Africa, as well as other parts of the world, on account of its strong constitution, shapely, close-grained, good quality mutton, and its early maturity combined with a fair weight of useful medium to strong quality wool. In

New South Wales the romney is a favourite for crossing with large-framed merinos in the Tablelands and south-east districts to produce dual-purpose sheep and mothers for second-cross lambs. Romney ewes crossed with the southdown ram give a very desirable type of lamb for the English trade and this cross is the basis of New Zealand's fat lamb industry.

Of the long-wools, the romney is only surpassed by the border leicester for early maturity. It is an excellent mother with a plentiful supply of milk; hence its suitability as a fat lamb dam.

Points

Head	The broad hornless head is carried rather low and has a strong masculine appearance. It is level between the eyes, with a wide forehead.
Face	Moderate to short in length, fleshy and broad above the nose. White hair on face with square, broad muzzle, black in colour (although a mottled nose will not disqualify from the show ring). Thick, fleshy, open nostrils. The lower jaw is deep and strong.
Eyes	Should be large, bright and fairly prominent, with fearless expression, but not wild.
Forehead	Broad and carrying a good, thick top-knot of wool.
Ears	Large, thick, prominent, low set, covered with white hair. Frequently take blueish tinge as sheep ages.
Neck	Is rather short, thick and muscular, with a well-developed strong scarf and low withers.
Chest	Wide, deep and well-rounded.
Back	Is broad and level but not as long as lincoln, leicester or border. The barrel is deep but ribs are not as well-sprung as the border leicester.
Rump	Has good length and is wide and heavy. Thick, heavy tail, set high; well-developed and well let-down thighs. The twist is well-arched.
Legs	Are short, heavy boned with wool carried well down, otherwise covered with white hair. The hoofs, are large, particularly sound, well-shaped and black in colour, although a streaky hoof will not disqualify in the show ring. Brown or rusty hair on legs is objectionable.
Covering	The wool is slightly finer than that of the border leicester but the fleece weight is somewhat similar. The wool is bulky and denser than the border leicester. The length of the wool, depending on quality numbers, is about 152 to 229 mm (6 to 9 inches) for twelve months growth. The wool is of a softer shade than the border leicester and is lustrous, but lacks the golden tinge of the other long-wools. The character is not as good as that of the border leicester. It is not as soft as the lincoln, English or border leicester. The staple is bulky, with the bulk of the fleece of even quality, between 50/44s, but there is often a running away to as low as 40s on the breech. A hairy breech is a common fault, but in view of the fact that hairiness in the romney is rather strongly inherited, careful breeding can do much to eliminate this.

Faults in a Romney Marsh Sheep
(From Romney Sheep Breeders Association)

Head	Scurs or horns; papery ears; narrowness of head; lack of depth in jaw.
Face	Lack of width in the nostrils; pink nose; brown spots on face; excessive black spots; harsh, coarse hairs.
Eyes	Narrow set and dull eyes.

Neck	Weak scrag is certain indication of poor conformation and constitution.
Shoulders	Devil's grip or high wither and shoulder depression.
Carriage	Badly balanced; a droopy, sloping sheep.
Chest	Weak brisket.
Back	Swampy or roach back.
Ribs	Narrowness of frame invariably denotes weak constitution.
Rump	Goose rump and cruiser stern — common faults which detract from essential mutton quality.
Legs	Thin bone, cow-hocked; down in the pastern; white hooves.
Skin	Dark and coarse, hard skin.
Fleece	Lack of uniformity; lack of bulk particularly on the back and running to fineness or wastiness on the extremities. Presence of coarse fibres or hair-kemp fibres.

Romney Marsh

English Leicester

This is the oldest improved British breed and in 1775 the pioneer breeder, Robert Bakewell, made history by improving the 'old dishley' breed of Leicestershire. He selected for carcass quality, but paid little attention to the wool, although the hair growth of the original breed was largely lost. He recognised the importance of giving his stock good nutritional conditions and he also accelerated the rate of improvement by hiring out sires to other farmers, recalling for his own use those that showed most promise. Due to Bakewell's sound breeding practice, the improved leicester was most prepotent and at some time or other, it was used in the improvement of all other long-wools. Indeed, it is sometimes contended that the English leicester was introduced even into the merino.

The English leicester is active and a good traveller, is a heavy feeder and puts on flesh quickly. The mutton, however, is not regarded as prime, being coarse-grained with excessive fat, particularly over the rump. It is finer boned than the lincoln and the carcass, while being more shapely, is of lighter weight. The wool is full lustre, bulk 40s, but is shorter and finer than lincoln wool. The staple is more rounded and tapers to the tip, which is characteristically spiralled. In Australia it crosses well with the large-framed merino but it is not as popular as either the border leicester or the romney for this purpose. The English leicester is most popular in Victoria and Tasmania.

Points

Head	Is carried well, not too high; neat, clearly chiselled, wide between the ears, slightly tapering to the nostrils. No sign of horns and carrying a light forelock.
Face	In general form, wedge-shaped and covered with short, white hairs, face tending to blue. Lips and nostrils black; black specks on face and ears, not objectionable.
Eyes	Full and bright.
Ears	Rather thin and long.
Neck	Short to medium length, thick and level with the back. Tapering from full developed shoulders to the fine cut head.
Shoulders	Upright and wide over the withers.
Chest	Breast should be deep, wide and prominent.
Back	Wide and level filled up behind the shoulders, giving a great girth, showing thickness through the heart, and carrying firm, even flesh.

English Leicester

Ribs	Well-sprung.
Hindquarters	Full-sized and square, showing good legs of mutton. Tail well set-on, almost level with the back.
Legs and feet	Legs straight, well set-on and wide apart. Short cannon bone, good pasterns, not sloping, and free of rust-coloured hairs on legs. Hooves black.
Skin	Healthy pink colour.
Carriage	Free and active with alert appearance.
Covering	The covering of the leicester is, comparatively speaking, hardly up to the standard of the lincoln. The staple is approximately 203 mm (8 inches) long, not nearly as long as lincoln, is round in formation with two distinct curls at the tip. The colour is lighter, being of a more silvery appearance; it is a full lustre, with well-defined, regular crimp, finer in fibre and lighter in condition than lincoln wool. The texture is soft and silky. The quality of a well-nourished, well-bred leicester is approximately 40/46s.

The Lincoln

As far back as can be traced, this breed has inhabited the rich lowlands of Lincolnshire and the east coast of England, so it is not surprising that it requires the richest of agricultural land with an abundance of nutritious feed to give of its best. Given suitable conditions, it is the largest and heaviest of all breeds and carries the heaviest fleece of the strongest and longest stapled wool. It is the slowest maturing of the British breeds. While it will not stand continuous adverse conditions, it will hold condition under temporary checks longer than other British breeds used in Australia. The fleece, even more than the carcass, is less affected by temporary checks.

From a wool point of view, this breed crosses well with large-framed merinos and, prior to the corriedale gaining popularity, was extensively used in Australia for crossbreeding purposes. Unfortunately, the half-bred, although a good dual-purpose sheep, is rather too slow of maturity to be a mother of fat lambs. The lincoln, with its heavy head, rather long back and neck, heavy bone, coarse-grained flesh, and inclination to run to fat when mature, is not the best for mutton purposes. It is not a good traveller or fossicker and is more liable to worms and footrot than the border or romney. Not a great number of lincolns are used at present and of these many are not up to the English standard.

Points	
Head	Although in proportion to the body, the head is of a massive appearance due to the broadness between the ears and eyes.
Forehead	Is covered with a well-developed topknot, or forelock, hanging down the face well below the eyes.
Face	Is not too long, but broad, white in colour, often with a bluish tint from the black spots showing through the soft white hair covering. Lips are thick and nostrils wide open; both are fleshy and black in colour. Pink muzzle is considered objectionable.
Eyes	Are of a docile expression and set well apart.
Ears	Are rather large, thick and fleshy, and set well back on the side of the head. They are frequently black and blue spotted and this is not discriminated against, but pink ears are objectionable.
Poll	Is arched.

Neck	Is inclined to be long and tapering from shoulders to head. The back is broad and straight. The old lincoln, before its improvement by the infusion of leicester blood, was described as swampy-backed, but this cannot be said of good English lincolns today.
Body	Is rather long with firm handling flesh on the back.
Ribs	Well-sprung from the spine, but rather flat on the sides.
Hindquarters	Are thick and well-rounded with a reasonably well-developed twist.
Chest	Is deep, a little flat at the sides.
Skin	Is pink.
Underline	Is straight.
Legs	Are of reasonable length, straight, heavy-boned with black hooves.
Covering	A fleece of 36/32s quality number is desirable. It is very heavy and even in quality, free of heterotypic fibres about the points and breech. The thick, bulky, flat staples grow up to 381 mm (15 inches) in length with large crimps of about 25.4 mm. It is highly lustrous and is of a rich golden colour and handles soft and silky. The tip is not wasty or pointed.

Lincoln

Short-woolled Group: Dorset Horn, Southdown, Ryeland, Shropshire Down, Suffolk Down, Hampshire Down, Dorset Down, South Suffolk

The Dorset Horn
The dorset horn is a breed which in recent years has sprung into prominence in Australia as being most valuable for the production of fat lambs under the prevailing

environmental conditions. Today the ram is most popular for crossing with half-bred ewes for the production of high-grade export lambs.

The dorset horn originated in Dorsetshire and its development is obscure. However, it would appear that long-wool, down and even merino blood was used in its evolution. It possesses the big bone, clarity of colour in the wool and white face and legs of the long-wools, the early maturity, fecundity and short, harsh wool typical of the downs, and horn, hooves, muzzle colour and the breeding habits of the merino. It has proved to be a hardy, prolific, thrifty, active breed of sheep in Australia, the ewes being heavy milkers, and excellent mothers. In recent years the conformation of the dorset horn has been considerably improved. It has a very long breeding season, most strains mating throughout the year, and this fact, together with the dorset being early maturing, makes the breed invaluable to farmers in early areas where the short period of good food is in winter–early spring, necessitating a summer mating.

The dorset is larger framed and coarser in the bone than the other down breeds and the quality of the mutton, although very good, is not up to the high standard of the southdown. It 'nicks' very well with the merino and the resultant lamb is the best of the short-wool-merino crosses, but unfortunately the wool is unsuitable for a carry-over and the crossbred lambs must be sold and not retained for breeding purposes. A feature of the breed is that there are no dark fibres in the skirtings and some of its popularity is due to this fact.

Points

Head	*Ram* — the ram has a large, strong-boned head with massive horns, well apart at the crown, which curl, or curve, downwards and forwards, as close to the face as may be, without necessitating cutting. They are not so corrugated and are of a lighter colour than those of the merino.
	Ewe — The head is lighter in build, with a smaller and more delicate curved, but not curled, growth of horns. Large broad nostrils are full open. Is well covered with wool from brow to poll.
Face	Long, broad, with strong bold muzzle. Face covered with soft white hair — often the merino creamy white rather than the pure white of the long-wools. Pink nose and lips.
Ears	Medium size and thin.
Neck	Thick, short and round. Well-sprung from the shoulders, very muscular across the top portion; no depression at the collar.
Chest	Full, rounded and deep.
Forequarters	Full, heavy, with no depression behind the shoulders; big girth, deep prominent brisket.
Shoulders	Arched, wide, with plenty of thickness through the heart.
Back	Broad, long and straight.
Ribs	Well-sprung from the spine, showing a good rounded deep side but there is a tendency toward being a little slab-sided towards the flank.
Hindquarters	Full, broad, well-rounded rump, and deep, with flesh extended to the hocks.
Tail	Well set-up in line with the back; wide, firm and fleshy.
Legs	Medium length, well-placed at the four corners, straight between the joints, stout bone; woolled to knees and hocks. Covered with white hair from knees to hocks.

Dorset Horn

Covering   The Dorset Horn Flock Book describes the desired wool as 'of good staple and character, dense and firm to the touch', the quality number being 56s, and a well-grown dorset will approach a 130 mm (5 inches) staple. The wool shows a better staple formation and character than typical down wool. It is doubtful if it is correct to state that this wool is chalky white in colour, for many dorset wools are now showing a softer colour with the sheep of almost a second demi-lustre. It is of a harsh texture, reasonably dense with a square tip, but the presence of coarse wasty fibres in the staple detracts from its appearance. The fleece is free of any dark fibres.

Off-type point   Dark spots on skin, black markings on horns, coarse hair on legs and
objections   tendency of horns to grow backwards.

The Southdown

The southdown is one of the oldest and purest of British breeds and was the first downs breed to be improved. It is regarded as the leading short-woolled breed and the other down breeds have had southdown blood introduced at one time or another. They were originally bred on the chalky downs in the south of England: well-drained lands producing an even, ample supply of nutritious, short, sweet food. The breed is renowned for its prime mutton which is shapely, succulent, tender, fine-grained, of good flavour, with an even fat distribution and a high percentage of muscle to the bone and fat. While the southdown does not gain weight faster than other breeds, it can be marketed profitably at an earlier age and weight.

In Australia the southdown gives best results in cool districts on well-drained paddocks, carrying short, sweet, improved pastures. In New South Wales the use of the southdown is more restricted to the cooler districts where the ram is mated with half-bred ewes for late-season, first-grade export lambs. In the warmer areas it is difficult to breed from downs rams sufficiently early for the lamb rearing or suckling period to coincide with the period of heaviest supply of suitable, soft food, and for this reason the dorset horn is more popular than the southdown.

In conformation the southdown is a contrast to the heavy long-woolled breeds. It is small-framed, yet the carcass weight is much heavier than the size would indicate. It is closely coupled, with good width, well-sprung ribs, low set-on, short, well-boned legs. The hindquarters are thick and heavy, are particularly well filled, and are always used as a standard of excellence for mutton breeds. The fleece is short, spongy, light in weight, with a quality number of 58/60s.

Points

Head	Is small, level between the ears and narrow between the eyes. Arched poll, well covered with wool; no sign of scurs, horns or dark poll; nostrils full and wide; woolled on forehead and cheeks but no wool round the eyes or across the bridge of the nose.
Face	Full, not too long and of an even mousy brown colour, not too dark or speckled; under jaw lighter in colour.
Eyes	Large, bright and prominent.
Ears	Of medium size and covered with short wool.
Neck	Short to moderate length, deep and full at the shoulders and tapering to the head; throat clear.
Shoulders	Well-set and top level with the back; no depression behind the blades.
Chest	Wide and deep with a bold rounded front.

Southdown

Back	Level and wide and well-covered with firm flesh.
Ribs	Well-sprung, giving a well-rounded barrel. The body is short, compact, thick through the heart with fore and hind flanks fully developed.
Hindquarters	Rump wide, long and well-turned; tail broad and set level. Thighs wide and well let-down with a deep, wide twist which gives an ideally shaped joint.
Skin	A healthy bright pink, free from any blueness.
Flesh	Even and firm to handle.
Legs	Short, straight and set on outside the body, woolled to the knees and well down the hind legs; hair covering of a mousy-brown colour.
Covering	The southdown carries the finest-fibred wool of all British breeds, the recognised quality being 58/60s. The fleece of 1.4 to 1.6 kg (three to three and a half pounds) has a wasty, open, mushy tip. The wool is harsh, short-stapled and has an undesirable flat-white, or chalky-white colour. The wool is absorbed in the hosiery trade.

Stud breeders are not so concerned about improving the wool, but merely look for a fleece true to the types as set down in the *Flock Book* — 'of fine texture, great density, and of sufficient length of staple, covering the whole body down to the knees, and below the hocks and up to the cheeks, with a full foretop but no wool round the eyes or across the bridge of the nose; no black hairs on body'.

The Ryeland

The old ryeland was a small-framed, well-shaped, light-boned animal growing a light fleece of about 1 kg (2 pounds) and was thought to be related to the Spanish merino. The introduction of English leicester blood altered the breed completely, and the present day ryeland has taken its place among the best breeds for the production of fat lambs. The ryeland was introduced into Australia by the New South Wales Department of Agriculture in 1918. It quickly gained favour in Victoria and South Australia where it is now used for crossing with long-wool × merino half-breds. In New South Wales the breed at first made rapid progress, but in recent years threre appears to have been little expansion in their popularity.

Although larger in frame and heavier in bone, the ryeland resembles the shropshire in conformation. The wool is longer and more stylish than down wool, there being a distinct staple formation. These sheep are hardy, of compact build and produce good quality mutton. They are quicker growing than the southdown, but are slower to mature. They can withstand harder conditions than the downs breeds proper and are footrot resistant, but they are not as fertile.

Points	
Head	Medium size, broad, top covered with wool, no trace of horns.
Face	Medium length (long, narrow face objectionable) showing character and length, dull white colour (not grey, blue or china-white); white hair on dark skin around eyes.
Eyes	Bright and clear of wool; dark skin around nose, nostrils not contracted, broad muzzle, narrow strips of wool down each side of face permissible, but present fashion is for clean faces.
Ears	Medium size, set low on side of head, carried slightly back (not heavy or hanging down) dark colour — either brown shade without wool or hair, or with white hair sometimes spotted with black or covered with short wool. Ears free of wool preferred.

Ryeland

Neck	Strong, broad, set so as to give head a bridled appearance.
Shoulders	Well set, wide, firm and full — no depression behind.
Chest	Broad and fairly deep. Brisket carried well forward.
Back	Straight, level from base of neck to tail setting, which should be broad and set on fairly level with the back.
Ribs	Well-sprung, body well set down giving a good bottom line.
Legs	Full-fleshed and meat carried well down. Dull white colour, single black spot on legs not objectionable; legs placed well apart, woolled well-down. Feet firmly set, white hooves objectionable.
Skin	Healthy pink colour.
Flesh	Even and firm handling all over.
Carriage	Smart appearance, head held high
Covering	Fine quality of 58/60 with a staple length of about 102 mm (4 inches). Densely covered throughout, including belly and purse; free from coarseness, firm handling, close level appearance, free of tip and kemp. (Any trace of grey is a decided objection.) Harsh handle, flat-white colour, very similar to shropshire in style.

The Shropshire

The shropshire was evolved by the introduction of cotswold, southdown and long-wool (possible English leicester) blood into the old more common breed and it now resembles the southdown in appearance and is esteemed for its hardiness, thriftiness and quick maturity. It is larger in the body, deeper, not so compact, with a broader, larger head than the southdown and shows more variation in type. Pastures most suitable for southdown are also best for the shropshire, although the latter is more adaptable to conditions which are less than ideal. The shropshire is a very early maturing breed and will gain weight quicker than the southdown, but the carcass is not as symmetrical as that of the southdown and the quality of mutton, although excellent, is not considered equal to that of the southdown. It carries more wool on the legs and face, and the hair on the face and ears is of a dark brown (nearly black) colour. The fleece is slightly heavier, the staple longer and the fibre stronger than in the southdown.

Points

Head	In proportion to size to body, larger and longer than that of the southdown; the forehead is broad, well-rounded and thickly covered with wool which extends well on the face, although an open face is preferred; broad between the ears and no sign of horns.
Face	Broader and with bolder muzzle than the southdown; soft-black or dark-brown muzzle.
Eyes	Full, medium-sized and bright, showing clear of wool.
Ears	Somewhat short and thick and set on the side of the head, covered with dark brown or nearly black coloured hair.
Neck	Thick, not too long, wide at the base, well-developed scrag, throat clean.
Shoulders	Well set and top level with back; no depression behind blades; fleshy forearms; thick through heart.
Chest	Well rounded, full chest, prominent brisket; width and depth is desirable.
Back	Straight, broad from base of scrag to top of tail; wide over loins; covered

Shropshire

	with firm flesh; longer barrel than southdown, good girth measurement.
Ribs	Well arched, giving good top and well-sprung sides.
Hindquarters	Hindquarters are square and show good width from loin to tail, good full fleshy leg, broad tail, well set on almost level with the back. The quarters are not as well developed as those of the southdown, being shorter and not as wide and inclined to narrow towards the hocks.
Legs	Legs short, straight, strong bone, covered with short wool below the hocks; fairly straight hocks; legs well-apart; hair on legs dark brown, matching face and ears.
Skin	Of a healthy, bright pink, not inclined to blueness.
Flesh	Even and firm handling all over.
Covering	The staple is longer, shows more character and better colour than that of the southdown, reaching 102 to 130 mm (4 to 5 inches) for the twelve months' growth. The fleece is also heavier by approximately 0.5 kg (1 lb), but it is not as heavy as that of the oxford down. It is harsh to handle, the colour is chalky-white with a slight golden tint and the recognised quality number is 56/58s. The fleece is open and spongy with little staple formation and should be free of black fibres except in the extreme skirtings.

The Suffolk

The suffolk was evolved about a hundred years ago by crossing the old norfolk breed with the southdown. The norfolks were a large-framed, strong-constitutioned, prolific breed with black points, but the carcass lacked shapeliness. The modern suffolk reflects some of the good features of both the southdown and norfolk; the southdown accounting, no doubt, for the improvement in shape and mutton qualities, and the norfolk for size, length of legs and neck, hardiness, activity, fertility and distinctive black head and legs.

The suffolk requires good nutritive conditions for best results but they are good foragers, prolific, and a very fast-growing breed. The ewes are excellent mothers and heavy milk producers. Suffolk × half-bred lambs may be marketed at a very early age but as export sucker lambs they are not rated very highly, for at light weights they lack shape, depth of fleshing and the flesh and fat are badly coloured. They mature as relatively heavy weights when they yield a long-bodied, economical carcass. When the lamb is about 34 kg (75 lb) it increases weight at a faster rate than most other breeds and in New South Wales is often carried to heavier weights for the local lamb trade.

The suffolk is a hardy, large-framed (one of the largest downs), very quick-growing breed. In proportion to its body weight, this breed carries a very light fleece (2.5 kg, approximately 5 lb) of harsh, mushy, short wool often with dark fibres in the fleece wool. It has a large black head with large loppy ears free of wool, a long barrel and stands rather high on jet black legs.

Points

Head	Large, hornless, carried high and bare of wool; jet black colour.
Face	Roman nose, long face, moderately fine, especially in ewes.
Ears	Large loppy ears, jet black and covered with hair of fine texture.
Eyes	Bright and full.
Neck	Fair length, full, well set, deep at shoulders.
Shoulders	Broad, inclined to be low at withers.
Chest	Deep, wide and prominent.

Suffolk

Back and loins	The longest back of downs, level — ideally with a good flesh covering.
Tail	Broad and high.
Ribs	Long and well-sprung, with a full flank.
Legs and feet	Straight and jet black in colour, with fine, flat bone; woolled to kness and hocks, clean below; legs all set apart; hind legs well filled.
Belly and scrotum	Well covered with wool.
Skin	Fine, soft and pink.
Flesh	Even and firm handling all over.
Covering	The fleece is light — 2.5 kg (approximately 5 lb). Not as fine as south-down, the quality number being 58/60s, but character and style are similar to the southdown; staple 76 mm (about three inches); there is an absence of staple formation; harsh texture and chalky-white colour. Dark fibres are present to a certain extent in the fleece but breeders try to avoid this fault.

Hampshire Down

The hampshire down dates its origin from the crossing of the old wiltshire horned sheep and the old berkshire knot with southdowns early in the nineteenth century. The first volume of the breed *Flock Book* was published in England in 1890. The breed was introduced to Australia in 1888.

Points

Head	Face and ears of a rich dark brown — approaching to black, well-covered with wool over the poll and forehead. Intelligent bright full eye. Ears well set on, fairly long and slightly curved. In rams, a bold masculine head is an essential feature.

Neck and shoulders	Neck of strong muscular growth, not too long, and well-placed on gradually sloping and closely fitting shoulders.
Carcass	Deep and symmetrical, with the ribs well-sprung, broad straight back, flat loins, full dock, wide rump, deep and heavily developed legs of mutton and breast.
Legs and feet	Strongly jointed and powerful legs of the same colour as face, set well apart, the hocks and knees not bending towards each other. Feet sound and short in the hoof.
Wool	White, of moderate length, close and fine texture, extending over the forehead and belly, the scrotum of rams being well-covered. Suggested wool count 56/58s.
Skin	Pink and flexible.
Objections	Snigs, white specks on face, ears and legs, thick coarse ears, black wool, coarse wool on breeches, protruding or short under jaw, excessive strength and loose skin over neck.

Hampshire Down

Dorset Down

The dorset down breed can be traced back to about the 1820s by crossing berkshire, hampshire or wiltshire ewes with southdowns. The breed was introduced to Australia in 1937.

Points

Type and appearance	Symmetrical, combining size and scope. Carriage and action in walking, gay and vigorous.
Head	Forehead fairly wide, eyes prominent and lively, muzzle moderately

Dorset Down

fine. Throat clearly defined. Ears of medium length, thin, pointed, of fine texture, brown in colour, as is the face, and carried well above the level of the eyes. (Ram masculine in character.)

Neck	Strongly set on broad shoulders and of moderate length.
Chest	Wide, full and deep.
Legs and feet	Wide-apart in front and hind legs straight and strong with flat, clean bone, standing well up on hoof.
Back and flesh	Level with no unevenness behind the shoulders and covered with firm, well-developed flesh with entire absence of unevenness or cushions. Loin long, level and well covered with meat.
Hindquarters	Wide and long, filled with flesh showing a good leg of mutton. Flank both deep and full.
Tail	Large and well set-on, almost level with chine.
Fleece	Of fine texture, dense, covering the whole of the body, down to the hocks and knees, round the cheeks, between the ears and on the forehead, but should not have wool under the eyes, across the bridge of the nose or on the ears. Skin of a delicate bright pink. Suggested wool count 58/56s.
Objections	Horns or evidence of their presence, dark poll, blue skin, speckled face, ears and legs, bad wool, overshort or undershot jaws.

South Suffolk

South suffolk sheep owe their origin to the late Mr George Gould, of Canterbury, New Zealand, who was the first importer of suffolk sheep into New Zealand and who also owned a southdown stud. In 1929 Mr Gould began crossing these breeds and then started to interbreed their progeny. Once the breed was established and was being sought after by prime lamb producers, Mr Gould approached the New Zealand Sheepbreeders' Association for recognition of the breed and in 1940 the council of that body agreed to admit the now accepted breed into their *New Zealand Flock Book*. South suffolks were first imported into Australia in 1946. The South Suffolk Sheep Society was founded and published its first *Flock Book* in 1959. In 1973 the Australian South Suffolk Sheep Society amalgamated with the Australian Society of Breeders of British Sheep and the history of these flocks was first printed in the *Australian Society of Breeders of British Sheep Flock Book* in 1973.

Points

Head	Medium length, with moderately strong muzzle. Hornless. Face colour chocolate, with a topknot of white wool on the forehead, face clean. Ears dark and of fine texture. Eyes bright and full.
Neck	Short and well set in the shoulders.
Shoulders	Broad and oblique.
Chest	Deep and wide.
Back and loins	Back of good length and level. Ribs well-sprung with no hardness on shoulder. Tail broad and well set on.
Legs and feet	Forelegs set well apart — hind legs well filled with meat. Legs chocolate, clean, short and flat boned. Hooves strong and black.
Belly	Well-covered with wool.
Fleece	Short to medium length fleece of medium fine 'downs' type wool, even throughout.
Skin	Fine, soft and pink.
Carriage	Bold and active.

South Suffolk

Disqualifications Horns, or any evidence of horns, blue skin, defective jaw, bad wool and black fibres.

Mountain Breeds: Cheviot

The Cheviot

The cheviot is the most popular mountain breed in the United Kingdom and it originated in the bleak Cheviot Hills and hilly parts about the north of England and south of Scotland, where it remains popular as a dual-purpose breed. While the quality of cheviot mutton is good, it is not up to the standard of the downs breeds, but the flavour is better than that of long-wool mutton.

Cheviots mature later than the border leicester and are very hardy and well adapted to cold conditions, the wool being in demand for the manufacture of cheviot cloth, or for blending in the production of Scottish tweeds. In Great Britain young ewes are bought by farmers on the richer lowlands for cross-breeding purposes, a popular cross being with the border leicester ram. The hairy kempy nature of the wool is a disadvantage in crossing although it must be pointed out that some strains are relatively free from kemp. The cheviot is not as large as the long-wool breeds and is square, compact and low set for a mountain breed.

Point	
Head	Hornless, medium-size, wide forehead, free of wool, covered with fine white hair, sometimes an odd black spot.
Face	Strong Roman nose, wide black nostrils.
Ears	Carried erect, covered with white hair, sometimes an odd black spot.

Cheviot

Neck	Strong, deep, wide shoulders.
Back	Straight and fleshy.
Ribs	Well sprung.
Hindquarters	Good.
Legs	Short, white; free from wool with black hooves. Very active, alert appearance.
Covering	The fleece is 56/50s quality number, full handling and of good staple length, lacking in character. Long fibres, often hairy, protrude from the staple giving a rough appearance, particularly about the breech. Australian breeders are particularly anxious to eliminate hairy and kemp fibres but in the United Kingdom cheviot wools are readily absorbed.

All photographs in Chapter 3 are courtesy of the *Australian Society of Breeders of British Sheep.*

4

AUSTRALIAN AND NEW ZEALAND BREEDS

The Corriedale

The Corriedale was evolved in New Zealand, and stands as a living monument to a few early studmasters. Realising the need for a distinct breed intermediate between the English long-wools and the merino, they started crossing, inbreeding and interbreeding to evolve a dual-purpose breed. In the early 1860s, Mr James Little, manager of the *Corriedale* estate in North Otago, found that he was unable to get satisfactory returns from pure romneys, and he turned to 'inbreeding' half-bred romney–merinos. The results created much interest and no little criticism, but Mr Little was not satisfied, so later, in 1878, on his own property, *Allandale*, North Canterbury, he crossed lincoln rams on to large-framed merino ewes. Unfortunately the type he aimed at was of too high a standard, both in carcass conformation and particularly in fineness of wool. The actual result was again not encouraging as there was a tendency to faults such as droop at the tail, weak shoulders, rough breech and hind legs too close together. Although South Australian merino ewes were introduced to correct these merino carcass faults, the wool standard had been set too high, for he had to throw more emphasis on wool to the detriment of carcass.

Prior to Mr Little's second attempt at *Allandale*, Mr W. S. Davidson, manager of *The Levels*, owned by the New Zealand Australian Land Company, near Timaru, South Canterbury, in 1874 started what may be considered as the foundation flock of the present-day corriedale. One thousand medium-woolled merino ewes were crossed with lincoln rams and the progeny interbred, with heavy culling. In eight years the breed was brought by careful selection to a type somewhat similar to our present-day corriedale. Many corriedale stud breeders in New Zealand and Australia to this day advertise that their stock is pure '*Levels*' blood. They know that their sheep are not pure lincoln–merino, yet these breeders are producing excellent top-grade corriedales. In some cases an infusion of border leicester has been made to improve the rate of maturity, whereas in others romney marsh blood has been judiciously introduced to improve conformation. In 1882 Mr H. A. Corbett of Victoria, and in 1888 Mr A. M. MacKinnon of Tasmania established studs on the same lines as at *The Levels*. But it was not until 1924 that the New Zealand Sheepbreeders' Association accepted the corriedale, and the first *Corriedale Flock Book* was issued.

In 1911 Mr Donald Macfarlane introduced *The Levels* corriedale to Australia and started a stud at Narrabri. The corriedale was quickly recognised as the long-wanted dual-purpose sheep and became very popular; so much so that numerous breeders produced sheep of doubtful blood and inferior type that did much to condemn the breed. However, it had to come, and its popularity has increased until now it is the second most popular breed in Australia. It is also found, still holding to type, in many other parts of the world, including the United States, the Argentine, Kenya, the north of England, Manchuria and Japan. Australia has proved the corriedale an excellent dual-purpose breed and a good mother for fat lamb breeding.

Many of the one and a half million 'corriedales' in New South Wales are sheep that should not be described as such by flock owners, and the majority would be more correctly described as comeback, a common mistake being to continue breeding from stock carrying wool too fine or finer than 56s. These finer-woolled sheep are inclined to revert too much to the merino type of carcass.

Although it must be recognised that there are strains within the corriedale breed, just as there are within the merino breed, it is noticeable that one strain with fine, soft wool will suit certain districts whereas the more robust type, with stronger wool, will

Corriedale

suit other conditions. Where the fine wool type is satisfactory, stud breeders should, and the majority do, take advantage of its excellent sale value and sell to flock or fat lamb breeders.

Being hardy, the corriedale is adaptable to a great range of climates and pastures and although not to be compared with the border leicester or the downs, will throw a high percentage of lambs, twins being common. The corriedale ewe is an excellent mother for lamb breeding, throwing a heavy-weight lamb, which will make up to a good hogget if required. It meets favour with the farmer requiring a dual-purpose sheep, particularly on top-dressed Slopes country or in the marginal areas, as it eliminates the necessity for carrying two distinct breeds.

Points
These are set down in the *Australian Corriedale Sheep Breeders' Association Flock Book*. In general appearance the corriedale should give the impression of being a well-woolled evenly balanced sheep with a remarkably hardy constitution. Being a dual-purpose sheep, due consideration should be given to both wool and carcass.

Head	Hornless, broad, strong, well-woolled, but free from wool blindness. Black or blue spots on the ears are no defect but black or brown spots on the face are, as are brown spots on the ears. Wide-open nostrils, black preferred.
Neck	Broad and forming a good scrag.
Back	From neck to rump, long, level and broad.
Brisket	Deep and wide.
Ribs	Well-sprung and deep.
Hindquarters	Well-apart, deep and broad and well let down towards the hocks.
Legs	Moderate length with good bone, set straight and well-apart. Black markings are considered a defect and brown a serious defect on the wool or hair of the legs.
Hooves	A fair size, well-formed and black in colour preferred.
Wool	The corriedale should carry a heavy, even fleece of good length, dense staple, pronounced crimp and even tip. The quality desired is a long-stapled, dense, bulky 50/56s, but a somewhat lower spinning quality, especially in a ram, is not to be discriminated against. A characteristic of the pure corriedale sheep is the remarkable evenness in length, density and quality of the fleece throughout. In the males, the purse should be covered with wool, which should not be coarse or hairy.

The Polwarth

After the foundation of the dual-purpose corriedale, the polwarth, a fixed come-back type of sheep, resembling a very plain-bodied, plain-fronted, extra long staple-woolled merino, was evolved in Victoria in 1880 to meet the requirements of light wool-growing localities which were too wet and cold for the pure merino. About this time the comeback was becoming increasingly popular with small landholders, particularly in Victoria and Tasmania: this type was more profitable on mixed farms than pure merino, while an even line was difficult to breed and costly to purchase.

In the 1870s Mr Richard Dennis of *Tarndwarncoort*, Victoria, was meeting with success from comebacks founded on lincoln–merino blood, so decided to try to establish a breed carrying the same good qualities of both wool and carcass. In 1880 Mr Dennis fixed a desirable type by mating lincoln–merino ewes with merino rams and line-breeding the progeny. Alexander Dennis of *Eeyeuk*, following his brother's example, also bred a similar type on the same lines. About 1887 Holford Wettenhall of *Carrs Plains* near Stawell fixed a type on exactly the same lines. Mr Wettenhall's sheep became well known in Victoria, Tasmania, New South Wales and South Australia as 'Ideals'. Each of these three studs and others, striving to produce the same type, used *Carrs Plains* merino rams. This, no doubt, was an advantage when it came to the interchange of strains or stud types. In 1919, at a meeting in Melbourne, breeders of this fixed type of comeback decided to adopt the name 'polwarth' as the first experiments were carried out in the County of Polwarth, on the property of *Tarndwarncoort*. The Polwarth Sheep Breeders' Association of Australia lays down the breeding of the polwarth as shown below.

With three-quarters merino blood, it is natural that this breed more closely resembles the merino both in covering and frame than the corriedales. The aim is a higher class of finer wool and consequently less attention was paid to conformation. As a result, the polwarth is entirely different from the corriedale in build, being smaller and not so thick, later maturing and not so heavy. The corriedale is also a better mother. The polwarth carcass is of better quality and size than that of the merino; they are good foragers and light fences will hold them.

Polwarth Sheep Breeders' Association of Australia — Breeding

LINCOLN   ×   MERINO

             Ewe progeny   ×   Merino ram
             Ram progeny   ×   Ewe progeny
                     Progeny

Eligible for flock book — first generation   Ram × Ewe
Eligible for competitive exhibition — third generation   Ram × Ewe
Eligible for inclusion in stud book — fifth generation   Ram × Ewe

Polwarth

Points

The ram and ewe may be horned or hornless, but it is considered inadvisable to cross these two distinct types, for weak horn growth is not favoured. The hornless type is now the most popular.

Long narrow faces are faulty. Fine noses and light hooves are preferred, although black-spotted or mottled nose and dark horns and hooves (from lincoln ancestry) are permissible. Freedom of wool about the eyes is a strong feature. A soft face, free of kemp, whether white or a darker shade, is desirable. Fairly level back and reasonably well-sprung ribs. Drooping rump and narrow or crooked legs are cull points. Length of leg should be moderate but not lanky.

The polwarth has a good top-knot, has less wool about the legs than the merino, and has an open face making wigging unnecessary. It should be free of skin folds. Heavy neck folds are not desirable but the front should be full and well-covered.

Covering is reasonably dense for comeback type. The quality is best kept about 58/60s. The weight of fleece compares favourably with good merino flock sheep. The fleece is even in quality, although a little roughness below the tail is tolerated in otherwise good sheep. The staple is approximately 130 mm (5 inches) long (not less than 102 mm [4 inches] for *Stud Book*) and is full-bodied. The wool shows very strong merino characteristics. It is of good character, soft texture, of a bright colour and is high-yielding. Compared with the merino, it is longer, brighter, lighter in condition but not so dense. In value the fleece is little short of that of the merino.

The breed does particularly well in merino country of a light nature, especially in cold districts with a good rainfall — 508 mm (over 20 inches). They still remain most popular in Victoria and give excellent results in the Mallee country. The breed has a good following in Tasmania and a fair number are used in New South Wales along the Victorian border and South Western Slopes.

The Poll Dorset

The poll dorset is the most popular and widely used breed in Australia for the production of prime lambs. The breed evolved in Australia through the introduction of the poll gene into dorset horn flocks from other poll breeds (corriedale, ryeland). Australian studmasters commenced development of the breed with initial crosses in 1937 and, after a planned upgrading period, the breed was officially registered and the Poll Dorset Association formed in 1954.

Three studmasters who pioneered the poll dorset breed were W. J. Dawkins of *Newbold*, Gawler River, South Australia, T. A. Stuart of *Valma*, Whitemore, Tasmania; and R. H. Wilson of *Kismet*, Howlong, New South Wales.

The poll dorset is a short-woolled, meat-producing sheep. It has a long, well-fleshed body, a pink skin, a dense down type of wool of 30 micron mean fibre diameter, and a white, wool-free face. Fleeces from breeding ewes average 3 kg of fine down-type dense wool.

The breed is noted for many outstanding characteristics. Poll dorsets will mate at any time of the year. They are an early maturing, prolific breed with percentage of lambs tagged to ewes mated being commonly over 100 per cent under good stud management. The ewes have good milking and mothering ability. Lambing in the pure breed is usually concentrated between March and June.

Rams start to work at three to four months of age, but are usually eight to fifteen months of age before they are used for flock joining.

The appropriate average mature liveweight of rams is 90 to 100 kg with ewes in the order of 70 to 80 kg. Purebred dorset have very fast growth rates with yearling rams

Poll Dorset

commonly attaining in excess of 110 kg. Poll dorset cross lambs on good grazing can reach 14 kg carcass weight in nine weeks and up to 22 kg at thirteen weeks. They can produce a young lean lightweight, or go on to higher weights without becoming grossly overfat.

About 75 per cent of all short-wool rams in Australia are dorset and a similar proportion of crossbred prime lambs is sired by rams of this breed. In New South Wales, of the annual drop of short-wool rams, approximately 85 per cent are dorset (17 per cent dorset horn, 68 per cent poll dorset).

The poll dorset is by far the preferred sire for the production of second-cross prime lambs. Poll dorsets are being increasingly used to sire first-cross prime lambs for slaughter and provide a segment of first-cross breeding ewe requirements.

Because of the adaptability of the breed, poll dorset studs have been established in the drier pastoral areas of Australia through to wetter colder or irrigated areas of improved pastures. In New South Wales, the major studs are located in Tablelands, Western Slopes and Riverina areas.

Perendale

The perendale was introduced by Massey Agricultural College of New Zealand (now Massey University) in response to the demands for a sheep that would give a higher lambing percentage on the drier hill country on the east coast of the North Island. Work on the problem was started in the late 1930s and, except for the period of the last war when critical trials were not possible, was continued through the late 1940s and 1950s. Comparisons of the New Zealand romney, the south country cheviot and the cheviot–romney cross were carried out under hard, hill-country conditions in the North Island. Everything considered, the cheviot–romney cross proved to be a very satisfactory answer to the problem. Incidentally, the south country cheviot was used in preference to the north country cheviot since it is very much more of an active hill sheep, fitted as a result of natural selection over a great number of years to utilise the unimproved natural grazing of the border hills between England and Scotland. By 1960 a number of farmers had started to breed the cheviot–romney cross. It was clear that something must be done to standardise type and so the sheep was given the name of 'perendale', and a society was formed and duly registered. The perendale is particularly hardy, very active and easy to muster, labour-saving and particularly adapted to doing well on poor, stemmy feed. It performs equally well on improved pastures at high stocking rates.

Points

General appearance	A medium-sized, white-faced sheep showing unmistakable cheviot ancestry in its open face, the carriage of its head and neck, and its ease of movement and resemblance to its other parent, the romney, in its type of fleece and, to a lesser extent, in its size.
Head	Strong without being over-large, wide between the eyes and broad in the crown, medium top-knot desirable but not essential. Horns not permissible.
Face	Open, white and free of wool below the eyes. Black spots not permissible unless very minute. Nose should be broad, black and with good, open nostrils.
Eyes	Dark, bold and intelligent, full of character.
Ears	Medium length, somewhat erect giving an alert appearance, soft to handle. Black spots, though permissible, are to be discouraged.
Neck	Strong and should hold head well up.
Shoulders	Not too heavy, but well-set. Withers can be a little higher than shoulder blades.
Chest	Wide, with plenty of heart room.

Perendale

Back	Because of the withers being a little higher there is usually a slight slope to the tail-head; this is acceptable.
Ribs	Well-sprung, with a good-length loin.
Hindquarters	Firm and deep, well-set to legs.
Legs	Medium length, clean and bone not too heavy. Rust not permissible.
Feet	Medium size and black.
Skin	Healthy pink colour.
Carriage	Well-balanced, free-moving and stylish.
Fleece	Quality range 50/56s, 48s permissible in older ram, staple should be regular and well-defined, crimp should be round and from butt to tip, showing limited lustre, handle should be full, springy and fairly soft. Free of hair, kemp and black fibres.
Avoid	Abnormally broad and long heads. Heavy or excessive width shoulders and brisket. Short dumpy type of sheep. Pink ears, white or pink nose. Excessively heavy bone and very short cannon bone. Down type wool. Hairy breech. Weak wool on back. Variation in count throughout fleece. Down on pastern.

The Gromark

Mr A. C. Godlee began developing the gromark in 1965 at *Marengo* near Tamworth. It is a fixed half-bred of border leicester and corriedale. It is a large-framed, polled, plain-

Gromark

bodied breed that carries a stylish white fleece with an average fibre diameter of 30 to 33 microns. Annual greasy fleece weights are in the range 4.5 to 5 kg.

The gromark is a dual-purpose breed mainly used for meat production, and is increasingly being used as a self-replacing prime lamb breed. The lamb carcass is longer and has a slightly flatter leg than traditional types, and butchers have noted a lack of waste in the carcass even at a heavier weight. Average realistic production levels for the breed under good conditions would be a weaning weight of 45 kg and a yearling weight of 75 kg.

Because the gromark is predominantly a British long-wool, the breeding season is restricted to January to July. Earlier-mating strains are expected to develop as the breed spreads into areas that require this characteristic. The fertility of the breed is high and average lamb marking performances of 125 per cent are common.

Carpet Wool Breeds

Wool carpets are made from the very coarsest wools and until recently, little or no wool of this type was produced in Australia. New Zealand, by contrast, is a relatively large producer of carpet wool, as its sheep industry is primarily based on the coarse-woolled romney marsh.

In Australia two breeds of sheep are beginning to attract a following for their ability to produce carpet type wool. These are the tukidale and drysdale, both genetic mutants or 'sports' derived from the New Zealand romney. Both types are stocky in appearance and produce an extremely long-stapled fleece of very coarse wool.

The Drysdale

Drysdales have a robust romney-like frame but grow harsh, medullated, chalky-white wool with a fibre diameter of 35 to 40 microns. As the wool grows rapidly, sheep need

Drysdale

to be shorn twice a year. Ewes average 7 to 9 kg a year; rams average to 12 to 13 kg a year. Most males and females are horned. Drysdales are very hardy and have a medium fertility of 100 to 110 per cent. Lambs have an excellent 'meat' carcass.

Drysdale sheep originated more than 50 years ago in New Zealand. Dr Dry, after whom the breed is named, discovered that the very hairy fleece produced by some romney sheep was due to the action of a single major dominant gene.

A feature of the drysdale breed is that it is possible to distinguish the halfbred from the purebred phenotypically. Registered drysdales are all homozygous and carry the dominant carpet wool gene (NN). So on the first cross (drysdale ram over romney ewe) progeny will all be carpet wool-producing sheep (Nn). An advantage of this dominant carpet wool gene is that the heterozygous sheep (Nn) can be distinguished from the homozygous sheep (NN) in the second cross. The identification is known to drysdale breeders as the 'shoulder patch' because in the first weeks of life the heterozygous lamb has a patch of crimpy wool behind the shoulder. This crimped wool can be readily distinguished from the long, straight birthcoat of the homozygous (NN) drysdale lamb.

Numbers of drysdales in Australia have increased steadily. There are now 10 000 romney and drysdale ewes mated to drysdale rams. The breed has spread to New South Wales, Victoria, Tasmania, South Australia and Western Australia. In Australia there are 118 registered stud breeders plus over 200 commercial breeders of drysdale sheep.

Tukidale

The Tukidale

Tukidales have a very robust and hardy romney-style conformation, and both males and females are horned. The breed is of medium fertility (100–110 per cent). Its wool is highly medullated, harsh to handle and chalky white. These qualities provide the necessary resilience and hard-wearing capability needed for carpets. The wool has the added advantage of accepting dyes readily. As with other carpet wool breeds, the wool grows quickly and needs to be shorn twice a year.

The breed originated from a mutant romney marsh ram born in 1966 on *Tuki Tuki*, the property of Mr M. Coop of New Zealand. The ram displayed the medullated, or hairy, characteristics distinctive of carpet wool sheep. In subsequent matings this characteristic proved dominant, and the gene responsible for the medullation is now known as the T gene.

In Australia most tukidales are in New South Wales, especially on the Tablelands, although there are numbers in Victoria, Tasmania, South Australia and Western Australia. As with other carpet wool breeds it is recommended that these sheep be run under reliable feed conditions and where vegetable fault in the wool is unlikely. Such areas have improved pastures and are suitable for prime lamb production.

5

SHEEP AREAS OF AUSTRALIA

I n New South Wales sheep are raised under wide variations of climate and pasture. The state may be divided into five main divisions:
1. Coastal Districts;
2. Tablelands;
3. Western Slopes;
4. Near Western Plains;
5. Far Western Plains.

In the following notes these districts are further subdivided, for within these areas there are large tracts of country which are totally different in nature. High and low altitudes, 'burry and free' grass and herbage, sour and sweet areas may, in numerous instances, lie adjacent to each other. This fact must be borne in mind when an attempt is made to generalise about types of sheep and wool: it is wrong to imagine that one type of wool is common to any district. Within the areas mentioned, particularly the Tablelands and Western Slopes, these is a big variation in climate, this being reflected in a marked variation in the nutritive value of grasses and herbage. Here it must be remembered that the value of pasture as a feed is not so much a question of the quantity of production as the quality measured in terms of digestible nutrients (i.e. the relative amounts of digestible carbohydrate, protein, fat, vitamins and minerals). For instance, the winter pastures of the New England and North West Slopes areas are often abundant but lacking in protein with drastic results for the health and breeding potentiality of sheep.

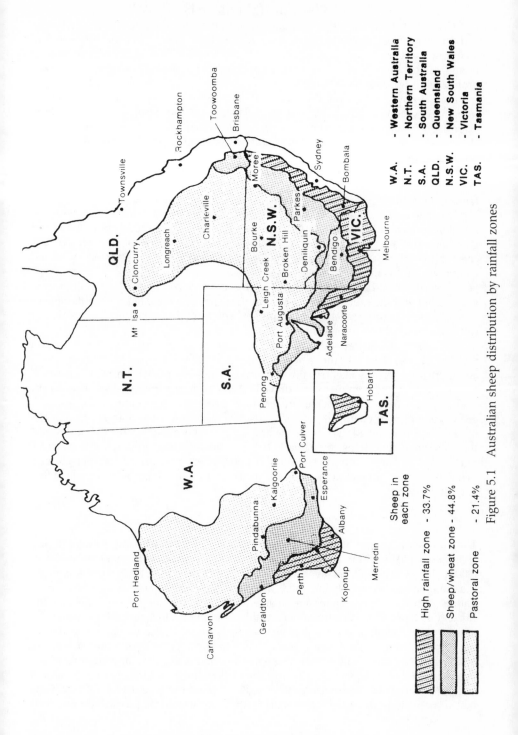

Figure 5.1   Australian sheep distribution by rainfall zones

The Coastal Districts

The Coastal Districts comprise that narrow strip between the Great Dividing Range and the coast. The rainfall is heavy, over 1016 mm (40 inches) and the soil is fertile, except in the County of Cumberland, and this high soil fertility combined with a humid atmosphere leads to a rank growth of grass. Cattle are generally more profitable under such conditions.

Since the days of the early foundation flocks, the coastal region has never been a recognised sheep farming area, but at one time wool-producing sheep were more extensively raised than at present. The luxuriant grasslands stood heavy stocking for a short time, but under heavy rainfall conditions the region proved unsuitable for the economical production of good wool and healthy sheep. Not only did the land quickly become 'sheep sick' but the stimulus to dairy production, largely as a result of metropolitan demands, made dairying more profitable. Now, after many years, sheep farming for mutton production with wool as an important sideline, is increasing. This is due to the demand for mutton from increasing populations in the large towns and a better understanding of the effects of pasture on the health and development of the sheep and advances in the control of disease. While the wool produced on the coast is free of burr and vegetable matter, it frequently shows the effect of humid conditions, and is mixed in type.

The North Coast is particularly rich country of heavy soil type. The conditions are very humid and the rainfall the heaviest in the state. The bush tick is troublesome to sheep in the far north, practically excluding sheep from this area, with the result that cattle are generally more profitable.

The County of Cumberland is the birthplace of the sheep industry, but, except on the alluvial river flats, the land is of poor quality. Very few sheep are depastured here, and there appears little likelihood of expansion. The swampy flats and poor pastures, together with humid summer conditions, lead to a high incidence of disease in sheep.

The South Coast is very fertile and produces better sheep pastures, although rank growth is a difficulty and, in general, cattle are more profitable than sheep. On the foothills and plateaux of the Great Dividing Range on both the North and South Coast there are good conditions, particularly for the British breeds, although disease gives much trouble. The country is made up of poor, steep, hilly areas interspersed with rich flats. A great number of merinos are grazed on the steeper parts and excellent fine wool is produced.

The Tablelands

The Tablelands extend along the Great Dividing Range which runs parallel to and about 30 to 130 km (20 to 80 miles) from the coast. This area is from 600 m to 1500 m (2000 to 5000 feet) above sea level, and enjoys regular seasons of fairly equable climate, the summer particularly being mild, although the winters are rather cold. Under natural conditions the grasses are sparse and short, and as a result, the carrying capacity is light. This country is very good in summer, but generally in winter the grasses are deficient in protein and vitamin A. For this reason, supplementary feeding, top dressing and/or pasture improvement have given spectacular results in building up the sheep's resistance to disease, and in making breeding possible where previously only dry sheep could be run. In parts of the Tablelands with provision of adequate fodder, fat lamb raising has rapidly advanced in recent years. Many areas that respond to improvement (particularly those in reasonable proximity to market)

are jumping in value for the increased carrying capacity following improved management practice. This has more than doubled the return per hectare.

With regular seasons, mild summer heat, cool winter conditions and a good grass
cover throughout the year, there is a freedom from dust, burr and vegetable matter.
Thus the Tablelands are famous for the production of superfine merino wool of good
style, soft and of good colour. It has been found that under the prevailing conditions
the stronger types of fleece do not give sufficient extra weight to compensate for their
lower price per kilogram, and moreover, the relatively moist conditions seem to favour
the development of an undesirable amount of colour. Wools from the cooler region
frequently show a darker tip which sometimes becomes beady, black and may be
difficult to scour.

The Tablelands may be divided into Northern or New England, Central and
Southern, which includes the Monaro area.

The Northern Tablelands

The Northern Tablelands, which are better known as New England, extend from the
heights overlooking the Hunter Valley about Murrurundi and follow the range and
good wool districts through Walcha, Kentucky, Uralla, Armidale, Guyra, Ben Lomond
and Glen Innes through to the Queensland border. The climate is rather severe in
winter, although snow conditions are experienced only occasionally and in this respect the region differs from the Southern Alps. The soils are heavy and inclined to
become sour and hard owing to the severity and length of the winter, but they respond to cultivation. The rainfall is over 762 mm (30 inches) with heavy storms in
summer and light winter rainfall. The bulk of the country is suitable only for grazing,
but lends itself to pasture improvement and this, in conjunction with farming, will
increase its carrying capacity of both merinos and British breeds. The winter feed is
hard and sheep require more than the dry, frostbitten or watery natural grasses. The
poverty of the winter grazing weakens the sheep, lowers their disease resistance, and
affects both lambing percentages and lamb vigour.

Some of the state's best free superfine to fine merino wool of 70s and better is
produced in the New England district around Walcha, Uralla, Armidale and Guyra,
with the quantity of good wool thinning out as we move north to Glen Innes. With the
exception of Glen Innes, the area grazed by these superfine to fine sheep extends well
east of the centres mentioned. Following the mountains north of Glen Innes on the
coastal side, the sheep population quickly thins out, for this is typical cattle country.
Tenterfield and east of Tenterfield is also cattle country, although sheep country
occurs again to the west of the mountains. Wool from the mountain spur out towards
Barraba is similar in type to that described on the New England range. Although much
free wool is produced in the New England district, it is wrong to imagine that it is all
free of vegetable faults, for wool of light vegetable fault is as common as free wool.
Tamworth, Werris Creek and Quirindi should not be included with New England.

The Upper Hunter River Valley is important from a sheep producer's point of view.
This is a comparatively small area of land, within the basin of the Hunter catchment
area, nearly surrounded by mountains, taking in Scone, Muswellbrook, Denman,
Sandy Hollow and Merriwa. It includes a wide variety of pasture land, mostly rich.
The summer temperature is higher, and the rainfall lower, than on the surrounding
Tablelands. Along the creeks, Bathurst and trefoil burr, thistles, and a wide variety of
weed growth is forever present, and in suitable years Bathurst burr and black thistles
extend over the hills. Small areas of trefoil will be found widely distributed throughout

the whole region. The season makes a big difference in the quantity of vegetable matter present in the wool from this district. On the hills round Cassilis and Merriwa, an excellent type of mostly free, fine merino wool is grown. The rich fertile flats round and east of Merriwa are first-class fat lamb country and throughout the whole Hunter River basin cattle are fattened. The dairy industry also extends up from the coast.

The Central Tablelands

The Central Tablelands extend from the west of the Hunter River Valley to a little south of Sydney. They are very wide in parts and scattered in nature, extending well into the drier western areas as the Warrumbungle spur, out to Coonabarabran. Again it is difficult to separate the high country around Bathurst, Orange and Tuena from the Central Tablelands.

The elevation of the Central Tablelands is not as high as either the Northern or Southern Tablelands and, with the exception of some of the more easterly side, which is of no interest to the sheep farmer, this division is less rough and the climate is milder than either to the north or to the south.

Throughout the wide area of the Central Tablelands there is a great variation in soils, although they are generally of a light nature. The contour of the country is also variable, ranging from virgin wooded peaks to ideal undulating grazing hills, interspersed with small rich flats. As there is a great mixture of free and burry, rich and poor country at different elevations, it is little wonder that here we find a great diversity of merino wool types, fat lamb and dual-purpose breeds of sheep. The rainfall, varying from 634 to 762 mm (25 to 30 inches) comes from all quarters evenly throughout the year, and is regular from year to year, making the whole admirably suitable for sheep grazing. Throughout the hills near Cassilis, Coolah, Gulgong and Mudgee, a large percentage of the clip is high grade superfine to fine free merino wool, a little shorter than the Goulburn–Yass wools of the same quality. This type is to be found sprinkled throughout the higher areas of the whole of the Central Tablelands, clips of fine high-grade wool being dispatched from the hills round Rockley. The balance of the merino wool from the Central Tablelands is mostly 64/70s, carrying light vegetable fault.

The Southern Tablelands

The Southern Tablelands extend along the highlands of the southern chain of mountains from north of Taralga, passing through the Crookwell, Goulburn, Queanbeyan, Captain's Flat, Cooma and Bombala districts to the Victorian border. The region also includes the western spurs of snow lease country around Mt Kosciusko. To the north there is light carrying country including much quartz, with a little sandy country interspersed with rich flats around the Taralga–Goulburn area. To the south, there is a wide variation of soils, some being black and heavy and others light and in places very stony. Much of the Southern Alps is heavily timbered, rugged and the rainfall is comparatively light — approximately 660 mm (26 inches), the bulk of which falls in the winter months. The country west of Goulburn, including the famous Yass district, is excellent superfine to fine wool-producing country, whereas from the Australian Capital Territory southwards a broader wool of 64/70s is found. The bulk of the wool from the whole area is free, or nearly free, of vegetable fault, with an increase of burr content in the extreme south or Cooma area. The richer pastures of the northern area of the Southern Tablelands — Taralga, Crookwell, Goulburn and west thereof — and the Australian Capital Territory, particularly when they have been improved, are suitable for fat lamb raising. A small number of late season fat lambs are also produced around Bombala to the south.

The Western Slopes

The Western Slopes, which are of varying width, extend from the Queensland border to the Victorian border, running parallel to the mountain range. This strip, although varying in character and rainfall, embodies the best and safest wheat lands, and the bulk of the first-class lamb-producing country in the state. As the area is a rich agricultural one, with hot summer temperatures, the wool from the slopes contains varying degrees of grass seed, burr and dust contamination. The slopes are usually divided into three sections:
1. North-Western Slopes;
2. Central-Western Slopes; and
3. South-Western Slopes.

The North-Western Slopes

The North-Western Slopes, bounded by the Nandewar Range to the south and Pallamallawa to the west, and taking Inverell as the centre, is a pocket of rich agricultural and grass land. This area includes Warialda, Bundarra, Bingara and Emmaville, where mixed sheep and cereal farming may be profitably undertaken. The rainfall is high for Slopes country, being appoximately 711 mm (28 inches), much of which falls in heavy summer storms.

The bulk of the merino wool is 64/70s of light-to-medium vegetable fault, mostly grass seed. Around and to the west of Inverell and Glen Innes is first-class fat lamb country. The border leicester–merino cross is popular in the North-West and the use of this cross extends far west of the area under review. The Liverpool Plains between the Nandewar and Liverpool Ranges form a rich basin of agricultural and fat lamb country surrounded by excellent fine wool-growing country. Quirindi, Werris Creek, Tamworth, Manilla and Gunnedah are all situated in first-grade fat lamb country and this class of country, including the best of lucerne flats, extends even further west along the Namoi River into the marginal area towards Wee Waa. With extensive fat lamb breeding, much crossbred wool is dispatched from this area, but the bulk is 64/70s merino with free, finer wool from the surrounding higher levels. Barraba, although not on the low levels, is on the northern outskirts of the Liverpool Plains area and this district has always been famous for its free, high superfine to fine merino wool. Wool from the Liverpool Plains as a whole carries much vegetable fault ranging from medium grass seed to heavy burr.

Central Western Slopes

This area is bounded by Dubbo, Parkes and Forbes on the west, Grenfell and Cowra to the south and follows a line through Carcoar and west of Blayney, Orange, Gulgong and Dunedoo on the eastern side. The section is typical of the Slopes in contour, being undulating to hilly, mostly of a rich red soil, with extensive river lucerne flats and consists of excellent sheep/wheat and fat lamb country. The rainfall is regular, approximately 584 to 635 mm (24 to 25 inches) and serious droughts are rare as it catches both the summer rains from the north and the winter rains from the south, making it one of the safest wheat-growing districts in the state.

There is a little nearly-free fine wool from this area, mostly from east of Wellington, and the hills about Manildra, but the big bulk of the merino is 64/70s, carrying seed and medium burr with a fair amount of red dust from the agricultural paddocks. There is a substantial production of crossbred wool. Practically the whole of this area is noted for its first-grade fat lambs, especially Cowra, Canowindra, Eugowra and Forbes along the Lachlan, while other centres away from the river flats market good quality lambs. Crossbreds and corriedales are popular and English rams such as border leicester, romney marsh, dorset horn and southdown are extensively used. To a lesser extent, ryeland and polwarth are found in this territory.

South-Western Slopes
This area is taken as a strip south of Young to the Victorian border, with a western boundary roughly on a line from Grenfell to Corowa, the eastern boundary being the Southern Tablelands. This territory experiences southern or winter rains with a total annual fall of 482 to 660 mm (19 to 26 inches) which is very suitable for wheat and fat lamb raising, even in the lighter rainfall areas.

The eastern portion, east of Young and Harden and including Boorowa and a portion of the famous Yass district, Gundagai and Tumut, produce free, well-nourished merino wools, although as we move south from Yass to the richer and more broken country the vegetable fault increases. The merino wool from the western side, by comparison, is of broader type 64/70s and leaning towards the medium 64s, mostly free, although dusty in years of light rainfall. The western side of the area is typical rich wheat country and particularly suited to crossbreeding in conjunction with wheat farming. The district referred to includes Cootamundra, Junee, Wagga, Henty, Culcairn and Corowa. Fat lamb breeding is a big industry on the whole of the South-West Slopes. The border leicester, romney marsh, corriedale and even polwarth are popular in the wheat areas and large numbers of second-cross lambs sired by dorset horn, southdown or ryelands come from this district, particularly near the rivers and from the eastern portion.

The Near Western Plains or Marginal Country
West of the Slopes districts running from north to south is an area that may be described as the Mid-West, Central West, Near West or marginal country. This extends approximately to a line extending southwards from Garah in the north, through Burren Junction, Coonamble, Warren, Tottenham, Condobolin, Hillston, Griffith, Deniliquin, and includes the Wakool Irrigation District on the Victorian border near Moulamein.

Throughout practically the whole of this area wheat is grown, although it is a much more precarious undertaking than in the Slopes districts. Large-framed wanganella-type merinos carrying 60/64s wool suit this area and a large percentage of this type of ewe is mated with border leicester rams to produce fat lambs or half-bred ewes for fat lamb breeders in Victoria, and the better fat lamb districts of New South Wales. In the north and central sections the wool clip carries medium to heavy trefoil burr and in the south light to medium burr. With the dry, hot summer conditions, especially where the paddocks are cultivated, the wool is dusty but seldom is it necessary when skirting to remove the back wool of fleeces on this account.

Between the marginal districts and the Far West is country too good to be classed as dry, dusty, far west plains. This area extends west from the marginal districts to a line approximately running through Brewarrina to east of Byrock and west of Nyngan, east of Nymagee, to Hillston and down to Balranald. This country, with a rainfall of approximately 381 mm (15 inches) and a carrying capacity of one sheep to one to two hectares, is the home of the wanganella type of merino. A wide variety of native grasses and herbage plants and edible trees flourish in this area. Pine grows freely on the lighter soils but on the better soils are found yarran, ruddah, myall, belah, wilga, box and white gum. Red soils predominate but, particularly towards the north and about the rivers, the red is mixed with heavier black soils.

The summer is hot and dusty with a mild autumn, winter and early spring. The grass and herbage in this area are remarkably nutritious and the stock very healthy. Although the rainfall is light, with judicious stocking and care at the time of grass seeding, a fair cover can be maintained. After autumn and winter showers, the light red soils quickly cover with crowsfoot and trefoil and the faster growing grasses, but this is quickly burnt off by the hot windy weather of late spring. The heavier black soils are slower to respond and require the heavier rains to produce the more permanent and coarser grasses, and these, together with dry trefoil burr, last through the summer and autumn. In this area every means of conservation of fodder and edible trees and the prevention of sheep erosion demands close attention. The bulk of the clip from this area is of 60/64s quality number, carrying medium to heavy burr and a fair amount of dust. Many of the leading medium to strong-woolled merino studs are located here, for it has proved very suitable for the development and maintenance of healthy large-framed merinos of this type.

The Far Western Plains
This large expanse of the state could be roughly divided into areas east and west of the Darling River, for the aridity increases in the westerly direction, with a corresponding drop in stock-carrying capacity. In the Far West, while the natural herbage is erratic in growth, it is very nutritious. In dry periods when herbage and coarse grass is not available, stock are dependent to a great extent on saltbush and other edible shrubs.

On the eastern side of this area, a fair balance of native grasses, herbage and, in parts, saltbush in its many varieties, permits a light constant stocking up to two hectares per sheep, but even here the stocking is seldom constant. Towards the western border the stocking is so irregular that a stocking rate is difficult to quote, possibly down as low as twenty hectares per sheep. The difficulty in the more arid areas is that leaseholders stock up in good years, and as dry times approach the sheep eat out and kill the more permanent shrubs and open the way for erosion, so lowering the carrying capacity for future years. The northern portion benefits from the Queensland summer showers and so the carrying capacity in this area is better.

The great proportion of the land west of Nyngan is Crown Lease, which was held originally in large areas, but which now is slowly being broken into maintenance areas, ranging from 2025 hectares (5000 acres) east of the Darling to 20 250 hectares (50 000 acres) in the more western areas.

Far West is merino wool country, with the bulk of the clip falling into the 64/60s group. The wool on account of the hot dry conditions is dusty or sandy and the density over fineness has to be closely watched to avoid as much as possible an open perished tip. Trefoil burr is seldom found, but the clip as a whole can be classed as of light vegetable fault on account of the seed and other larger vegetable matter. The sheep generally are not as large in frame as those on the eastern side of the Western Plains but all sheep are good travellers and are particularly free from disease and parasites. Lessees in the Far West are regular buyers of wanganella and South Australian strains from the more favoured stud farm areas of the Western Plains.

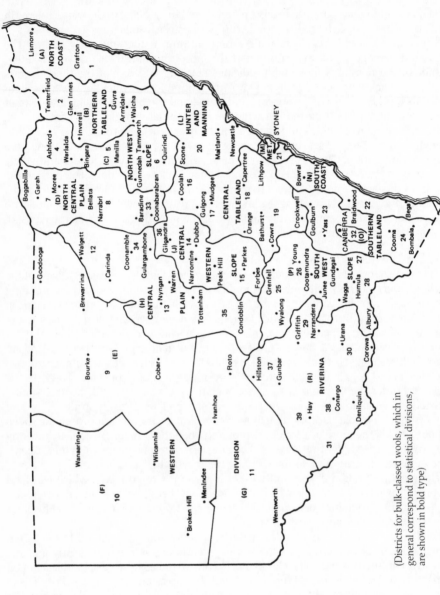

Figure 5.2   AWC wool areas of New South Wales

(Districts for bulk-classed wools, which in
general correspond to statistical divisions,
are shown in bold type)

Key to AWC wool areas
New South Wales

Wool area	Bulk code	Division	Shires and larger municipalities within statistical divisions	
1	A	North Coast	Shires:	Bellingen, Byron, Copmanhurst, Gundurimba, Kyogle, Macleay, Nambucca, Nymboida, Terania, Tintenbar, Tomki, Tweed, Woodburn, Coffs Harbour, Maclean, Ulmarra
2	B	Northern Tablelands	Shires:	Macintyre (part), Severn, Tenterfield
3	B	Northern Tablelands	Shires:	Walcha, Dumaresq, Guyra, Uralla
4	C	North-West Slopes	Shires:	Ashford, Bingara, Macintyre (part), Yallaroi
5	C	North-West Slopes	Shires:	Barraba, Cockburn, Mandowa
6	C	North-West Slopes	Shires:	Liverpool Plains, Peel, Nundle, Tamarang, Murrurundi
7	D	North Central Plains	Shires:	Boolooroo, Boomi
8	D	North Central Plains	Shires:	Namoi
9	E	Western Division (north-eastern portion) (part)	Counties:	Barrona, Booroondarra, Canbelego (part), Clyde (part), Cowper, Culgoa (part), Gunderbooka, Irrara, Landsborough, Mouramba, Rankin, Robinson, Woore, Yanda
10	F	Western Division (north-western portion)	Counties:	Delalah, Evelyn, Farnall, Fitzgerald, Killara, Livingstone, Menindee, Mootwingee, Poole, Tandora, Thoulcana, Tongowoko, Ularara, Werunda, Yancowinna, Yantata, Young, Yungnulgra
11	G	Western Division (southern portion)	Counties:	Blaxland, Caira (part), Franklin, Kilfera, Manara, Mossgiel, Perry, Taila, Tara, Walgeers, Wentworth, Windeyer
12	H	Central Plains	Shires:	Walgett (part)
13	H	Central Plains	Shires:	Bogan (part), Warren
14	J	Central Western Slopes	Shires:	Talbragar, Timbrebongie, Wellington (part), Coolah (part)
15	J	Central Western Slopes	Shires:	Molong, Goobang, Jemalong, Boree
16	K	Central Tablelands	Shires:	Coolah (part), Merriwa
17	K	Central Tablelands	Shires:	Wellington (part), Cudgegong, Rylstone
18	K	Central Tablelands	Municipalities: Shires:	City of Blue Mountains Canobolas, Turon, Blaxland, Colo

Key to AWC wool areas New South Wales (cont.)

19	K	Central Tablelands	*Shires*:	Waugoola, Oberon, Abercrombie, Crookwell, Lyndhurst
20	L	Hunter and Manning	*Municipalities*:	City of Greater Cessnock, City of Maitland
			Shires:	Scone, Gloucester, Manning, Hastings, Stroud, Port Stephens, Dungog, Patrick Plains, Muswellbrook, Lake Macquarie, Wyong, Gosford
21	M	Cumberland, including Metropolitan	Statistical Metropolitan Area of Sydney	
			Municipalities:	Campbelltown, Camden, Liverpool (part outside Metropolis), Penrith, Windsor
			Shires:	Baulkham Hills (part outside Metropolis), Blacktown (part outside Metropolis), Hornsby (part outside Metropolis)
22	N	South Coast	*Municipalities*:	City of Greater Wollongong
			Shires:	Eurobodalla, Imlay, Mittagong, Mumbulla, Shoalhaven, Tallaganda, Wingecarribee, Wollondilly
23	O	Southern Tablelands	*Shires*:	Goodradigbee, Gunning, Mulwaree, Yarrowlumla
24	O	Southern Tablelands	*Shires*:	Bibbenluke, Monaro, Snowy River
25	P	South-West Slopes	*Shires*:	Bland, Weddin
26	P	South-West Slopes	*Shires*:	Narraburra, Jindalee, Demondrille, Burrangong, Boorowa
27	P	South-West Slopes	*Shires*:	Gundagai, Illabo, Mitchell, Tumut
28	P	South-West Slopes	*Shires*:	Holbrook, Hume, Kyeamba, Tumbarumba
29	R	Riverina	*Shires*:	Coolamon, Leeton, Wade, Yanco
30	R	Riverina	*Shires*:	Berrigan, Corowa, Culcairn, Jerilderie, Lockhart, Urana
31	R	Riverina	*Shires*:	Murray, Wakool (part County Caira)
32	O	ACT	Australian Capital Territory	
33	D	North Central Plains	*Shires*:	Coonabarabran
34	H	Central Plains	*Shires*:	Coonamble
35	H	Central Plains	*Shires*:	Lachlan
36	J	Central Western Slopes	*Shires*:	Gilgandra
37	R	Riverina	*Shires*:	Carrathool
38	R	Riverina	*Shires*:	Conargo, Windouran
39	R	Riverina	*Shires*:	Waradgery, Murrumbidgee (part County Caira)
40	E	Western (north-east portion) (part)	*Counties*:	Culgoa (part), Finch, Narran
			Shires:	Brewarrina, Walgett (part)

Principal Sheep and Wool Types of Victoria, Queeensland, Tasmania, South Australia and Western Australia

Victoria

The sheep and wool industry in this state dates from about 1834. The merinos were founded on Macarthur stock. Victoria produces some of the choicest wool grown, particularly in the Western Districts, and usually commands the highest prices for greasy wool sold in Australia. Although the country and climatic conditions in this state vary considerably, they are more even than in New South Wales.

The Western Districts are eminently suited for the production of fine merino wool of very high spinner properties. Fineness of fibre is not the only outstanding feature of the wool; the colour, softness of texture and character, together with cleanliness and light conditions, are just as prominent. These latter properties are revealed in all types of wool — comeback and crossbred as well as merino. For these reasons, Victorian wool is very high yielding.

Crossbreeding has developed to a much greater extent in this state and quite a large proportion of the flocks are of this class of sheep. The clip consequently includes a large percentage of crossbred wool.

Queensland

This state possesses great potential sheep-breeding and wool-growing possibilities. Its rich pasture and sound conditions combine to rank it as one of the greatest pastoral states of Australia. The sheep and wool industry was established soon after 1829. When the Darling Downs was discovered by Cunningham, the squatters of New South Wales soon extended their activities into what is now Queenland. The flocks, consequently, are also founded on Macarthur blood. Although on the whole much warmer than New South Wales, the country is nevertheless particularly well-adapted for the merino. Given good seasons, a wonderful growth of excellent, indigenous grasses flourishes practically throughout the year. These grasses, particularly Mitchell and Flinders, provide excellent feed for merino sheep.

Except for the south-eastern corner (the Tablelands), most of the pastoral country is given over to wool growing — entirely merino. The wool generally is of a good useful type ranging from strong to fine with the bulk medium. The state produces few, if any, super clips comparable to the New England wools; nevertheless, the wool generally is held in very high esteem. It develops an excellent colour in the scoured product. Owing to the dry conditions, the wool frequently carries much foreign matter, with the result that it does not yield so well as the southern wools.

Tasmania

This state produces some of the finest wool in Australia. In many cases it compares favourably with the finest Mudgee types of New South Wales. The industry was established as far back as 1825 on Macarthur stock. Subsequent importations of the finest strains of Spanish merino, both of the negretti and saxony types, assisted the development of this superfine fibre. For many years Tasmania was regarded as the merino 'stud farm' of Australia. Quite a large proportion of the stud rams bred on the island found their way into the larger flocks of the mainland, especially those of New South Wales, Victoria and even Queensland. This outlet was held until the present type of plain-bodied sheep began to come into favour early in this century. In recent years the demand for Tasmanian stud merinos has declined, with the result that many of the best studs of Tasmania have disappeared and have been replaced by crossbreds.

Although quite a large proportion of the clip now consists of crossbred wool, some very high prices are still obtained for the finest merino.

South Australia

The sheep and wool industry of this state was founded on Macarthur blood about 1837. Although the pastoral industry is still the major one, this has been replaced in many districts by agriculture, especially wheat growing. The natural conditions prevailing throughout the pastoral areas are very similar to those of the western districts of New South Wales, but the rainfall is, if anything, lighter. These conditions favour the strong-woolled type of merino which predominates. South Australian breeders have for many years specialised in this particular type of merino, and some very high-class studs are to be found in this state. The type is more robust, both in conformation and covering than the New South Wales strong-wools. These studs are among the foremost of the Commonwealth. The wool of this state consists generally of a medium to strong and often stronger. Noted for its great substance, draught and body, its tensile strength is also regarded as very good. Owing to the dry conditions and the colour of the soil, the wool generally carries much red dust.

Western Australia

The industry commenced in this state about 1829, the sheep introduced being mostly Macarthur blood from New South Wales and Tasmania. The state is well suited for merino sheep. The conditions are comparatively dry and sound but vary with the result that the wool itself shows considerable variation. Many good clips are produced, especially in the south-western area, and the quality is generally of average to good medium type. Most of the wool carries much red dust which affects its greasy value.

6

BIOLOGY OF THE SKIN AND THE WOOL FIBRE

I n general, the fibres forming the covering or fleece of the sheep are similar to those of other hairy-coated mammals. They are composed of long, splinter-shaped cells closely cemented together, and are derived from special structures (follicles) in the skin. In a general sense they are all hair fibres, but those grown by the sheep are referred to as 'wool'. Some breeds of sheep in Asia and in Africa do not possess a fleece in the accepted sense, but have a hairy coat similar to that of cattle and horses. Most sheep of wool-growing importance have fleeces in which the fibres grow continuously as, for example, the merino and breeds derived from it. However, other breeds, such as the mountain long-wools of Great Britain, grow fleeces which are more akin to most animal species in being shed annually, either wholly or partly.

The follicles are located just beneath the outer skin surface of the sheep and are fed from the bloodstream. They may be classified into two main types: primary follicles, that is, follicles with a closely associated 'sweat' gland (sudoriferous gland), and usually a small follicle muscle (*Arrector Pili*), and secondary follicles without a closely associated sweat gland. Both types of follicles have a wax gland (sebaceous gland).

The Mature Non-Meduallated Wool Follicles

Figure 6.1 shows a vertical section of a mature wool follicle containing a non-medullated wool fibre. The figure is idealised in some respects because most wool follicles are not as straight as the one illustrated. It is most important to notice that the papilla is usually directed towards one side of the follicle, which thus has the shape of a golf club. Further, the 'shaft' of the follicle and fibre is usually curved in a longitudinal direction. It is possible that this orientation of the papilla and the shape of the follicle are connected with the formation of crimp in the fibre. The sebaceous gland

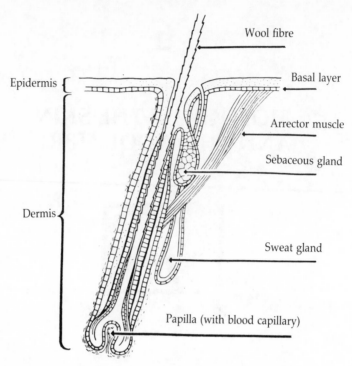

Figure 6.1   Diagram of a wool follicle. The figure shows a wool fibre growing out of its follicle.
Source: World Book of Wool by the International Wool Secretariat. Courtesy CSIRO.

normally has two lobes separated from one another in the horizontal plane, in such a way that only one lobe is seen in a vertical section as in Figure 6.1. Both lobes are visible in horizontal section through the skin, such as in Figure 6.2.

The sweat gland has, in most sheep breeds, the shape of a thin elongated bag, although in the merino it may be coiled at the lower end. It opens into the neck of the follicle just above the sebaceous gland. In this respect the structure of the wool follicle is different from that of the human hair follicle, in which the sweat glands open directly to the surface of the skin and appear to have no definite association with the hair follicles. In the sheep the sweat gland lies between the two main lobes of the sebaceous gland (see Figures 6.1 and 6.2).

In general the primary follicles tend to produce the medullated type or hairy fibre, and in some cases, kemp fibres. This medulla may vary from a series of small isolated cavities to a continuous canal. These fibres, though relatively always fewer, are often the more prominent fibres in the fleece. They tend to be less regularly crimped, are thicker, longer and less circular in cross-section. The kemp fibre is a notable exception, being usually very thick, very short, often somewhat flattened in cross-section with a tapering tip and a flat white colour. The secondary follicles tend to produce the non-medullated type of fibre. Generally these fibres are more numerous, are mostly well-crimped, fine, short, nearly circular in cross-section and in some breeds — such as the merino — are almost the only type present. In the merino and certain other breeds, non-medullated fibres of the secondary type tend to grow equally from both primary

Sebaceous gland (two lobes)

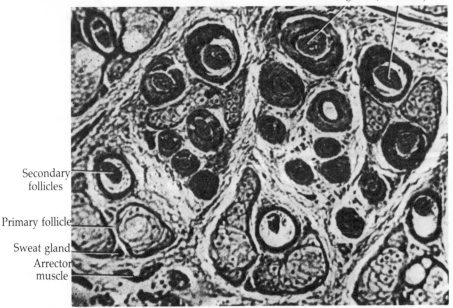

Secondary follicles

Primary follicle

Sweat gland

Arrector muscle

Figure 6.2   Horizontal section of skin showing primary and secondary follicles. Horizontal section magnified 200× of the skin of a romney lamb showing a follicle, accessory structures and associated secondary follicles.
Source: Wool Science Review by the International Wool Secretariat. Courtesy CSIRO.

and secondary follicles. The follicle group in sheep consists typically, but not exclusively, of a basic group of three primary follicles and a variable number of secondary follicles, these primary follicles being arranged in groups of three known as 'trios'.

All primary follicles are distinguished by certain accessory structures, namely:
1. a simple coiled sudoriferous or 'sweat' gland;
2. an involuntary muscle, the *Arrector Pili* muscle;
3. a sac-like sebaceous or wax gland.
Of these structures (1) and (2) belong exclusively to primary follicles and form a clear and definite means of distinguishing them.

The secondary follicles appear after the completion of the 'trio' group of primary follicles, and relatively late in the pre-natal development of the young lamb. The secondary follicles have only one type of accessory structure associated with them, that is, the sebaceous or 'wax' gland — which is usually smaller than those attached to primary follicles, and more simple in shape (see Figure 6.2). The average number of follicles in each cluster varies from one part of the skin to another. The cluster is numerically small at birth but increases rapidly during the first year and may not be mature until the animal is one or two years old, each cluster consisting of one primary follicle and its variable number of secondaries characteristic for the body region, individual sheep, breed or strain.

The smallest clusters are found in the skin of the groin where they usually consist of primary follicles alone. The largest clusters are generally found along the skin of the

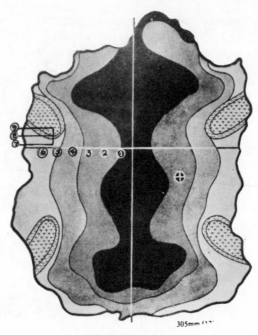

305mm (1 7'

Figure 6.3   Chart of a tanned merino skin showing the main variations in density of the wool follicle population. The denser the population, the darker the shading. The numbered points on the left side indicate the position at which the photomicrographs were taken. These are shown in Figure 6.4.
Source: Austin, *Merino, Past, Present and Probable.*

back, where, in merinos, up to 50 or more secondary follicles may be found beside each primary (Figure 6.3). Average clusters are usually found on the skin of the side. Comparing side regions of different breeds, the merino is found to have the largest clusters and such breeds as the lincoln and the ryeland are among the smallest. Most breeds and species of sheep appear to fall within the range of three to fifty secondary follicles to each primary follicle.

The primary follicles are the first developed in the embryo and this usually takes place between the thirty-fifth and fortieth day of gestation. As the embryo develops, secondary follicles appear and these are added to even after the lamb has been born, but at a decreasing rate as the lamb matures. The primary follicles produce fibres referred to as 'mother hair', these being of a heterotypic type when the lamb is born, but which are usually shed to be replaced by wool fibres.

Fibre Population

The least obvious difference between individual sheep and the most difficult to measure or estimate is the total population of fibres in the skin. The follicle groups, according to type and breed of sheep, vary greatly in number, but follow a system of order and regularity in the individual sheep. The merino lamb at birth may have fifteen to twenty million fibres in its fleece, (i.e. fibres which have emerged from the skin follicles). At twelve months of age, the number of fibres in the fleece may have increased four or five times, that is to sixty to 100 million. On the other hand, it should be

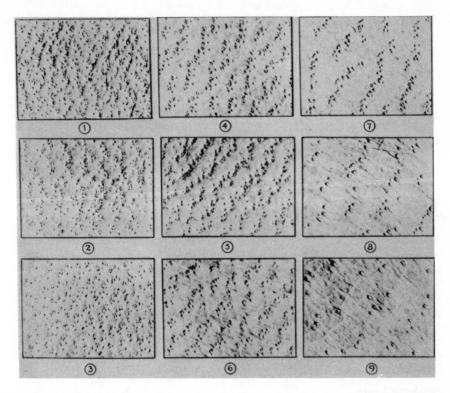

Figure 6.4   Photomicrographs demonstrating the main variations in the wool follicle populations on the tanned merino skin.
Source: Austin, *Merino, Past, Present and Probable.*

remembered that the number of fibres per square centimetre at birth is very much greater than at twelve months of age, values being reduced. For example, from between 10 000 and 16 000 per square centimetre to between 5500 and 10 000. This apparent anomaly is due to the wool-growing skin area expanding with growth at a greater rate than that at which new fibres are added to the fleece. Tests reveal that in merino sheep the variation in follicle population is from 2300 to 8500 and in some cases to as many as 13 000 follicles per square centimetre of skin. The general average for Australian merino would be about 5000 follicles per square centimetre of skin.

Environmental conditions during growth may affect the fibre population of the skin, but available evidence suggests that of all fleece's characteristics, it is probably the most stable. This means that complete suppression of follicle activity is apparently one of the last changes which can be produced in a sheep by external factors.

The Wool Fibre and its Structure

Under a purely practical examination, wool will be found to differ from all other fibres, including hair, in crimp formation, texture, elasticity and staple formation.

Crimp Formation

Wool, when viewed in staple formation, presents a wavy or undulating formation. These waves, or corrugations, are known as 'crimps'. They vary in size, formation and number (in a given length) according generally to the fineness and character of the wool. They are not found in the same regular form of development in other animal fibres, where in most cases the fibres are quite straight. Vegetable fibres show no crimpy development.

Texture
Wool is also readily distinguished by the texture or handle. Generally it handles soft, warm and full. Some of the finest fur fibres handle with a texture closely allied to some wool fibres, but most animal fibres handle with a comparatively hard, unyielding feel. Vegetable fibres have a lean, firm texture; they are also cold.

Elasticity
Wool, due to its chemical composition, possesses a higher degree of elasticity and resilience than any other fibre used for textile purposes. This is chiefly responsible for its fullness of handle. An equal degree of elasticity will be found in some of the finer fur fibres. Silk and nylon are also somewhat elastic, but with most other textile fibres, little or no elasticity is present.

Staple Formation
In mode of growth, wool is also quite distinct. The same characteristics are not found even in other animal fibres in this respect. An examination of a sample of wool will reveal the marked disposition of wool to grow in clusters or groups of fibres. These clusters are known as 'staples'. These staples are in turn held together by fibres which pass from one staple to another and are known as 'binders'. Through the agency of these fibres, it is possible to remove the whole fleece in an almost unbroken form. Other animal fibres do not group together but have a more independent growth, and fall apart when clipped.

The Wool Fibre
From the following illustrations it can be seen that the wool fibre consists of thousands of tiny units within units. A wool fibre is even finer than a strand of hair and it is easy to understand just how small are all these units which go together to make it.

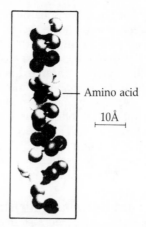

Figure 6.5   Alpha helix — Basically the wool fibre is a very complex unit. Research indicates that it is probably constructed as follows: chemically the wool fibre is made up of units which are linked together to form chains. These chains are not straight but form spirals like corkscrews. This is known as an alpha-helix.

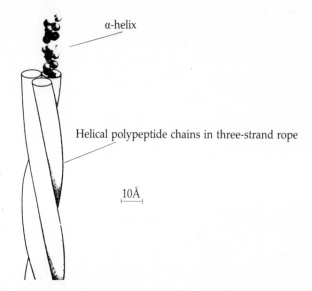

Figure 6.6   Protofibril — Three of these spiral chains wrap around one another to form another unit known as a protofibril.

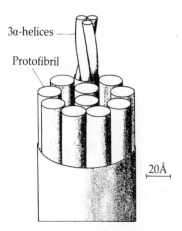

Figure 6.7   Microfibril — Eleven of these protofibrils then form another unit which is known as a microfibril.

Physical Structure of Wool

A microscopic examination reveals a very complex structure compared with other textile fibres, and important distinguishing features from allied fibres, such as fur and hair, may be readily detected. The most important features of the structure are the density and form of the composite cells, and the mechanical arrangement of the surface scales or flattened cells. In the latter instance, a means is not only provided for distinguishing wool from hair and other animal fibres, but moreover, different breeds

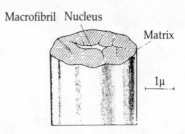

Figure 6.8   Cortical cell — Hundreds of microfibril units are cemented together with the matrix, to form larger units called macrofibrils. The cortical cell consists of a number of macrofibrils.

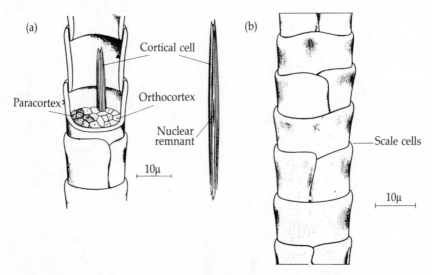

Figure 6.9   Wool fibre — (a) The cortical cells form the inner major part of the fibre which is known as the cortex. This is in two parts with different properties — the orthocortex and the paracortex.
(b) Finally all these different units are enclosed within an outer layer of scales to form the wool fibre.

and types of wool show considerable variations in this respect. This factor is partly responsible for the great variability of manufacturing quality existing in wool.

The true wool fibre is built up of innumerable minute cells. These are formed from substances fed from the bloodstream into the base of the papilla in the wool follicles. Microscopic examination of the cross-section of a wool fibre reveals the existence of these cells closely packed together, with two distinct layers. The larger central layer, or cortex, is made up of more or less round, long, tapering spindle-shaped cells, tightly packed together. Certain faulty wools, and most hairs, possess a central core of cells or medulla. Such wools are known as 'medullated' wools. The presence of a medulla detracts from wool's superiority in colour, softness and elasticity. The thinner outer layer of cuticle cells of the wool fibre are tightly compressed, flattened, or scale-like cells.

An examination of the surface structure of the fibre reveals one of wool's most valuable structural features. The surrounding scales, or cuticle cells, do not fit up flush, but the upper edge of each scale overlaps the scale immediately above it to a greater or lesser extent, thus forming innumerable minute projections and indentations along the whole surface of the fibre. These projections and indentations are known as serrations. The peculiar surface structure is found in most animal fibres, but the serrations are most highly developed in wool. Another important feature of these laminated cuticle cells worthy of note is that they always point in the same direction with the detached edge upwards and outwards towards the tip end of the fibre. Although they can only be seen by the use of a powerful microscope (for in a fine wool there are over 2000 to a centimetre) they may be felt if a staple is passed lightly through the thumb and forefinger from tip to base.

In the fine merino, these cuticle cells have a greater overlap and stand out a wider angle, thus forming better developed serrations. In the coarser wools, the cuticle cells are not so tightly packed and the overlap is less, giving the fibres a smoother surface. In merino wool with its great development of serrations, there is less light reflection, whereas the coarser English long-wools have larger cuticle cells lying close into the shaft, thereby causing a smooth surface which gives a higher reflective power, amounting to a sheen or lustre.

Chemical Composition of the Wool Fibre

Wool is composed of a complex protein named keratin. The chemical composition is not constant in different wools. A variation is even noticeable throughout the length of the fibre. The sulphur content varies from 2 to 6 per cent, whereas the nitrogen content is fairly constant. Wool's approximate elementary composition is: carbon 50 per cent; hydrogen 7 per cent; nitrogen 17 per cent; oxygen 20 per cent; sulphur 6 per cent.

Other substances, including horn, tortoise shell, feathers, fingernails, and various hair fibres are all composed of the common protein keratin. Wool is a protein — a group of chemical substances which includes meat and many other animal products. Proteins are built up of substances called amino acids. These are the fundamental 'building stones' of nearly all biological structures. There are about twenty or more known amino acids and the nature of the protein depends on the proportions in which these amino acids are present and upon the way they are joined together. Each amino acid is itself built up from atoms, the most fundamental chemical units which constitute all matter. Analysis of the proportion of the various elements present tells us nothing of the way in which the atoms of the elements are grouped together into amino acids, but it may give us some indication of which building stones are present in the substances we are studying.

In the case of wool and other fibrous proteins such as muscle fibre, the amino acids or building stones are grouped together in the form of long chains which run along the length of the fibre. These chains are not straight, but are wavy and, when a fibre is stretched, this process of elongation consists of stretching out all these waves in the fundamental internal structure of the protein substance. Such a pattern is characteristic of the animal fibres and has not as yet been imitated by the majority of the manmade fibres. Vegetable fibres such as cotton and linen do not possess this pattern and for this reason do not have elasticity. A further point about the long chains is that they are linked together by cross-linkages which tie them into sheets or bundles. Modification of these cross-linkages can be carried out chemically, and is responsible for some of the mothproofing treatments, some of the anti-shrink treatments for increasing the strength of fibres, and for increasing their resistance to attack by chemical agents such as alkaline substances.

The bundles of peptide chains, as the groups of building stones are called, constitute what are called crystallites — very minute structural units which are too small to be seen even with the most powerful microscope. The crystallites contain the more highly organised carefully arranged portion of the protein substances and are surrounded by a substantial amount of protein material which is apparently not arranged in any orderly manner into chains. Groups of crystallites comprise fibrils which can just be distinguished with very powerful microscopes and the fibrils, in turn, are grouped together to form long, narrow, cortical cells. These cells constitute the majority of the material inside the ring of scales which surrounds the fibre. Sometimes, however, in the case of so-called 'hairy' fibres, the material in the centre of the fibres does not consist of the solid cortical cells, but shows a spongy structure containing air, called a medulla. Such fibres are spoken of as medullated fibres.

7

PROPERTIES OF WOOL AND WOOLLEN GOODS

The fibres which comprise the fleece of the sheep have certain properties which distinguish wool from all other textile raw materials. These properties are common to all wools, whether coarse or fine, whether grown on wool breeds or on dual-purpose breeds, or even on sheep for which the fleece is a secondary consideration. The fundamental properties of wool vary to only a relatively slight degree with breed or type and, broadly, they are quite distinct from features which wool experts recognise as a basis for high and low commercial values.

These fundamental properties, which may be developed in particular sheep to a greater or lesser degree, according to breed, are:
1. the presence of scales giving felting capacity;
2. elasticity and resilience;
3. non-conductivity;
4. chemical stability and affinity for dyes;
5. hygroscopic power and heat of wetting;
6. non-inflammability.

Scales and Felting Capacity

The serrated formation due to cuticular scales is more highly developed in wool than in other animal fibres and is non-existent in vegetable, mineral and man-made fibres. Because these serrations all point in one direction along each fibre, they cause wool 'creep', entangle and felt when a mass of fibres, softened by moisture and other agents, is subjected to alternate application and release of pressure. Due to its fineness, crimpiness, softness and elasticity, the Australian superfine merino wool is renowned for its felting capacity. Coarse wools, because of their flatter scales and their thicker fibres, do not felt as readily.

Scales in Relation to Lustre
The increased scale covering with well-developed serrations of the fine wool tends to diffuse the light, giving less reflection and softer colour. Coarser wools, with larger scales lying closer to the shaft, have a smoother surface and so reflect more light, giving a sheen or lustre.

Scales in Relation to Spinning
Softer yarns can be produced from wool because of the frictional effect of the surface scales. Fine merino types are good examples and from them manufacturers are able to produce fine yarn less tightly spun, to manufacture high-grade, soft dress materials of light weight. With loosely spun yarn, elasticity is better able to come to expression and so give the required softness.

Felting Capacity in Relation to Felts
Felting capacity is very important in the manufacture of felted materials, a large quantity of which are used as components in industry as washers and gaskets, insulation of heat and sound, also for slippers, floor coverings and for millinery purposes. The very short and lower grades of wool can be absorbed in this branch.

Felting Capacity in Relation to Special Fabrics
Special fabrics, held in high regard, such as velours, face cloths, meltons and billiard cloths, are dependent for their production on the felting capacity of wool.

Scales in Relation to Shrinkage and Non-shrink Processes
Fine fibres which have most scales are also more elastic, making them liable to 'creep', and goods made from them require careful laundering. However, with recently discovered non-shrink methods of treatment, this is not such a handicap. Shrinkage and felting is the result of the frictional effect of the scales, the elasticity of wool fibres, their high wet strength, and high power of elastic recovery when wet.

Elasticity
Elasticity is the power to stretch under strain and revert to the original position when the strain is released. Wool has greater elasticity than any fibre with a damp wool fibre able to stretch up to 70 per cent of its original length. Elasticity is perhaps wool's most valuable asset, for so many of its other important features, such as fullness of handle, softness and felting capacity are to a great measure products of elasticity. The crimp formation is dependent to a large extent on wool's elasticity and these two features together give wool its wonderful heat insulating power.

The elasticity is derived from the fibre's chemical structure. The various structures of the fibre are built up from ribbon-like keratin molecules which are about 0.25 of a micron long, and lie lengthwise throughout the fibre. These molecules are built up of chains of smaller units which are not straight but are crimped or folded. When the fibre is stretched, the folds in these chains are straightened out and revert to their normal position when the strain is released.

Durability
The durability of wool clothing is closely connected with elasticity. Wool fibres may be repeatedly bent thousands of times before breaking, where other fibres become weakened and break much sooner. Wool clothing does not wear quickly on surfaces subjected to friction, or bending, such as elbows and knees. Partly because of this feature, clothes required for hard wear, such as uniforms, are always made from wool. Elasticity, combined with the durability and strength of the coarser wools, gives us valuable furnishing materials and luxurious pile carpets. Because of its durability,

wool fibres can be recovered from old materials and remanufactured for the chemical structure is not easily destroyed, and so long as the fibre has two ends, it can be used.

Felting
Under pressure and movement, a portion of the fibre is stretched without breaking and the scales grip adjacent fibres. When the pressure is released, the fibre reverts to its original length. By a repetition of this process, fibres become entangled and are drawn closer together, resulting in felting. The coarser and less elastic wools do not 'creep' so readily, for elasticity in conjunction with strength plays an important part in felting.

Crease Resistance
It is because of this property of elasticity that wool clothing crushes less easily than other fibres and if creased in packing, soon recovers its normal appearance, especially when hung in a damp atmosphere. Moisture aids the bent chains of the wool fibre molecules to return to their natural form.

Draping
Woollen dress fabrics, due to elasticity, are easily draped to graceful lines, and do not hang straight as do materials of cotton or linen fabrics. For the same reason, a tailor is able to make up woollen materials giving a smart cut in garments that will retain their appearance after wear in sunshine and rain.

Shape During Wetting
Wool's high power of elastic regain is not lost when wet, making it specially suitable for swimming costumes and underclothing; when wet, wool garments do not have the 'clammy' feel which dampness gives to other materials.

Elasticity-looped Stitch Structure
The elasticity of the wool fibre is particularly suitable to the looped stitch of knitted materials for form-fitting garments such as underwear, stockings and cardigans.

Non-Conductivity

Woollen material is renowned for its non-conductivity. This property is chiefly the result of the elastic, crimpy, serrated fibres always forming small air spaces in the fabric, no matter how closely felted. Keratin in a more solid state would conduct heat more readily than the same mass of wool fibres. Even in the closest felted woollen materials, innumerable air pockets are formed which build up non-conductivity, and the finer the raw wool, the more effective is the fabric in this respect. This feature of non-conductivity makes woollen wearing apparel embodying warmth and lightness particularly suitable for cold climates. Woollen blankets, too, are widely used on account of their lightness and insulating properties.

Chemical Stability and Affinity for Dyes

Wool's chemical stability is a very important property, making it possible to obtain positive results in many manufacturing processes. On this depends, to a great extent, wool's durability and fast holding of a multiplicity of colours. Wool, in comparison with vegetable fibres, is very much more resistant to acids and much less to alkalis. On this fact depends the process of carbonising and also the separation of mixtures of wool and other fibres. Wool clothing, on account of its resistance to acids, is specially useful to industrial workers coming in contact with weak acids. In laundering woollens, care must be taken to avoid strongly alkaline and very hot solutions, which cause felting and damage to the material.

Wool is chemically decomposed by light, heat and air, as will be noticed in the weathering of the tip of the staple on the sheep's back. This affected tip is of less tensile strength and lacks the ability to hold a permanent set in the finishing processes, such as crabbing and blowing. Perhaps more serious is the resulting uneven distribution of colour when the wool is dyed. This weathering effect is impeded by dyeing, particularly in the darker shades. Wool is of such a chemical structure that dye is bound more firmly than in other fibres. Due to this fastness of dyeing, and its comparative resistance to weathering agents, woollen materials are universally used for bunting. Again, many forms of dye chemicals are acceptable to wool and so a wide range of colours is possible. In dyeing, which often requires a boiling-acid liquid, it is the chemical stability of wool which makes this process more effective than for vegetable fibres. In acid-dyeing of wool it is the fundamental chemical nature of the fibre which allows the dye to be bound more firmly to the fibre than is the case with cotton, imparting fastness of dyeing.

In such finishing processes as crabbing and blowing, it is wool's chemical reactivity which makes these processes effective. It is the breaking of the links between the peptide chains by heat and the reformation of the links in a new position that provides the permanent set, the main function of crabbing and blowing. Wool is very resistant to most forms of mildew and bacteria, particularly if the surface structure is not damaged. Damage from the biting and enzymic digestion of the moth larvae can be prevented by impregnation of poisons or chemicals rendering the wool indigestible. One of the main reasons for felting or shrinking is the instability of the cross links between the keratin chains. Anti-shrink properties are imparted to the woollen materials by either destroying these sensitive cross links, particularly in the surface structure, or chemically modifying these cross links to a more stable linkage. A third method of shrink proofing is impregnation with synthetic resins (plastics) which interfere with the surface nature and/or the elastic properties of the fibre — the two main physical features involved in shrinkage.

Non-Flammability

In this respect, wool differs from most other textile fibres. Cotton, flax and rayons and other cellulose fibres are highly flammable, especially when dry. When great heat is applied to wool, it will smoulder but does not burn readily. This low flammability is due to its high nitrogen to oxygen ratio. Woollen clothing is very suitable for children in preference to the highly flammable cheap substitute, flannelette.

Hygroscopic Power and 'Heat of Wetting'

No other textile fibre will absorb moisture to the same extent as wool. Dry wool will absorb one-third of its weight before becoming saturated. The chemical composition of wool attracts moisture, swelling the fibres by approximately 10 per cent. The IWTO standard regain, recognised in Australia for scoured wool, is set at 17 per cent. The actual regain, however, is dependent on type of wool, atmospheric temperature and moisture. Greasy wool absorbs more atmospheric moisture on account of suint's high affinity for moisture.

During manufacture, moisture is added to wool to increase elasticity and eliminate much friction. Dry wools are not so satisfactory in the carding, combing, drawing and spinning and so moisture is added to reduce breakage of fibres and static electricity and to gain smoother yarn in the spinning. As wool's resistance to deformation is only slightly changed by wetting, a wool garment worn near the skin on becoming wet does not lose its open structure as other materials do — in other words, wool garments never become 'clammy'.

As wool absorbs moisture, heat is generated. This property adds to wool's usefulness for wearing apparel, for after doing strenuous work, the putting on of a woollen garment near the body prevents chills, because the body is actually warmed as the perspiration is being absorbed. In addition, the wool garment dries slowly, thereby reducing the possibility of chill which follows rapid evaporation. Again, when passing out into the cold outer air in woollen clothing, the wool is not only a good insulator, but takes up moisture from the air and generates heat.

Too much water in greasy wool when pressed or stacked will cause loss of colour, staining and mildew damage. Heating in slipe wool, however, is more the result of the chemical reaction of impurities left in the wool by the sliping process.

8

CHARACTERISTICS OF WOOL
IN RELATION TO CLASSING

In addition to the properties which are common to all types of wool, the practical wool expert and the classer must learn to distinguish various types which, because of their different characteristics are suitable for different manufacturing purposes or which differ in commercial values. The following are the physical characteristics by which various grades of wool may be distinguished:
- fineness;
- length;
- character;
- soundness;
- density;
- colour;
- softness;
- lustre.

Fineness

Fineness of fibre is much sought after, for fine fibres are necessary for spinning fine yarns, with which are made lightweight, attractive materials. As elasticity is increased by fineness, softness, which is dependent on elasticity, is a strong feature of materials made of fine wool. Spinning capacity is closely affected by fibre diameter; so much so that fineness for practical purposes is generally expressed by a series of 'quality numbers' which were originally determined by the number of hanks of 512 m (560 yards) which weighed 454 g (1 lb) when the wool was spun to as fine a yarn as possible.

Since the finer counts of yarn are now seldom produced, quality numbers have only a general relation to spinning limit and are usually determined by the appearance of the raw wool, and by mean fibre diameter in processed wool.

Table 8.1   Quality numbers

Merino brand	Description	Crossbred brand	Description
Superfine	74/80	Fine CBK	60s
Fine	70s	Strong CBK	58s
Medium	64	Fine XB	56s
Strong	60	Medium XB	50s
Extra Strong	below 60s	Strong XB	46s/44s
		Extra Strong XB	40s/36s

Table 8.2   Relationship between quality number and mean fibre thickness micron range

Mean micron	Range	Quality count numbers
19 μ	18.6–19.5	70s
20 μ	19.6–20.5	66s
21 μ	20.6–21.5	64s
22 μ	21.6–22.5	62s
23 μ	22.6–23.5	60s
24 μ	23.6–24.5	58/60s and/or 60/58s
25 μ	24.6–25.5	58s
26 μ	25.6–26.5	58/56s and/or 56/58s
27 μ	26.6–27.5	56s
28 μ	27.6–28.5	56/50
30 μ	28.6–31.5	50s
33 μ	31.6–34.5	46s
35 μ	34.6–36.0	44s
38 μ	36	40s

Source: Australian Wool Corporation

Fine fibres have a relatively high felting capacity on account of the increased notches or serrations on the surface of the fibre, and increased elasticity. This means greater care in laundering is necessary. The finer wool fabrics will not stand wear from friction as will materials made from stronger wool, but the finer-fibred materials are more pliable, will hold shape better and will not crease so readily, or if creased will come back to their original form more readily. Where delicate shades are desired, fine wool has an advantage as it absorbs dyes readily and reflects clear tints without sign of a harsher sheen. Materials made of fine fibres contain more tiny pockets of air, giving greater warmth.

Length

Twelve months is considered as natural growth of the wool staple at this stage; all of its component properties are likely to be developed to the best possible extent. There is a marked variation in length in the twelve months' growth between the different breeds and types even in this country ranging from 38 mm (1½ inch) staple from the old types of superfine merino to 305 mm (12 inch) staple from well-fed lincolns.

Cause of Variations

Length and fibre diameter variation is due to quite a number of causes, such as:

1. **Natural or inherited characteristics of the breed**. For example, a staple of polwarth wool should be longer than merino of the same fineness and grown under the same conditions.
2. **Individuality of strain**. Length can be increased within a breed by selective breeding.
3. **The relation existing between diameter and length**. As a general rule, the finer the fibre, the shorter the staple and vice versa.
4. **Nutrition and health**. This is a very big factor. An abundance of well-balanced feed stimulates growth whereas starved or diseased sheep grow shorter and finer wool. On a low plane of nutrition, the secondary fibres are frequently more quickly affected in length and fineness, leaving the primary fibres longer and stronger, and showing a wider variation between the two groups.
5. **Age of sheep**. As sheep age, the rate of growth slows down.
6. **Climate**. Naturally climate alters the nutrition and indirectly the health of the sheep and its covering; however, irrespective of nutrition and health, temperature has a slight effect on growth rate. Wool growth is increased by an increase in temperature.
7. **Pregnancy and lactation**. In the latter stages of pregnancy and when suckling a lamb, especially when nutrition is low, a slowing of growth results.

Length of Staple Affecting the System of Manufacture

In most cases, length will decide the purpose for which the wool can be most usefully employed and also to a very large measure the most suitable method of treatment to be adopted. With merino wool, to a great extent, length determines in the first instance whether a wool will be treated by the worsted or woollen system of manufacture. The worsted section (providing other necessary properties are present) absorbs the longer or 'combing' wools, and produces generally a higher-grade material, whereas the short-stapled wools are more profitably used in the woollen section. The woollen section produces some high-grade materials from short superfine wool, as well as absorbing the very lowest types of wool.

Using the Noble Comb or Bradford system, sound wools of 51 mm (2 inches) and over are profitably combed, whereas European and Asian manufacturers using the Heilmann comb will profitably comb wool 38 to 51 mm (1½ to 2 inches) long. Crossbred wools of about 178 mm (7 inches) and over in length undergo a preparing process before combing or in lieu of combing, whereas sound wools under that length are carded prior to combing.

Effect of Fibre Length on Strength of Yarn

In the first instance, length greatly determines whether a wool will be treated by the woollen or worsted system. Length is to no mean extent a factor in deciding for what particular purpose a wool will be used within its group and the part to be taken in the construction of the material.

In the cloth construction, it must be kept in mind that a stronger yarn is made from the longer fibres and for this reason the longer, sound-stapled wools are selected for the manufacture of warp yarn, which has to stand higher tensions and be more elastic during weaving.

Length in Relation to Evenness of Yarn

The longer-fibred wools also give a smoother yarn when spun on the worsted system, and with evenness of length a level, smooth yarn is produced — a yarn desirable for worsted purposes and particularly warp yarns. On the woollen system, on the other hand, the best and most even yarns are produced with shorter fibres, which are readily 'drafted' in spinning.

Character

Character of wool may be defined as strongly marked distinctive 'qualities' of breed or type; the inbred or characteristic traits of the wool. A well-bred wool of good character will generally show even pronounced crimp and clear-cut staple formation. Good crimp formation in wool is used as an indication that features valued by the manufacturer are present. A regular crimp formation is an indication that the sheep growing the wool has been on an even plane of nutrition throughout the year and the fibre diameter and serrations are regular from tip to base. Wool of good character would be free grown, giving more economical combing, with less fibre breakage and more even length of fibre in the top and yarn. Character is indicative of good breeding and growth under conditions most likely to bring out the good manufacturing features of the wool of any type of breed.

Soundness

The terms 'sound' or 'tender' refer to the tensile strength of the fibres and their ability to stand the tension placed on them during the combing process. The practical wool expert tests wool in the staple by applying a strain by hand, but no particular strain can be stated on account of the variation of the fibre bulk within the staple. There are varying degrees of tenderness and the classer, by applying a uniform tension on a staple of approximately the same bulk of fibres, is able to separate the comparatively sound group from tender.

Some wools may show a break near the tip, leaving sufficient percentage of fibre of even length to make the wool economical to comb, whereas in another instance the staple may show a definite break halfway down the staple, making this wool totally unsuitable for profitable combing. Frequently, wool is of a delicate growth, leaving the wool staple unsound. The percentage of 'noil' and unevenness of fibre length make this wool unsuitable for combing and more economically absorbed in the woollen section. The degree of soundness has much to do with the purpose for which the wool will be used; for instance, soundness is of great importance in 'warp' (or foundation) yarns, whereas wool of less tensile strength may be combed and used for 'weft' yarns; both of which are 'combing' wools. Length variation as a result of fibre breakages, as well as increasing noil, interferes with the spinning of a smooth yarn. Uniformity of fibre length is important in worsted yarns, and one of the main functions of the combing machine is to equalise length; but variation of fibre length is an asset in most woollen yarns. The longer fibres in the woollen process act as a core to the yarn binding in the shorter, softer fibres.

Cause of Break

Tenderness is generally caused by a sudden rise in the body temperature of the sheep, whereby the nutriments supplying the construction of the wool fibre, and the supply of yolk, are affected. This may be brought about by sickness, a likely example of which is milk fever, or by a sudden change in pasture, particularly from dry, possibly drought conditions to soft green feed. Sheep deprived of water, or with unsuitable water, or starved over a long period, will grow fibres of a delicate, frail structure of insufficient tensile strength for combing purposes.

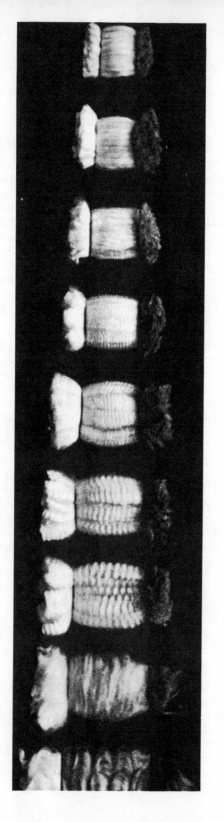

Superfine merino 80s

Fine merino 70s

Medium merino 64s

Strong merino 60s

Comeback 58s

Fine crossbred 56s

Medium crossbred 50s

Strong crossbred 44/46s

Figure 8.1

Lincoln 40s

Density

The term 'dense' is applied to wool with the fibres closely packed. Density, combined with length, is of great importance to the grower, for other features being equal, a sheep with more fibres carries a heavier fleece and, of greater importance to the wool expert, the fibres are better preserved from the damaging effect of sunlight, rain and dust. Thinly grown wool, on the other hand, is depreciated in market value by a wasty, perished tip.

In warp and shafty types, as well as good length for fineness, bulk of staple and density are necessary features in the production of a sound, high-yielding wool. Density is of particular importance in the growing of fine wools. Such wools are valued for their high felting and spinning capacity and these features in fine wools are much impaired by the effect of weathering.

Colour

On a close examination of various breeds and types of wool, it is revealed that there is considerable variation in colour, both in the grease and after scouring. Colour of the scoured wool in relation to its use is of great concern to the manufacturer, for a soft bright tone may be very desirable for the dyeing of delicate shades in such materials as dress goods, whereas a high, lustrous colour is more suited for furnishing drapery.

In merino wools the most desirable colour is a bright creamy white. Perhaps the best-coloured wools in Australia are produced in the Western Districts of Victoria, Yass, Barraba and parts of the New England districts. The wools are attractive in the greasy state and scour out to the snowy, pearly whiteness, a colour esteemed by the manufacturer for delicate dyeing. There are many forms of discoloration, some natural, some due to seasonal influences, others due to the presence of bacteria, foreign and vegetable matter, dust and parasites. These may amount to a permanent stain or merely give the greasy wool a dull or 'sad' appearance, scouring out to a dull white colour. Dusty wools, such as the 'pink' wools of Western Queensland and Western New South Wales often scour out to a surprisingly white colour, better than that of some other wools of more attractive appearance in the grease. For dyeing purposes, freedom from pigmentation is very desirable in all wools.

Softness

The manufacturer is prepared to pay higher prices for soft wool, for it possesses higher felting and spinning capacity, and makes softer-handling, more expensive cloth, whether it be material near the body or outside garments. Soft wool worsteds are warmer, will not crease so readily and hold shape better than harsh wool worsteds. Soft woollens, worn near the skin, conform more readily to the shape of the wearer, are warmer and give less irritation to sensitive skins than harsh woollens.

Lustre

To a large extent the size and set of the overlapping cuticle scales of the fibre regulate the tone of brightness, lustre or non-lustre, as this is largely brought about by the reflection of light from the fibre surface. With wool from sheep of the British long-wool group, various degrees of lustre are peculiar to each breed. Wool of the downs breeds lacks both lustre and brightness, and is often described as being of a chalky or flat white colour.

9

DEFECTS IN WOOL AND WOOL FAULTS

In gauging wool's usefulness, as well as the degree of development of good qualities, defects must be taken into consideration. The defects found in wool of concern to the grower, woolclasser, sorter, buyer and manufacturer, may be divided into two groups: those caused by inherited factors and those caused by external factors. Some of the more common defects that may be modified, if not eliminated, by careful breeding, are hairiness, unevenness of the fleece, harshness, coloured fibres, tenderness, cotted, webby and stringy wool.

Hairiness

Various types of fibres of a hairy nature contain, to greater or lesser degree, a medulla or centre core of air-filled cells, causing a dull colour. They are coarse, harsh-handling, lack strength and elasticity and grow to varying lengths, compared with the wool growth. Kemp fibres grow as single fibres anywhere throughout the fleece, although the breech, head and folds are the most likely location. The presence of kemp depreciates the value of wool and stud sheep showing kemp meet with little demand. The fibres are shed when of short length and are found floating in the fleece. They are coarse, straight, smooth (lacking serrations), harsh, brittle, tapering at both ends, flat white in colour and lacking pliability. These fibres destroy the smoothness of the yarn and are difficult to dye. Kemps are most harmful in the worsted and milled goods, so are confined to the lower grades in the woollen section. However, like most fibres they have a special use, for kemps are an asset in such materials as rough tweeds where they add to the rough flecked finish.

Heterotypic Fibres

Hairy fibres usually stand out longer than the wool staple length about the breech and folds and are sometimes referred to by the grazier as 'dog hair', whereas the same fibres which are generally of wool and hair structure may be described as 'heterotypic' fibres. Hairy fibres found in the fleece contain a medulla to a greater or lesser degree. These soft, spongy, air-filled cells may be only over a very short length of the fibre or may occupy the whole length. Again, the medulla may be thin or may be well developed, occupying a large area of the cross-section. In the latter case, the fibre will be coarse, dull, lacking strength and elasticity. Wool containing much of such fault would be best absorbed in carpet wools.

Unevenness in Fleeces and Staple

It is the aim of growers to produce wool sheep carrying an even fleece. It is natural for the fleece to vary in length, fineness, colour, condition and style, according to the position grown on the sheep. However, unnecessary variation is to be avoided. Wool grown on the breech, folds and wrinkles is naturally of a coarser fibre than the wool on other plain parts. Much has been done by breeders in eliminating coarse breech wool, and with the advent of the plain-bodied merino, a more even fleece in length and colour and fineness was popularised. Length again has been evened up by the reducing of unnecessary short wool about the face and legs. Wool on the face causes wool blindness and aggravates the grass-seed menace. Sheep with 'wool to the toes' are no longer popular. This wool gathers mud, vegetable matter, usually contains hair fibres and is no indication that the sheep is well covered elsewhere.

In colour and condition there have been marked changes. The belly is one part much improved in this respect. The belly is now well covered with a good type of wool falling little short of the side wool in length and character. The old-fashioned short, pasty, thin and even bare bellies are a memory of the past. This evening-up of the fleece in fibre length, fineness and type has improved the all-round quality of the fleece. The classer is better able to grade to particular types required by manufacturers, and sorting is mostly eliminated in the mills. Buyers are always seeking uniform lots of wool and are prepared to pay higher prices for regularity, particularly in fine and superfine wools.

Coloured Fibres

The presence of black or pigmented fibres in a white top, yarn or cloth is a very serious and expensive defect, and every care should be taken in the wool shed to prevent the odd black fleece or patch becoming mixed with good wools, either in fleece or skirting. It is a good practice to separate coloured wool from white on the shearing board and take care not to allow even black patches to pass on as far as the skirting table or piece-picking table before removal. Pigmented wool can enter the flock either as black or piebald sheep or through sheep having odd black fibres right through the fleece. The presence of a large amount of coloured wool is hereditary, but it is difficult to avoid odd-coloured lambs appearing in even well-bred flocks because there is no simple way of picking which ewes and rams are genetically pure for whiteness. To secure a pure-breeding white flock, special care must be taken to ensure that rams used are free from black fibres. If black was a dominant character, pure white would be easier to attain.

Coloured fibres are used extensively for blending with white to produce a grey finish known as 'natural colour'. Self-coloured garments are seldom made from naturally black wool for such garments would not stand the effect of light as well as artificially dyed wool. Again, it is impossible to produce a totally black wool as there is always a percentage of brown on the weathered tip of the darkest fleece. Coloured fibres can be dyed and used in materials where clarity of colour is of little consequence.

Tenderness

A thinning of the wool fibre at one point is often the result of a sudden change of feed, sickness, disease or any cause of a rise in body temperature of the sheep. The wool is left in a state of insufficient tensile strength to stand the strain of the combing process. The weak growth is not necessarily confined to one point of the staple; the whole staple length may be of a weak growth, due possibly to starvation or sickness. Fibre breakages in the cards and combs increase the 'tear' or proportion of 'top' to 'noil' and so tender wools are less profitable for combing purposes. The wool, if slightly tender, may be lowered to the 'topmaking' grade, whereas if showing a bad 'break' it is then unprofitable to comb and is confined to the 'woollen section'.

Cotted Wool

Wool partially felted or matted on the sheep's back is termed 'cotted' wool. This is largely due to environment rather than breeding, but one is tied up with the other, for if a breed or type is grown under unsuitable conditions, a cotted wool may result, whereas another suited type of breed will produce a 'freegrown' fleece. Thirty to forty years ago, cots were a more common fault, but breeders, in striving for a free-grown wool and by building up the general health of sheep, have checked this fault to a great extent. Cotting is often a result of some serious setback, such as starvation or sickness, causing fibres to break or shed. By movement of the sheep or by rubbing in the case of infested sheep, the fibres are then liable to creep, particularly under warm moist conditions and when the wool is not well nourished. Old ewes, sheep of poor constitution and strong-woolled sheep on poor country suffer most severely, and the incidence is highest when lush feed follows drought conditions. Cotted wool is unsuitable for worsted purposes on account of the fibre breakage resulting in processing, and therefore tends to be confined to the woollen section. This results in lower prices.

Webby Wool

This wool is of a light, weak growth with little or no staple or regular crimp formation. It lacks density and body and is often found at its worst on the back of old fine-woolled sheep. Sheep with this type of wool are light cutters and the wool, being open, is subject to weathering. The type is most unsuited for combing purposes because the fibres become weathered and weak, and in wool experts' terms are 'too wefty and wasty'. It is prevalent in weak-constitutioned sheep.

Stringy Wool

A thin staple formation is most undesirable, as more fibres are exposed to dust, rain and sunlight, resulting in a loss of colour and fibre breakages in combing. Sheep carrying this low-grade combing or topmaking type, as with webby wool, are drastically culled.

Discolorations and Stains

A discoloration that is not completely removed by the ordinary scouring process is commonly referred to as a 'stain'. However, it is often the degree of discoloration rather than the type that decides whether a wool is temporarily discoloured or permanently stained. For this reason, it is impossible to definitely divide all types of coloration faults into 'stained' or 'discoloured'. Stains in wool reduce the use of such wool and this must result in less competition.

Bacterial Coloration or Water Stain

Some can and some cannot be removed by normal scouring. The normal fleece contains many types of bacteria without any apparent ill-effect. Under certain conditions, such as dampness over a long period, particularly when the sheep are carrying six months' wool or more, these bacteria multiply. If many types of bacteria multiply, the result is a dirty yellow, light brown or grey discoloration that will generally scour out. Sometimes, however, a particular type of bacteria multiplies resulting in a brighter discoloration of yellow, brown, green, black, purple, pink, blue or red. The green, pink, brown and black colours are the more difficult to remove.

Mycotic Dermatitis and Fleece Rot

Mycotic dermatitis is a fungal growth, attacking the skin itself, whereas the micro-organism causing 'fleece rot' works more on the surface of the skin. Both are cultivated by damp conditions. Mycotic dermatitis multiplies chiefly deep in the wool follicles and gives rise to the heaping up of thick scabby material, which binds the wool fibres into a hard mass. This is known as 'lumpy wool'.

The organisms responsible for fleece rot grow mostly on the surface of the skin and in the overlying wool and hair. Fleece rot frequently causes a banding of the staple with some damage to the fibre and definite coloration, which reduces the market value of the wool.

Yolk or Canary Stain

If heavily discoloured, will not scour out. The wool is discoloured by a yellow pigment, lanaurin, excreted with the sweat. Canary-stained wool retains something after scouring which prevents uniform dyeing.

Bluestone

Will not scour out. When copper sulphate is used as a fly dressing the wool is stained a green colour.

Charcoal

Generally can be scoured out. Caused by sheep lying on burnt country or rubbing against burnt trees and logs. Some charcoal is difficult to remove.

Dusty

Generally can be scoured out. In hot dry areas the fleece is impregnated with dust and extra scouring is necessary.

Tick

If badly discoloured, will not scour out to a good colour. This staining is caused by the excreta of the sheep tick or ked, giving the wool a dull, dirty yellow colour.

Tar or Paint

Will not scour out. Caused by branding sheep with tar or paint. If it is found necessary to brand sheep only fluids such as Si-Ro-Mark, which will scour out, should be used.

Urine

Will not scour out. Caused by wool being constantly saturated with urine to a dark brown colour.

Fern or Vegetable

Cannot be thoroughly scoured out. Caused by drippings from wet trees or contact with wet bracken fern. Is usually light brown.

Dip
Cannot be scoured out. Slight brown stain can be caused by faulty dipping.

Mildew
If wool is pressed when damp it loses its bloom and mildew is cultivated which cannot be scoured out. Stained wool should be carefully dried, otherwise heating will result and the discoloration will spread.

10

VEGETABLE MATTER IN WOOL

Grazing sheep brush over plants and some parts of these plants adhere to the wool. These plant parts include leaves, twigs or, more frequently, the fruits. Leaves which are spiny, hairy or rolled readily adhere to the wool, but often even the smooth ones become entangled. Twigs are broken off shrubs or trees. Smooth twigs often stay enmeshed, but more frequently those which remain have prickles or spines, such as the rose or galvanised burr. The greater part of the 'burr' found in wool, however, is made up of the fruits of the plants. Naked seed is seldom found in wool, the seed being enclosed in protective structures. These vary from thin scaly structures, as in grass 'seeds', to hard, often prickly structures, as in the Noogoora burr. For example, trefoil 'burrs' are fruits (or pods) which unroll to expose the tiny yellow beans inside. Galvanised, Noogoora and Bathurst burrs have fruits which are made up of parts of the flowers which have become modified to form hard, prickly structures, known as 'burrs'.

'Burrs' assume varying degrees of importance, according to the angle from which they are viewed. In the manufacture of wool tops, the shive type of 'burr' (e.g. shive, black jack, wire grass, corkscrew, etc.) offers the greatest menace, as these burrs, lying parallel with the wool fibres, slip through the carding and combing machines and remain as blemishes in the manufactured product. Noogoora burr is a pest to the carder because the wool is usually wound very tightly around it, the whole being large and hard. During the carding process it catches in the teeth of the machine, either blocking or breaking them, because of its size and toughness. Bathurst burrs cause no inconvenience in the carding or combing processes, because the spines fall off very easily, allowing the burrs to fall away freely from the wool. Trefoil burrs or medics, however, unroll so readily that, in the 'monkey's eyebrow' state, they slip through the

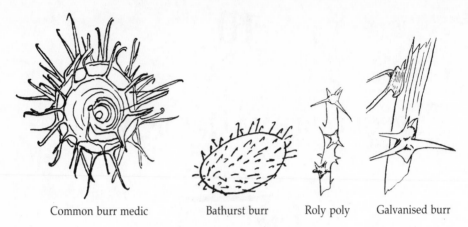

Common burr medic   Bathurst burr   Roly poly   Galvanised burr

Figure 10.1   Commonly-found burrs

carding machines with the wool. Carbonising of burrs in wool is something made difficult by the small burrs such as Bogan flea or the New England crusher which, because of their small size, can slip through the crushing rollers. They are therefore not powdered and remain to mar the finished product.

'Burrs' have another set of values to the wool grower. Although the presence of burr depreciates the value of wool, the plant from which the burr originates is often of great importance. Trefoil clover forms an excellent pasture, and in times of drought the burr is often the sole fodder for sheep. Crowsfoot (geranium), black jack, wire grass, barley grass and bromegrass are good fodder when young, but when pastured are a pest to sheep, causing injury to the face parts and to the skin. Shive is the best-known cause of this type of injury, because of its prolific seeding, especially as, being rather unpalatable, stock allow it to seed more than other grasses. There are numerous other grass seeds which occur in wool but which do not cause much trouble, such as wallaby grasses and mitchell grasses. These are excellent for grazing. The former are distributed over the whole of the grassland; the latter restricted to the zone of summer rainfall. Galvanised burrs, roly polys, Bogan fleas, Bathurst burrs and Noogoora burrs serve no useful purpose for sheep at any stage of their growth. The first two are far too spiny, the third is a small tough bush. Bathurst burr is very prickly, and Noogoora burr is said to be harmful when young.

The various burrs present a complex problem to the wool appraisers, especially when there are many different types present in the same sample. The greatest number of different kinds of burr present in any one sample recorded at the Testing House was twelve. The estimation of the vegetable matter in the wool after scouring by visual means is made difficult by such spiny burrs as trefoil or Noogoora, which provide centres around which the wool felts, thus hiding the burrs from view. In cases such as these, burr is more readily detected by touch. 'Burrs' of the Noogoora and Bathurst type weigh heavily, whilst shive and other grass seeds weigh lightly, though they are often present in great quantities. As well there can be variation between the weight of the same sort of burr from different places. To these difficulties may sometimes be added that of the extreme inconvenience of handling.

The burrs from wool examined at the Testing House have originated chiefly from grasses and other herbaceous plants. There have been very few trees represented in the burry material so far. Eucalypts (gums, boxes, etc.) are spread over most of New

South Wales, but the fruits, leaves or twigs occur rarely in wool. One or other of the species of casuarina (she oaks, belahs, bull oaks, river oaks, etc.) occur also over most of New South Wales, but parts of these are seldom found in the wool by the time it is scoured. There are other small seeds, such as those of many composites, Mexican poppy, or smooth grass seeds, which occur in the pastures on which sheep feed, but which fall out of the fleece or are destroyed before the sheep are shorn. In the remaining sections, a description of the more common 'burrs' which are found in wool will be given, together with notes on their distribution and importance.

Common burr medic (*Medicago polymorpha*). This is also known as trefoil and trefoil clover, found throughout the sheep country. This burr is very adhesive to wool because of its hooked spines. It also unrolls easily and in consequence slips through the machines in manufacture. Stock often exist in drought on these burrs only. The leaves make excellent fodder. Four other species of *Medicago* have also been noted, but they are not as common as the one described. They are cut leaf medic, small burr medic, barrel medic and *Medicago arabica* (no common name).

Noogoora burr (*Xanthium pungens*) This is also known as 'cockle' or 'clot burr' in many parts of Australia. The plant grows about 1.5 to 1.8 m (5 to 6 feet) in height. The burrs are borne in clusters and there may be as many as 10 000 per plant. They measure 19 to 25 mm (¾ to 1 inch) in length and are covered with very strong, hooked spines with two prominent spines at one end. These burrs are very tough and are detrimental to carding machinery. There are two seeds in each burr and one seed germinates in one season and the other seed in the next or some subsequent season which makes the burr very difficult to eradicate. It is not regarded as poisonous but in the very young stages can cause 'blown' or 'hoven' in sheep.

Bathurst burr (*Xanthium spinosum*) This plant is well known, as it has been a recognised pest for over 100 years. The burr is covered with fine, hooked spines which cling to the wool. These spines fall off easily in the carding process and in the process of scouring many of the burrs also fall out of the wool.

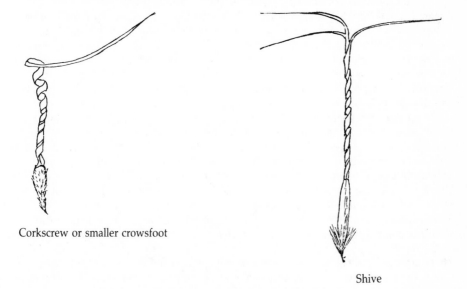

Corkscrew or smaller crowsfoot

Shive

Figure 10.2   Common burrs

Crowsfoot, corkscrew, or geranium (*Erodium cygnorum*) This larger species of crowsfoot is commonly called wild geranium, crowsfoot or corkscrew. The cork-screwed fruitlet which occurs in wool is one-fifth of the actual fruit, which splits up into fine crowned parts. The awn is first moist and straight, but on drying becomes curled. This process of curling and uncurling goes on each time the awn is wet and dried, so it is easy to imagine how thoroughly it can be wound in the wool after having been picked up by the animal and having been through a few storms. This winding action also enables it to penetrate the skin of the sheep and it has been reported that they sometimes collect in large quantities under the skin of the animal.

Corkscrew or smaller crowsfoot (*Erodium moschatum*). Similar to the last-named burr but smaller and seems to be more common in wool. The plant is good pasture when young but in many places is a noxious weed.

Shive (*Aristida* spp) This is a type of spear grass usually with three awns (tail part of the seed). They are annual or perennial grasses with long slender leaves, of little importance as foliage. Some when young are eaten, but when they mature are especially troublesome to stock for two reasons:
1. The seeds are produced in large numbers.
2. They are very sharp-pointed and are therefore troublesome to the face part of the animals as well as piercing the skin causing intense irritation and often inducing infection. They are summer grasses and occur all over sheep country.

Corkscrew grass (*Stipa* spp) These have a variety of names: corkscrew, needle grass or black jack. They are very similar to the other two corkscrews in that they penetrate the skin of the sheep. Generally valuable for forage.

Wild oat (*Avena fatua*) An annual grass recognised as a pest in crops and pastures. Flowers in October, November and December.

Barley grass (*Hordeum leporinum*) A grass common to grasslands all over the world and occurs in New South Wales throughout the sheep country with the exception of the Far Western division. An annual grass, good for fodder when young, but when mature the head breaks into sharp-pointed parts annoying the sheep.

Bogan flea and New England crusher (*Calotis* spp) These species comprise a number of different types. The plants are low, bushy and covered with rough hairs. The whole cluster of seed breaks up into many tiny, flea-like parts. When the cluster is picked up by the sheep, movement, together with drying of the cluster, causes the seeds to separate and cause matting of the wool.

Galvanised burr, roly poly and others (*Sclerolaena* spp) The bindii, the galvanised burr and the roly poly belong to this genus. The plants are usually small and shrubby with small, narrow, thickish leaves. The stems are covered with thorns. Stock do not touch these plants because of their spiny nature. The burrs, however, either singly or in groups on twigs, cling readily to the wool. Because of their spiny nature and because they become wound in the wool these burrs make the wool very unpleasant to handle.

Saffron thistle (*Carthamus latanus*) This weed is of the common thistle type with the leaves bearing sharp prickles on the edges, but smooth on the flat part of the leaf blade. The saffron thistle appears rapidly on recently cultivated areas, but when grass becomes sufficiently strong this weed, being a poor competitor, is driven out. The fruit, like a small shuttlecock, is about 6 mm long and both fruits and leaves are found in the wool.

These are the more common types of vegetable matter found in wool, but many more types of grass and seed have been noted in wool samples examined. However they are of minor importance and do not warrant discussion here.

11

PICKING UP, SKIRTING AND ROLLING

The duties of the picker-up are to pick up the fleeces when shorn, sweep the board, separate the pieces, locks, stains and bellies from the floor sweepings, and place each in baskets set on the board for the purpose, and dress wounds and fly-struck sheep as required. Any dags that may be present are freed of wool and thrown away. A picker-up will have no difficulty in performing these duties for four shearers.

The shearer, soon after starting to shear a sheep, will remove the belly wool and throw it out on the board away from the sheep being shorn. The picker-up immediately picks up this wool, removes any urine stain in the case of rams, or wethers' bellies and then places the wool in baskets or containers conveniently placed on the shearing board. The picker-up should be instructed not to allow large quantities of lower lines — pieces, locks, bellies and stains to accumulate on the board prior to 'smoko' or the finish of the day, as such a practice causes dissension in the wool room. Each Friday afternoon and at the cut-off of the shed, the shed hands are required to wash down the shearing board.

Picking-Up

Immediately the fleece is shorn, the next duty is the actual picking-up performed in the following manner:

Straighten out the neck end of the fleece and then, with toes and hands, push the fore part towards the breech. Invariably, one hind leg is tucked under, this must be straightened out, then grasp a breech in each hand, lift the breech end and give a quick shake to free it of any second cuts or sweat locks, then allow to fold down on the fore part; still retaining a grasp of the breeches, the fleece is picked up. When lifted up from the floor, no portions should be hanging down, for these hanging portions will catch on the end of the table when throwing out.

Throwing Out

The fleece is carried to the end of the wool-rolling table and thrown on the table. The neck end is allowed to go forward, and the breech end is held until the whole fleece gently falls flat on the table. The object is to spread the fleece tip upmost over the table in such a manner that it can be readily skirted and rolled into a loose ball with shoulder and side wool exposed. To assist the wool rollers in the skirting, the fleece is always spread over the table in the same position, for skirting is a most important duty and has a direct effect on the get-up of the clip. The head should be at one end and the breech at the other end of the table, with the line of the back occupying the centre line of the table. In this position, the edges of the fleece will be placed near the edges of the table, in which position the work of skirting can be more expeditiously performed. The table hands know exactly where to find the skirtings and can thoroughly trim the fleece in a minimum amount of time.

In throwing a fleece, care must be taken not to break the fleece or drape the wool over the edge of the table by throwing too hard. The type of wool being shorn greatly regulates the force required in throwing. For instance, old ewes' fleeces will break very easily and require careful handling, whereas more force can be used when throwing a well grown, bold type of young medium-to-strong merino or crossbred fleece.

Picking up of Lambs' Wool

On account of its shortness and lack of 'binders', a lamb's fleece will not hold together as will an adult sheep's fleece, therefore it is more convenient to pick up lambs' wool with two light boards about 61 cm × 15 cm (24 × 6 inches), hinged at one end with a piece of basil. To pick up and carry wool to the wool room, lambs' wool is squeezed between these lamb bats or clappers, as they are termed. Common mistakes a classer must watch for in the picking up of lambs' wool are the sweeping of two fleeces

Picking up the fleece *After the shearer has finished, the picker-up should stand right against the fleece and push it forward with his feet and legs so that it is compact. If the wool from the first hind leg is turned under, it should be straightened out. Then with each hand grasp a good handful of wool from each hind leg. This hold is most important and must be retained until the fleece has been thrown on the rolling table.*

The fleece is folded by lifting the wool shorn from the hind part of the sheep and putting it down on top of the remainder of the fleece, the picker-up being careful to retain the original grip.

By bringing the hands around in a semi-circular movement and pressing well towards the feet and legs, the fleece can then be picked up cleanly.

The fleece ready to be thrown on to the rolling table.

Throwing out fleece
Source: Department of Information

together and the dropping of single staples along the board and allowing them to accumulate before bringing them to the wool room. Much work in sorting will be saved if the mixing of length in merino and the mixing of wool of different fiteness in crossbred, is carefully avoided when placing the fleeces on the table.

The wool rolling table is generally used for the sorting of lambs' wool and is for this purpose covered (for preference) with duck sheeting. Jute material such as wool packs should be avoided, owing to the deleterious effect of the jute fibres becoming entangled in the lambs' wool. The picker-up merely opens his lamb bats and spills the lambs' wool on the covered table without unduly breaking up the fleece. This assists the sorters in locating the short wool from the points and belly. The picker-up should not place a second fleece on top of unsorted wool, but should keep each fleece separate.

As lambs' wool is light, there is a danger of a draught from the open delivery chute blowing it towards the back of the board, as it is shorn. The 'broomie', or board boy, should keep the wool pushed up to the lamb being shorn.

Skirting

A clip may consist of anything from 15 per cent to over 25 per cent lower lines (skirtings, bellies locks, etc.). With this wide variation, the why and wherefore of skirting should be clearly understood. Although very important, this operation is often carried out in a slovenly manner. The 'get-up' of a clip can be easily impaired by faulty skirting, yet it is common to see wool similar to the bulk of the fleece removed, and sometimes the fleeces are left 'skirty'. The classer should instruct the table hands how to skirt and roll the fleeces and throughout the shearing watch and advise on the

Skirting fleeces in a shearing shed
Source: Courtesy AWC

skirting for different fleeces requiring different treatment. It must be remembered that careless skirting will reflect the ability of the classer who carries this responsibility.

Object of Skirting
The aim of skirting is to:
1. remove the portion below the average standard of the fleece;
2. by doing so, leave the fleece more even in all respects
Remove as little wool as possible, but remember that the fleece must be as even as possible in vegetable fault, length and type generally. Over-skirting is a very common mistake. Careful light skirting is much more difficult than heavy skirting and, unless watched, the wool rollers will gradually dip deeper into the fleece. As an example of skirting, take the treatment of a well-grown merino clip on clean Tablelands country. These fleeces generally require the removal of the short, sweaty, heavy-conditioned, discoloured and stained wool around the edge which in this case may amount to very little (possibly 10 per cent). This contrasts with the treatment of fleeces from lightly burred Western Slopes country, particularly in a wet year. In this case, heavier skirting will be necessary to remove clumpy burr from the points of the fleece.

In the Far Western area, in dry seasons, it is frequently necessary to remove the 'backs' as they will most likely contain an excess of dust and are often thin and wasty.

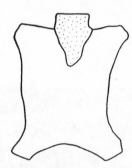

Fleeces otherwise free of vegetable fault frequently contain heavy thistle in the neck region. Skirt very lightly to remove the sweaty fribs from the points and edges of the fleece. Remove the neck wool.

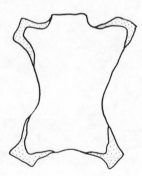

When the bulk of the fleece is free to nearly free, but contains light burr on the points and edges, heavier skirting is required. Remove the burry portions with the aim of leaving the fleece free to nearly free.

In the case of a lightly burred fleece with heavier burr on the points, again skirt heavily enough to remove the clumpy burr and so lift the fleece wool into a lightly burred category.

Heavily burred fleeces should only be lightly trimmed, as no improvement can be gained by heavy skirting.

☐ *Free of vegetable fault*

▨ *B fault (1.1–3% burr)*

▨ *C fault (3.1–7% burr)*

Figure 11.1   Skirting the fleece in relation to vegetable fault

The edges of the fleece are trimmed to remove any wool with excessive vegetable fault. The extent of vegetable fault varies considerably from year to year.

Advantages of Skirting
Both fleece wool and skirtings can be graded into uniform lines. This suits the requirements of buyers who have orders for definite types, and the wool can be more accurately valued, resulting in keener competition and a higher price for the clip. Skirting is generally affected by:
1. whether merino or crossbred;
2. whether burry or free;
3. whether of good or inferior standard.

Explanation

1. Merino fleeces require heavier skirting than crossbreeds, as they are better covered on the points and have large neck folds. The wool grown on the points will be shorter, duller and heavier in condition than the bulk of the fleece, whilst the wool grown on the neck folds will be irregular in length and fineness. All wool inferior to the overall quality of the fleece must be removed. Crossbred fleeces require lighter skirting than merino, as they are more uniform in length and condition to the extreme edges of the fleece. This is because crossbred sheep are cleaner or grow less wool on the points, and are straighter, or plain-necked, so that the wool grown here will be very similar to the rest of the fleece. With good quality, free crossbred fleeces, there will be only a small margin of short or sweaty wool from around the head and forearms to be removed. However, it is frequently necessary to skirt deeply across the breech, as the wool grown here is generally much coarser than the rest of the fleece and must be removed. It should be noted that the heaviest skirting in the case of the merino comes from the forepart, due to neck folds, whereas in the case of the crossbred, the heaviest skirting comes from the breech end due to the coarse breech wool.

2. The standard of preparation of burry wool clips is largely dependent upon careful skirting in relation to the variation of vegetable fault in the fleece. Skirting should always be as light as possible, but where the wool on the points or edges of the fleece is heavier with burr, it should be removed. This will leave the remainder of the fleece of a uniform burr content and at the same time lift it into a higher category for yield. Similarly in the case of dusty clips from dry inland districts, the backs should be removed, if they are thoroughly penetrated with dust and inferior to the rest of the fleece.

3. With both merino and crossbred fleeces, the standard of quality affects the skirting. Inferior grade fleeces, such as tick-stained, yolk-stained, cotted, etc. are not sufficiently improved in commercial value to warrant removing much wool from the fleece. Light skirting in this case is a better proposition, removing only urine-stained and short sweaty edges. As part of the job of skirting, table hands should give the fleece on the table a shaking to get rid of second cuts and small sweaty ends.

Rolling

After careful skirting, to reduce the fleece to a convenient form for handling in the shed, the fleece is folded and rolled into a loose ball and then placed on the classer's table. The wool roller aims at rolling the fleece in such a manner that the shoulder and side wool are left exposed. The more irregular parts are within. It is of considerable assistance to the classer to have the shoulder wool in each instance exposed. By the shoulder wool he/she is able to appraise, with a reasonable amount of assurance, the quality of the fleece throughout; however, it is not advisable for the classer to view only the shoulder wool.

The usual method of rolling, combining speed with efficiency, is as follows:

1. Fold the neck end into the shoulder point.
2. Fold the breech end well in, to reduce the length of the fleece.
3. Fold in the far side of the fleece, one-third of the width of the fleece.
4. The same side is again folded to meet the edge near the roller.
5. Turn in any unevenness on the near side.
6. Roll from breech end to shoulder.

Construction of the Wool Table

Dimensions: 3 m (10 ft) long, 1.5 m (5 ft) and 0.9 m (3 ft) high. The frame should be formed of stout timber and strongly braced.

The top is formed of dressed softwood rollers 38 mm (1½ inches) in diameter, spaced 16 mm (⅝ inch) apart, and fitted crosswise. A good selection of tubular steel, conventional or circular, wool-rolling tables are now available from shearing suppliers.

Shed Labour Required

The shed labour, as shown, is only approximate, for the number of hands required in a shed is greatly dependent on the layout of the shed, the type of sheep and the district. Under certain circumstances the duties of shed labour could be varied with equally satisfactory results.

Four Stands
1 picker-up
1 wool rooller
1 shed hand wool-presser (to pen up)
1 classer/expert/overseer/bookkeeper

Six Stands
2 pickers-up
2 wool rollers
1 classer/expert/overseer/bookkeeper
1 wool presser (bale branding and
weighing) — piece work rates

Eight Stands
2 pickers-up
2 wool rollers
1 penner-up
1 classer/overseer/bookkeeper
1 expert
1 wool presser (bale branding and
weighing) — piece work rates

Ten Stands
3 pickers-up
3 wool rollers
1 penner-up
1 classer/overseer/bookkeeper
1 wool presser — piece work rates
1 expert

Twelve Stands
3 pickers-up
4 wool rollers
1 penner-up
1 woolclasser
1 overseer/bookkeeper/expert
2 wool pressers

By award the classer is not permitted to expert in sheds of over six stands or roll wool or act as overseer in sheds of over twelve stands.

12

AUSTRALIAN CLIP PREPARATION STANDARDS

Aims of Clip Preparation

The aims of clip preparation are to prepare wool for sale to provide a textile fibre which processors may use with confidence (i.e. the wool must conform to their technical requirements); and to maximise the net return to the woolgrower. This can best be achieved by paying careful attention to:

1. producing as few lines as possible from the clip while maintaining an appropriate degree of uniformity within each line; and
2. eliminating contamination of the clip by removing stained and pigmented fibres and by keeping out all foreign material.

Definitions

Mob   Sheep of the same breed which have run together under similar environmental conditions since the previous shearing.

Note: No distinction for either age or sex is made for adult sheep.

Flock   All the sheep run on a property

Clip   All the wool from the flock

The following clip preparation standards must be read as an integral part of the code of practice for the preparation of Australian wool clips.

Skirting

All fleeces must be carefully skirted so that only inferior wool is removed and all good fleece wool remains with the fleece. The following list outlines the inferior wool which must be removed and indicates the lines into which it should be placed.

Inferior Wool		Appropriate Description
Stain	Any wool containing permanent discoloration that cannot be removed by commerical scouring, and includes flyblown wool, urine, dung, water and blood stains.	STN
	Stains must be carefully removed and branded STN. If pieces are not picked for stain, and stain is present, they should be branded STN PCS.	
	A severe price discount is applied if stains are allowed to remain in fleece and pieces, since the end product cannot be dyed to pastel shades.	
	Wet stains should be dried before pressing to avoid further deterioration and the danger of spontaneous combustion.	
Sweats and fribs	Remove and place in the pieces line.	PCS
Topknots	Remove and place in the pieces line if long enough.	PCS
Jowls and cotted edges	Remove and place in the pieces line unless heavily matted and seedy, in which case they should be kept separate.	PCS
Clumpy vegetable matter	Heavy clumpy burr (e.g. burr-on-burr) and/or seed should be removed only if the rest of the fleece is relatively free. As a general rule, fleeces containing light vegetable matter should be skirted to remove the clumpy portions and the inferior wool. Fleeces with heavy vegetable matter should be skirted lightly to remove only the inferior wool.	PCS
Shanks	If shanks are matted, they should be kept separate and consigned for bulk-classing (and so noted on the classer specification). Small, thin shanks, if present, may be dropped under the table into the locks (LKS). Pieces lines containing shanks will be discounted because medullated fibres from shanks cause spinning and dyeing problems.	BC
Short crutchings and topknots	Very short wool from the crutch, and short topknots, should be placed in the locks line.	LKS
Skin pieces	These should be removed and kept separate since they cause processing and dyeing problems. A container should be located near the wool table for them.	SKN PCS

Inferior Wool		Appropriate Description
Hairy breech wool	This should be removed and placed in a separate pieces line since it generally contains medullated fibres.	Separate PCS
Necks	These should be removed only if heavily contaminated with vegetable matter or water-stained in comparison with the rest of the fleece.	NKS
Backs	These should be removed only if very dusty, wasty and tender in comparison with the rest of the fleece.	BKS
Other Lines		
Bellies	Bellies should be picked for any urine, water or heavy mud stains. The clean portion should be placed in the BLS line and the stains in the STN BLS or STN line. If bellies are not picked for stain, and stain is present, they should be branded STN BLS.	BLS
Locks	Locks should always be kept separate from other lines.	LKS
Crutchings	Good-length clean crutchings should be picked for stain and branded CRT. If crutchings are not picked for stain, and stain is present, they should be branded STN CRT.	CRT
Dags	Wool from the breech area which is coated with dung should be kept separate.	DAG

It is stressed that if any of the above inferior wool is present on fleeces, it must be removed and placed in an appropriate line. Where extreme variations occur within a clip, more than one line may be necessary.

Bulkclassing

If there is an insufficient quantity of any line to press a bale, the wool should be branded BC and consigned for bulkclassing with other leftover lines. Various lines within a BC bale should be separated by paper, and the order of contents should be clearly stated on the classer specification. Fleeces consigned for bulkclassing should be carefully skirted before pressing.

Classing Merino Wool Clips

For Merino clips which meet the mob criteria, a maximum of three adjacent quality numbers should be placed together in each line. Approved standard bale markings and line contents are as follows:

Fleece

Description	Contents
AAAM	Main bulk line containing a maximum of three adjacent visual qualities. For example, the AAAM line may be bulk 64s with some 70s and 60s.
	Do not include any fourth adjacent quality number fleeces, or distinctly short or cast fleeces.
BBB	Coarse line containing any wool of the fourth quality number. For example, if the AAAM line is bulk 64s, then any 58s merino fleeces, if present, should be placed in the BBB line. Exclude any distinctly short or cast fleeces.
AAM	Distinctly short fleeces. A short line should not normally be required in clips which meet the mob criteria.

Possible cast lines which may be necessary, depending on the clip, include:

FLC	Tender wool, which may be broken relatively easily by hand when staples of about pencil thickness are tested. In some clips, the majority of fleeces may be relatively tender, in which case only very tender fleeces should be removed and placed in the FLC line. The relative tensile strengths should be stated on the classer specification (e.g. slightly tender, very tender, etc.).
COL	Unscourable coloured wool, including canary yellow, fleece rot, water stains or any other form of unscourable colour in fleeces.
COT	Cotted wool which is felted or matter together so that the fleece will not tear apart freely when handled.
DGY	Doggy wool which exhibits a distinct lack of crimp definition.

In each of the above cast lines, a maximum of three adjacent quality numbers should be placed together to form a large line. Normal standards of length apply. If there are not enough fleeces to make up a bale of any cast line, they should be consigned for bulkclassing.

Other Lines

Description	Contents	
WNS	Weaners' wool from sheep of either sex over six months of age which are being shorn for the first time (e.g. AAAWNS).	
LMS	Lambs' wool from young sheep up to about six months of age. Up to three lines may be necessary, depending on the variation in average length of the wool and size of the mob. Lambs lines may include:	
	AAALMS	The longest, lightest and brightest wool, with all trimmings removed.
	AALMS	Shorter than the AAALMS line, but still light and bright wool and with all trimmings removed.
	ALMS	Any very short wool and lamb trimmings (with skin pieces and stains removed).
RMS	Rams' wool should be placed in corresponding lines of fleece wool if it matches the flock wool in all characteristics.	
PRM	Prematurely shorn wool. After skirting, combing and carding, length wool should be kept separate.	

Description	Contents
OG	Overgrown fleeces which are distinctly more than twelve months' growth.
DBL	Fleeces of two or more years' growth.
DER	Fleeces or parts of fleeces affected by mycotic dermatitis.
PLK	Plucked wool removed from the carcasses of dead sheep as soon as the fibres become loose from the skin.
DD	Dead wool collected in the paddock, from sheep which have been dead for some time.
BLACK	Transfer of black fibres to white sheep will occur if coloured sheep are left in a white mob. Thus coloured sheep should be slaughtered to eliminate any form of dark fibre contamination. However, if black and spotted sheep remain, they should be shorn last.

1. The entire fleece should be removed, including skirtings and locks from the boards, and placed into lines branded BLACK.
2. Black wool should not be handled or pressed with white wool.
3. Where practical, consign wool to a rehandling factility to enabled the wool to be made into large lots and sold as BLACK.

Preparation of Superfine Merino Wool Clips

These recommendations apply only to superfine, speciality wool types. The definition of these wools is as follows: 'Wools of 74s quality and finer which are of spinners, superior or choice styles according to the Australian Wool Corporation type list.' Processors' requirements and market price premiums may warrant special preparation procedures. Classers should be aware of changes in seasonal conditions which may affect the production of these types.

Important Fleece Characteristics

Quality	If there is sufficient wool available to produce a bale, fleeces should be classed to individual quality numbers, (i.e. 100s, 90s, 80s and 74s). If there is not sufficient wool to make a bale of single-quality number fleeces, two quality numbers should be combined. Softness of handle is important.
Style	Fleeces should be of best style (i.e. spinners, superior or choice).
Length	Strict classing to length should be adhered to in each line according to the A, B and C length categories of the Corporation type list.
Soundness	Fleeces must be completely sound.
Colour	Fleeces must be of very good colour in accordance with the requirements of the Corporation type list.
Condition	Only light condition fleeces should be considered. In addition, only a very light dust content is permitted.
Vegetable matter	All fleeces must be free or nearly free of vegetable matter.

Skirting

The aim in skirting superfine fleeces is to leave the fleece lines as even as possible for all classing characteristics. Skirting should remove all vegetable matter, stains, fribs, shanks, skin pieces and short edges.

Backs

Backs should be removed if they do not meet the superfine requirements for style.

Necks

All seedy necks should be removed to leave the fleece free of vegetable fault. Neck runners should be removed to leave the fleece line as even as possible for quality number, length and style.

Note: For preparation of skirtings and other lower lines, refer to the section on skirting in the Clip Preparation Standards.

Superfine Lambs

For superfine lamb preparation, the separation should be for length. Lines should also be completely free of vegetable matter and stain and should be of good colour.

Bale Descriptions

The prefix SUP may be used where appropriate.

Classing Comeback, Crossbred and British Breed Long-Wool Clips

If fleeces are bulk 60s and finer, a maximum of three adjacent quality numbers should be placed together in a line, but if fleeces are bulk 58s or coarser, a maximum of two adjacent quality numbers is permited in each line.

Description	Contents
AAACBK	Main bulk line, containing a maximum of three adjacent quality numbers if bulk 60s with some 64s and 58s fleeces. If bulk 58s a maximum of two adjacent quality numbers may be included. Exclude coarse, distinctly short and cast fleeces.
AAAFX	Fine crossbred fleeces containing a maximum of two adjacent quality numbers (e.g. 58s/56s). Exclude distinctly short and cast fleeces.
AAAMX	Medium crossbred fleeces containing a maximum of two adjacent quality numbers (e.g. 50s/46s). Exclude distinctly short and cast fleeces.
AAACX	Coarse crossbred fleeces containing a maximum of two adjacent quality numbers. Exclude distinctly short and cast fleeces.

Not all the above lines may be necessary in every crossbred clip. *Note*: The adjacent quality numbers may also vary from the above examples.

The same descriptions are used for lower lines and distinctly short and cast fleeces as for merino wool, except that each of these lines are prefaced/suffixed by the appropriate quality number description. For example, in a fine crossbred clip, lines may include AAFX, FXFLC, FXPCS, FXBLS, etc.

Descriptions used for lambs' wool may include AAAFXL, AAFXL, etc.

Classing British Breed Short-Wool Clips

Wool from these breeds should be kept separate from that of long-wool breeds and can be identified by the brand DNS.

If there is enough wool to make a main line, it is preferable that it be branded with an abbreviation of the breed name (e.g. AAASFK (suffolk), AAADOR (dorset), AAASD (southdown) after fleeces have been skirted and cast fleeces removed.

Lower lines can be preceded by the breed abbreviation (e.g. SFKPCS).

Classing Carpet Wool Clips

Skirting

All stain must be removed. Stain in otherwise clean wool will lead to heavy price discounts. Remove all dags, fribs, sweats and clumpy burr or seed.

Description	Contents
ACWP&B	Pieces and bellies (combined or separate)
	These skirtings, etc. should be open, free of vegetable matter, contain no stain or dag, a low percentage of sweats, and display high bulk. Skirtings, etc. displaying these characteristics from both adult sheep and lambs may be blended together in this line.
	Length — Average 85 mm (3.5 inches), minimum 50 mm (2 inches), maximum 125 mm (5 inches).
	Colour — Good to average
CWP&B	Secondary pieces, bellies (combined or separate)
	Skirtings, etc. which may be shorter and inferior in quality and/or colour, carrying light vegetable matter, a higher proportion of sweats and being of lower bulk should be included in this line. *Note*: Cotted pieces, bellies, stains and dags, and wool containing excessive vegetable fault should be placed in the cast lines.
	Length — Minimum 50 mm (2 inches)
	Note: If pieces and bellies are separated they should be marked ACW PCS (or CW PCS) and ACW BLS (or CW BLS) to prevent confusion with the secondary fleece line.

Classing

Description	Contents
AAACW	Main fleece line
	Fleeces in this line should be open and contain a high percentage of medullated fibres, sound in staple, free or nearly free from vegetable matter and have a minimum amount of kemp. They should also be harsh in handle and be of high bulk.
	Length — Average 100 mm (4 inches) to 150 mm (6 inches)
	Fibre — Straight to almost straight
	Colour — Good/chalky white
	Lustre — Non-lustrous
AACWB	Secondary fleece line
	Fleeces in this line should be open, medullated, sound in staple, and free or nearly free of vegetable matter. Fleeces in this line should be softer in handle and lower in medullation and bulk than those in the top line.
	Length — Average 75 mm (3 inches) to 150 mm (6 inches)
	Fibre — May include some crimp
	Colour — Good/average
	Lustre — Low/average

Description	Contents
ACWS	Short fleece line (including premature shorn)

Fleeces displaying the characteristics of the above two lines, but lacking in length, should be included in this line.

Length — Average less than 75 mm (3 inches)

CWL	Lambs

Lamb fleeces in this line should be open, medullated, sound in staple, and free or nearly free of vegetable matter. They should be kept separate from the adult fleece wools as they differ in style and are usually softer.

Length — Average 75 mm (3 inches) to 150 mm (6 inches)

Fibre — May include some crimp

Colour — Good

CWSL	Short lambs (including premature shorn).

Similar to the lamb fleece line, but shorter in length.

Length — Average less than 75 mm (3 inches)

CW LKS	All locks and short inferior pieces, bellies and crutchings.

Cast Descriptions

All non-matching wool (e.g. cotted, permanently discoloured, tender, lustrous and fleeces containing excessive vegetable matter) must be kept separate. Small quantities of cast fleeces should be branded BC and consigned for bulkclassing.

Description	Contents
CW FLC	Wool which may be broken relatively easily by hand when tension is applied.
CW COT	Wool which has become partially felted or matted so that the fleece will not tear apart freely when being handled.
CW COL	Wool containing non-scourable colour (e.g. canary yellow, non-scourable brands, fleece rot).

Objective Measurement Interlotting Matching Tolerances

The matching of components to form an interlot is on the basis of both technically and commercially acceptable ranges for the measured characteristics and visual compatibility of the measured and unmeasured characteristics.

When matching components it is recommended that this procedure be followed:

1. Initially appraise components for type and approximately batch samples.
2. Apply objective measurement criteria.
3. Examine components for visual compatibility of the measured characteristics of fibre diameter, vegetable matter and yield.
4. Appraise components for compatibility of non-measured characteristics which include length, strength, colour and style.

It has been agreed by the industry that the following ranges are technically and commercially acceptable and thus should be used when matching components:

Range of Yield (All Descriptions):

1. 12 per cent, where the mean of the sale lot is 0–59.9 per cent;
2. 8 per cent, where the mean of sale lot is 60–72.9 per cent;
3. 6 per cent, where the mean of the sale lot is 73 per cent or greater.

Range of Mean Fibre Diameter (All Descriptions):

1. 0.8 micron, where mean of sale lot is between 19.0 microns or finer;
2. 1.0 micron, where mean of sale lot is between 19.1–19.5 microns;
3. 1.5 microns, where mean of sale lot is between 19.6–22.0 microns;
4. 2 microns, where mean of sale lot is between 22.1–32.0 microns;
5. 4 microns, where mean of sale lot is 32.1 microns and coarser.

For wool which is bulk 60s and finer, up to three adjacent qualities may be included in the sale lot. For wool which is bulk 58s and coarser, up to two adjacent qualities may be included.

Range of Vegetable Matter Base

Different vegetable matter types (e.g. burr and seed) should not be lotted together.

Fleece
1. 0.8 per cent, where mean of sale lot is 0.5 per cent or less;
2. 1 per cent, where mean of sale is greater than 0.6 per cent and less than 1 per cent;
3. 2 per cent, where mean of sale lot is greater than 1.1 per cent and less than 6 per cent;
4. 3 per cent, where mean of sale lot is 6.1 per cent or greater.

Skirtings
1. 1.5 per cent, where mean of sale lot is 2 per cent or less;
2. 3 per cent, where mean of sale lot is greater than 2 per cent and less than 6 per cent;
3. 5 per cent, where mean of sale lot is 6 per cent or greater.

Cardings
1. 1 per cent, where mean of sale lot is 1 per cent or less;
2. 2 per cent, where mean of sale lot is greater than 1 per cent and less than 5 per cent;
3. 5 per cent, where mean of sale lot is 5 per cent or greater.

Source: Clip Preparation Standards 1989, Australian Wool Corporation

13

CLASSER'S DUTIES

The Woolclassers and Shearing Staff Employees' Award sets out the duties of the woolclasser as follows: 'A classer shall carry out his duties in accordance with the directions and orders of the owner or his nominated representative. His duties include:

1. to classify the wool and advise and report generally as a wool expert, regarding all matters pertaining to the clip and the getting up and preparing of such for sale;
2. to instruct the woolrollers and supervise the skirting and rolling of the fleece;
3. to instruct and supervise the piece-pickers, the pickers-up as far as concerns their duty in picking up the fleeces, and all other men engaged in handling of the wool;
4. to instruct the wool pressers and exercise a general supervision over the pressing, weighing and branding of the bales;
5. to keep the shed wool book, or see that it is kept by the woolpresser or wool weigher, to the satisfaction of the employer, and where required to write up the station permanent wool and weight book daily (one copy only).

Any other duties required of the classer should be specified at the time of his engagement. If the classer is engaged to perform, and does perform, any of the additional (combined) duties covered by the award (i.e. overseeing, waybills and/or bookkeeping, woolrolling or other shedhand's work), then he must be paid the prescribed extra rates for those duties. It is generally advisable for the classer to arrive at the shed the day before shearing is to commence. This gives him the opportunity to see that the wool room is suitably prepared and all requirements are on hand for a prompt start at 7.30 a.m. the following morning. The following matters require checking prior to the commencement of shearing:

The Woolclasser's Checklist

On the day before Commencement of Shearing

1. See that the wool tables are in good repair and conveniently positioned ready for work.
2. Remove any crutchings, dead wool, skins or super bags and sweep out the shearing board and wool room.
3. Have the following items on hand in the shed in good condition and ready for use: brooms, wool baskets, lambs' bats, packs, bale fasteners, stencils, branding brush and ink.
4. Have the wool weight book and pencil ready for the presser.
5. See that the wool press is in working order with spare cable or parts on hand in case of breakdown.
6. Check the scales.
7. Ask the owner for a description of the sheep and the order in which flocks will be shorn. Request a copy of the previous season's wool testing certificates and classer's report or account sales. This should give you a general idea of the lines to be made in the clip.
8. Make a check of the AWC Code of Practice requirements.

At the Commencement of Shearing

1. Place the shed hands in their respective positions and instruct them how to carry out their duties, for example:
   - treatment of fleece, bellies, locks and stains by the board boy;
   - the extent of skirting necessary for fleeces.
2. Allow about twenty fleeces to accumulate to obtain a clear indication of the type of wool in the clip before setting the lines.
3. Clearly mark the bale descriptions on each bin for the presser.
4. Advise the wool presser of the bale weights required and of any special branding requirements, e.g. classer's registration, or destination brands. Insist on uniform and neatly branded bales.

During Shearing

1. Supervise the shed hands' work and be sure the wool preparation is up to standard.
2. Supervise the wool pressing to avoid mixed lines and branding errors. Description and bale number should be branded on the bale immediately after pressing. Check the wool weight book entries from time to time.
3. Find out when flock cut-outs are approaching. Regulate pressing and bale weights to avoid star lots, butts or mixed bales.
4. Rule off the wool book at the end of each flock and record a brief description of the sheep and wool in the remarks column.
5. Keep the shearing board and wool room well swept during shearing. At the end of each run sweep up thoroughly and place all loose wool in the bins.
6. Each evening write up the station copy of the wool weight book and progressively make out the classer's report. The bale totals in the wool weight book and classer's specification should correspond.

At the Conclusion of Shearing

1. Post the completed classer's report or specification to the wool broker immediately at the conclusion of shearing.
2. Complete 'Request for Inspection of Wool Clip' form, enclose a copy of the classer's

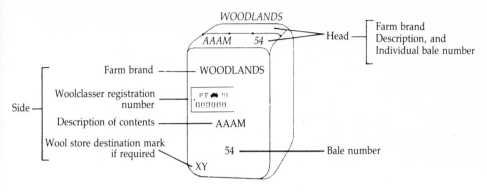

Figure 13.1 Bale markings

specification and forward them to: Clip Inspection Service, Australian Wool Corporation, 369 Royal Parade, Parkville, Vic, 3052

Pressing and Branding

A bale may be described as a parcel of wool in an approved wool pack weighing not less than 110 kg (242 lb) and not more than 204 kg (450 lb) gross.

The presser should be instructed to keep the weight of bales as even as possible in each line. Care must be exercised to see that no string, synthetic fibres or jute becomes mixed with the wool, and woolpacks should be opened up outside the shed and all loose fibres removed. When a bale is pressed, it should be branded as soon as it leaves the press. Branding of bales should be, whenever possible, of the following-sized lettering and in this order:

Station brand	10 cm (4 inch) letters
District	7.6 cm (3 inch) letters
Description	10 cm (4 inch) letters
Bale Number	10 cm (4 inch) letters

Brand marks should be placed on one side of the bale only: the side corresponding with the last flap to be fastened. The owner's brand or an abbreviated form of it, together with the description and bale number should also be placed on the head of the bale. As soon as the bale is branded, it should be weighed and the particulars entered in the weight book. Bales are normally branded as shown in Figure 13.1

Weight Book

This should consist of a book ruled as shown in Figure 13.2 in five columns, the first for the bale number, second for description or class of wool, the third for description of sheep, the fourth for the weight in kilograms, and the fifth for any remarks the

No. of bale	Description of wool	Description of sheep	Weight (in kg)	Remarks

Figure 13.2 Weight book

STANDARD CLASSER SPECIFICATION

IT IS ESSENTIAL THAT THIS SPECIFICATION BE
FORWARDED IMMEDIATELY SHEARING IS COMPLETED

			PACK	METRIO
BRAND	IONA		POLY ✓	COMEBACK
			JUTE	CROSSBRED ✓

BROKER NAME SUPPLIED

SALE CENTRE SUPPLIED   ROAD ☒ SEND WITH EACH CONSIGNMENT   MAIL ☐ POST IMMEDIATELY FOR EACH CONSIGNMENT

OWNERS TRADING NAME NAME SUPPLIED

POSTAL ADDRESS ADDRESS SUPPLIED

PROCEEDS INSTRUCTIONS   SUPPLIED

BANK: SUPPLIED

A/C No.: SUPPLIED

PHONE No. (___) SUPPLIED   POSTCODE 9999   DATE 1.7.1989

OWNER/MANAGER (Signature) SUPPLIED   SIGNED

OFFICE USE ONLY

CLASSERS REGISTERED No. **PⅡ** 8 6 1 2 9 6

CLASSERS NAME NAME SUPPLIED

POSTAL ADDRESS ADDRESS SUPPLIED

POSTCODE 9999   PHONE No. (___) SUPPLIED   DATE COMPLETED 1.7.89

IS PRESENT SHEARING COMPLETE? YES   TOTAL THIS SHEARING 71   BALES.   CLASSER (Signature) SIGNED

OFFICE USE	MOB No.	ADD. BREAK.	No. of BALES	DESCRIPTION					INDIVIDUAL BALE NUMBERS										
	1	✓	14	AAAM	1	2	3	4	5	6	7	8	9	10	11	12	14	15	
	2	✓	13	AAAM	18	20	21	22	23	24	25	26	27	28	30	32	33		
	3	✓	6	AAAM	36	37	38	39	40	41									
	4	✓	13	AAAM	47	48	49	50	51	52	53	54	55	56	57	60	61		
	1-4	✓	4	BBB	17	35	42	63											
	1-4	✓	4	FLC	16	34	43	62											
	1-4	✓	5	PCS	13	29	45	59	64										
	1-4	✓	4	BLS	19	31	44	58											
	1-4		2	LKS	46	65													
	5		4	AAALMS	66	67	68	69											
	5		1	AALMS	70														
			1	B/C	71														

Please tick Appropriate boxes (✓/Tick)	MOB NO	MOB/DESC		MOB BREAKS		QUALITY			WOOL GROWTH MONTHS	BURR/VEG				STAIN FREE PROC.	MONTH CRUTCHED	
Use Code		AGE	SEX*	FROM	TO	Fine	Med ium	Str ong		L H O U R	Light	Med ium	Hea vy			
E = Ewes	1	2-4	E	1	17		✓		12		✓			✓	NOV	
W = Wethers	2	2-4	W	18	35		✓		12		✓			✓	NOV	
M = Mixed E/W	3	5-6	E	36	46	✓			12		✓			✓	NOV	
H = Hoggets	4	H	M	47	65	✓	✓		12	✓	✓			✓	NOV	
L = Lambs	5	L	M	66	70	✓			4-5		✓			✓	NOV	
WN = Weaners	6															

NOTE: LINES OF THE SAME DESCRIPTION FROM DIFFERENT MOBS MAY BE LOTTED TOGETHER AT THE BROKERS DISCRETION UNLESS SPECIFIED HERE WITH REASONS FOR SEPARATE LOTTING.

BALE Nº 71 - ALMS/cot/COL/STN/SKN PCS

* Mobs 1-3 can be combined

classer sees fit to make on the wool of each flock. The wool weight book is most important and should be kept clean and tidy and the wool from each flock should be noted in it. The brokers require this information for cataloguing purposes as they can then decide which flocks will be sold together or separately. It also enables the grower to find the average cut of each flock. It is from this book that the classer will get the necessary information when making his report on the clip.

Classer's Report

The classer's work of preparing the clip is not complete until a report has been forwarded to the wool-selling broker, providing full particulars of the type of wool and lines made in the clip. This information enables the wool valuer at the brokers' store to arrange the clip into suitable sale lots by grouping together bales of the same description from similar mobs. That is to say, the top line of the clip may be branded AAA M but there may be a slight variation in type, burr or dust content between AAA M lines from various mobs. The main points required by the broker are:

1. name of clip and owner's address;
2. number of bales in the clip and individual numbers in each line;
3. brands and description of wool in each line;
4. where mob cut-outs occur;
5. any information on changes in type of wool such as, any burry mobs, dusty, etc.
6. The classer may also suggest that the wool from certain mobs is similar and could be matched together.

Figure 13.3 gives an example of the Australian Wool Corporation's standard classer specification. This document must be signed by both the grower and the classer.

14

MARKETING

Methods of Offering Wool for Sale at Auction

Wool received in the auction system may be offered for sale in one of three ways. Evenly classed lines may be sold as *individual growers' brand lots*, either with test certificates or without test certificates. Alternatively they may be combined with lines of similar wool from other growers to form an *interlot*. With interlotting, the wool is sold in its original packs and the grower is paid according to the quantity of his wool in the interlot. Unevenly classed lines and lines of less than one bale may be *bulk-classed*, whereby wool is removed from the bale, classed and repacked.

The proportion of the clip offered at auction by these methods has varied over time in conjunction with changes in the marketing system.

In the years immediately prior to the 1970–71 selling season, lines of more than four bales were sold as individual lots in the main sale room. Lines of four bales or less were either sold as individual lots in a second sale room or were subjected to rehandling procedures such as interlotting and bulkclassing.

The Australian Wool Board's wool marketing report (AWB 1976) had a major impact on small lot policy in the early 1970s. In that report, it was recommended that, wherever possible, all lines of one, two and three bales be eliminated by interlotting. Subsequently, a price-averaging plan was introduced in the 1970–71 selling season, under which all lines of one, two and three bales were pooled. The pooled wool was rehandled to form larger lots, and growers received payment from the pool for each wool type. Four-bale lines were sold as individual lots in the main sale room along with larger lots.

When the price-averaging plan was discontinued in the 1973–74 season, four-bale lines continued to be sold as individual lots. The system of handling lines of one, two and three bales generally reverted to that which existed prior to 1970–71, although

Selling the modern way—No. 1 Auction Room, The Wool Exchange, Melbourne. Photo courtesy AWC.

sale lots of all sizes were sold in the one room, with merino wool and crossbred wool sold in separate rooms. This change was associated with an increase from 77 per cent to 82 per cent in the percentage of wool offered as growers' brand lots. In the following year, brokers in some centres established joint rehandling facilities to reduce costs, and the percentage of wool offered as grower's brand lots fell back to 75 per cent.

The introduction of pre-sale testing of wool in 1972–73 permitted a major change in marketing methods. Under sale by sample and objective measurement, a core sample (of at least 0.8 kg) and a grab sample (4–5 kg) are taken from each sale lot. Objective tests on fibre diameter yield and vegetable matter are conducted on the core samples, and the results are displayed with the grab sample for buyer appraisal. Previously whole bales were displayed, which entailed moving these bales onto the show floor before sale, then off the show floor into storage after sale. In the 1987–88 selling season, of all the wool offered for auction, 99.21 per cent was sold by sample. Sale by sample and objective measurement also enabled brokers to interlot on the basis of a sample from each bale, reducing their handling costs. The proportion of wool interlotted has increased markedly since this system was introduced in 1975–76. In 1977, the Limited Offer to Purchase Scheme was introduced by the Australian Wool Corporation with the aim of demonstrating areas of cost saving in wool handling. Under the Limited Offer to Purchase Scheme, the minimum size of lots sold by sample and objective measurement was progressively reduced from six to three bales. The scheme ceased operating in 1980 and was succeeded by the Wool Marketing Service. The Wool Marketing Scheme adopted a policy of selling three-bale lines

Dumping bales ready for shipping.

as tested growers' brand lots, with only one-bale and two-bale lines interlotted or bulkclassed.

Commercial brokers currently offer most farm lines of greater than three bales as tested growers' brand lots, although a small proportion continues to be sold untested. Uniformly classed lines of three bales or less are interlotted where possible; these lines are sold separately only if the wool is difficult to match, is a specialty type, or by grower request. Brokers have indicated, however, that decisions on method of marketing are usually left to the broker.

Sale by Sample

With the exception of a very small percentage, all wool sold by auction throughout Australia is sold by sample, and the test results printed in the sale catalogue. In the extract from such a catalogue shown in Figure 14.1, the following information is provided as a guide for the buyer.

BSH	These are the percentages of burr (B), seed (S) or hardheads (H) which make up the gross vegetable matter base.
ACY	Australian carbonising yield
JSCY	Japanese clean scoured yield
SCD	Scoured yield with an allowance for 17 per cent moisture regain
SCH DRY	Schlumberger dry combing yield
VMB/GROSS	Gross vegetable matter base percentage
MIC	Mean fibre diameter in microns

S/L	Staple length
MM	Mean staple length in millimetres
CV%	Coefficient of variation for staple length
S/S	Staple strength expressed as newtons per kilotex
POB	Position of break. Indicates the percentage of staples that broke at the top (T), middle (M) or base (B) regions
COL	A measure of clean colour
LOT No.	Sale lot number
BS	Number of bales in the lot

The station brand ('CALLUBRI') and the classing line description (AAA M) are included for each lot together with a number to indicate the region of origin as shown in the map in the front of the catalogue (e.g. N35 where 'N' represents New South Wales and '35' the Condobolin/Tottenham region).

Sale by Separation

With sale by sample, it is now unnecessary for the display sample to be in the same centre as the stored bales, as the sample is accepted by the industry as being representative of the contents of the bales in the sale lot. This provides for a method of selling called *sale by separation*. Wool stored at minor selling centres may be offered for sale at the larger centres. About 22 per cent of the total offering was sold by this method in the 1987–88 season.

The Reserve Price Scheme

In 1970, legislation was passed to establish the Australian Wool Commission. It had a number of responsibilities for improving marketing of the clip, the most important being to operate a flexible reserve price scheme at auctions. The scheme was further improved in 1973 when the government agreed to amalgamate the Wool Board and the Wool Commission into the Australian Wool Corporation.

Prior to 1970 the Australian wool market operated free of control except during war years. Wool prices were influenced by many external factors such as currency fluctuations, aspirations of dealers within the market, and textile processors' perceptions of future demand. In the first instance, the Australian Wool Commission was capable of arresting any slide in buyers' confidence by entering the market and offering prices on a flexible basis which provided a foundation on which values could be established. This scheme was further improved in 1974 by the Australian Wool Corporation when a minimum value or *floor price* was determined for each wool type.

A market indicator is set and represents an approximate average value across the entire wool clip. During the sale, any lot which fails to attract bids equal to or in excess of this valuation is purchased by the Corporation at the 'floor price' valuation irrespective of the level of the losing bid. Wool purchased by the Corporation is held until demand improves and it can be sold at an acceptable price. At times the Wool Corporation has purchased 30 per cent or more of offerings at weekly auctions.

All wool growers contribute 4 per cent of their gross wool income each year to finance the scheme. At 30 June 1988, the Market Support Fund had a balance of almost $225 million during the year and interest on investments totalled around $175 million. It is the policy of the Wool Council of Australia to have sufficient reserves to purchase one-third of the clip for the coming year. Contributions by growers are progressively returned and in the 1980–81 and 1981–82 seasons the total amount was $175 million.

BSH	ACY	JSCY	SCD 17%	SCH DRY	VMB CROSS	MIC	MM	CVX%	N/KT	T	M	B	COL	LOT No.	BS
70 30 0	66.0 2878	69.6 3035	73.4 3200	67.7 2952	2.2 4406	21.0	95	15 CALLUBRI N35 AAAM	46		71	29		S2194	23
70 30 0	66.7 3303	70.4 3486	73.4 3635	68.8 3407	1.5 5004	22.0		AAA M						S2198	26
55 45 0	66.3 1508	70.0 1593	73.0 1661	68.3 1554	1.6 2299	20.7	96	11 AAA M	42		98	2		S2199	12
60 40 0	62.1 3447	65.8 3653	69.6 3863	63.7 3536	2.3 5609	21.7	92	15 AAA M	44		57	43		S2200	29
65 35 0	67.1 746	70.8 787	74.1 824	69.0 767	1.8 1124	24.0		BBB						S2202	6
75 25 0	63.9 1436	67.2 1511	73.8 1659	64.5 1450	4.6 2272	18.3	72	13 AAA WNS	48	36	64			S2203	12

Figure 14.1 Extract from sale catalogue

Intervention Price

The Australian Wool Corporation also operates a second level of reserve prices known as the intervention price. It is only introduced in exceptional circumstances when the prevailing market conditions are significantly abnormal. The intervention price is reversible, may be altered when warranted, and is not always automatically in force. It is quite different, therefore, to the minimum reserve price.

Sale with Additional Measurement

Sale with additional measurement was introduced in Australia in 1985. Staple length and staple strength are now available as routine measurements, together with the associated values of staple length variability and the position of break in the staple. The measurement of average yellowness of wool is now also commercially available.

Sale by Description

The next phase in the progression towards 'sale by description' is under way. It involves the provision of information on the non-measured characteristics, along with the new measurements. In the near future trials will be conducted with unmeasured characteristics being independently appraised. Feedback of these results will help guide the industry along the path towards 'sale by description'.

Wool Marketing Act

The *Wool Marketing Act* 1987 bound the Australia Wool Corporation (AWC) to perform certain functions as a statutory authority. Part IV of the Act is of particular significance to the auction system. Functions of the Corporation in relation to wool marketing are:
1. to operate a reserve price scheme in respect of wool offered for sale at auction or by electronic means;
2. to formulate, secure the observance of, and implement standards of preparation of wool for submissions for sale at auction or otherwise, including standards in respect of wool packs;
3. to formulate, and secure the adoption of, terms and conditions governing the acceptance of wool for sale, and the sale of wool, at auction or otherwise;
4. to facilitate the adoption of aids to the efficient marketing of wool.

Computer Selling

A system of computer marketing has recently been introduced in Australia by a major wool selling broker. After receipt at the warehousing centre, wool consigned for sale by this method is grab sampled and cored for testing. The cores are sent to AWTA and the sample forwarded to the central show floor in Sydney, Melbourne or Fermantle for buyers to inspect. Before the nominated sale, a special catalogue is made available with the traditional auction catalogue. At the same time, details of the offering, plus growers' reserves and the broker's valuations, are fed electronically to an independent computer system.

Once buyers have valued the wool, they enter their bids into the system from their own offices using a computer terminal or telex. Bids can be based on greasy or clean yield prices and buyers have the opportunity to specify various bidding strategies.

The flexibility of the system allows buyers to modify their bids right up to the moment of sale. The broker can also vary valuations up to an hour before the sale starts. At sale time the central computer system in Sydney will match buyers' bids against the wool offered. The lot will be sold to the highest bidder at one cent above the second-highest bidder, providing the highest bid exceeds the grower's reserve and/or the broker's valuation.

A section of the show-floor at the Yennora Wool Centre showing sale by sample with the traditional show bales in the background.

Immediately after the sale, results are transmitted by computer to buyers and the brokers' warehousing centres. Account sales to growers and invoices to buyers are then issued. Buyers can also gain direct access to the system to receive reports on their purchases, market summaries regarding price trends for various types and information about other buyers in the sale.

Sale by Private Treaty

Growers can negotiate a direct sale with private treaty wool merchants. The most common type of sale by this method is on the farm to country wool and skin dealers, or to private wool buyers who are often representatives of the larger wool-buying companies. For the woolgrower, the advantage of private selling is the quick payment for the wool clip. However, payment to the woolgrower might be lower as there may be less competition between buyers than there would be under the auction system. A few clips are sold by private treaty direct to wool processors, and this number may increase with the growth of early-stage processing of wool in Australia.

A third alternative is sale by private commission. The clip is delivered to a private treaty show floor where sale offerings are continuous. Purchases are negotiated between buyers and commission wool merchants acting on behalf of their clients, the woolgrowers. This system has the advantage of quick disposal of the clip, lower handling charges and early payment.

Wool Futures

In order to take advantage of high prices which may prevail at the time, growers who plan to sell their clip some time in the future can take advantage of the futures market. 'Futures' are sold prior to the wool being physically available, and the 'futures' are then repurchased at the time of the actual delivery of the wool. Any fall in the price of the wool is offset by the cheaper price of futures to buy back. The profit on the futures usually compensates for the loss on the wool sold. Futures provide a form of insurance which can make planning and budgeting more certain. The use of wool futures in Australia has declined since the introduction of the AWC Reserve Price Scheme.

15

TYPES AND YIELD

Types

Wool is classified into grades or types, governed greatly by mean fibre diameter, style and length. For instance, the type of wool required for making a high-grade dress material is entirely different from the type needed to produce blankets, tweeds and so on. The wool, therefore, must be graded as nearly as possible to the requirements of the different sections of the wool textile industry. 'Type', as applied to wool, refers to a class, kind or description which is governed by the development of all wool characteristics together with the state and condition of the wool. The chief characteristics concerned are mean fibre diameter, style, length, colour, tensile strength, yield and vegetable matter content.

Combing and Carding Types

All wool, irrespective of breed or quality, may be classified as combing or carding according to its manufacturing usefulness.

Combing

A combing wool is one having the necessary qualities to render it suitable to undergo the combing process in the worsted system of manufacture.

Requirements are:

1. *Minimum length* — approx. 35 mm for Heilman or French combing, approx. 51 mm for Noble Combing.
2. *Soundness* — not tender.
3. *Free grown* — not cotted or matted.

Carding

Wool which falls short of any of the above requirements is classified as a carding type and is used in the woollen system of manufacture.

Merino Section

The chief types in the merino section are:

- spinners' wools;
- topmakers' wools;
- French combing;
- carding types;
- carbo types; and
- various lower grades.

Spinners' Wools

A spinner's wool must be of a good even length; regular in quality and of good colour; sound; free grown and free to nearly free of burr, seed and/or dust. This section is divided into three main lengths, namely warp, warp and half and half-warp. The Australian Wool Corporation table of types describes these as lengths 'A', 'B', and 'C' respectively.

Topmakers

Topmakers are combing wools possessing faults which exclude them from the spinners section. The chief faults are: burr or seed, heavy dust; thin and wasty; irregularity; poor colour; lack of soundness, etc.

Topmakers are divided into the same lengths as spinners.

French Combing

These wools are too short for Noble combing, but are suitable for the Heilman or French comb. Both free and burry types are included in this section.

Carding

This embraces very tender or cotted wools which are unsuitable for combing. Also included are lambs' wool, locks, crutchings and other similar types, which suit the requirements of the carding trade.

Carbo

Carbo wools are those which cannot have the burr or seed economically removed mechanically (by carding and combing). This necessitates their removal by the carbonising process. Fleece wool, before falling into this category, would, generally speaking, carry over 7 per cent of burr. The most common types are short, heavily burred wools, such as locks, lambs, crutchings, thin or short bellies or prematurely shorn fleeces.

Lower Lines of the Clip

Under this heading are included such types as black wool, overgrown fleece, dead wool, stains, etc.

Crossbred Section

In the crossbred section the types are more limited, being classified into only two sections, namely comeback and crossbred. Each of these groups would include combing, carding and carbo types.

The Australian Wool Corporation Type List

The Australian Wool Corporation type list has evolved over many years and is a standardised system used to type or describe individual lots of wool. The type has a number of components, each describing a characteristic of the wool, and all wool appraised by the AWC as a certain 'type' will be allocated the same value in the reserve price scheme.

Prior to the introduction of presale testing and sale by sample in 1972, the 'type' was determined completely by subjective appraisal. However, with the development of objective measurement, the type list now incorporates measured 'micron' and 'VM' results, enabling a more accurate and equitable appraisal of the wool. Objective measurements for greasy staple length (mm) and staple tensile strength (newtons per kilotex) were made commercially available in January 1985. Since the start of the 1985–86 season, the AWC has incorporated these results into the type list and reserve price schedule.

Other characteristics being examined for possible objective measurement include clean colour, fibre diameter variability and resistance to compression. The type list may be altered in the future as these and more raw wool characteristics are measured and accepted commercially, allowing them to be incorporated into the system of 'describing' or 'typing' lots of wool.

The chief wool characteristics for assessment to determine the AWC type are:

1. **Mean fibre diameter** Wool is inherently variable in fibre diameter, but the average mean fibre diameter of any sale lot is by far the most important characteristic in terms of processing value (hence price received at auction). Mean fibre diameter is measured commercially by the Airflow machine and expressed in micrometers (or microns): one millionth of a meter.

   Merino fleece wools which are coarser than 24.5 microns are typed as crossbred with the addition of prefix M which denotes merino wool.

2. **Style** Style is a general term which covers a number of features of the wool related to general appearance. These include the general thickness of the staple and the uniformity of crimp throughout its length, the degree of wastiness and dust in the staple tip, the vegetable matter content, staple tensile strength, greasy colour, softness and fibre density of the fleece. Style is a characteristic of wool which has traditionally indicated the 'end-use' purpose (e.g. spinners, best topmaking, good topmaking, etc.).

   The principal AWC style grade categories for merino and crossbred fleece are as follows:

Merino Combing Fleece

Choice style	— Free of vegetable fault
Superior style	— Free of vegetable fault
Spinners	— Free to nearly free to B vegetable fault
Best topmakers	— F/NF–B–C vegetable fault
Good topmakers	— F/NF–B–C–D vegetable fault
Average topmakers	— F/NF–B–C–D vegetable fault
Inferior topmakers	— F/NF–B–C–D vegetable fault

Merino French Combing Fleece

Best style and length	— Free–B–C vegetable fault
Good average length and style	— F/NF–B–C–D vegetable fault
Short	— F/NF–B–C–D vegetable fault
Very short	— F/NF–B–C–D vegetable fault

Crossbred Combing Fleece

Choice style and length	— Free to nearly free of vegetable fault
Super style and length	— Free to nearly free of vegetable fault
Best style and length	— Free–B–C vegetable fault
Good style and length	— F/NF–B–C–D vegetable fault
Average style and length	— F/NF–B–C–D vegetable fault

Crossbred Combing Fleece (cont.)

Inferior style average length	— F/NF–B–C–D vegetable fault
Good style medium length	— F/NF–B–C–D vegetable fault
Average/short length	— F/NF–B–C–D vegetable fault
Short	— F/NF–B–C–D vegetable fault
Very short	— F/NF–B–C–D vegetable fault

3. **Length** For subjective appraisal, merino fleece wools are classified into three categories, namely A, B or C length:

A length (warp) A warp wool is a well-grown, bulky wool, absolutely sound and regular. It must be of recognised length, the minimum lengths being:

Mean micron range	Approximate quality number	Minimum length
17.5–17.9	80s	86 mm
18.0–18.5	74s	87 mm
18.6–19.5	70s	89 mm
19.6–20.5	66s	91 mm
20.6–21.5	64s	93 mm
21.6–22.5	62s	94 mm
22.6–23.5	60s	95 mm
23.6–24.5	60/58s	96 mm

B Length (warp and half) Medium length or wool of uneven length, but with approximately 60 per cent of the bulk being warp length.

C Length (half-warp) A wool possessing all the characteristics of a warp wool, but shorter than the minimum required length.

For combing wool types offered for sale with additional measurement, see the AWC table of lengths in millimetres for greasy wool.

For the subjective appraisal of crossbred combing wools, the length category is incorporated with the style grading, for example good style and length, average style and length, etc.

4. **Colour** Greasy colour of wool is often not a good indicator of the scoured product. Most Australian fleece wools scour to a white, bright clean colour. However, when unscourable colour is present, the following type prefixes may be used:

H^1 = light unscourable colour;
H^2 = medium unscourable colour;
H^3 = heavy unscourable colour.

5. **Tensile strength** Grades of staple strength (or tenderness) are indicated by the following type prefixes:

W^1 = part tender;
W^2 = tender;
V = very tender (rotten).

6. **Yield and V.M. content** These characteristics are usually objectively measured prior to sale by sample. The test results are taken into account for typing the wool, although the yield is expressed separately to the type (e.g. TYPE 62 at 70 per cent yield).

Vegetable fault (V.M.) content is indicated by suffixes. No suffix indicates free or nearly free (FNF) wool; other suffixes are as follows:

Combine wool   For burr fault:
                                  B = 1.1 to 3%
                                  C = 3.1 to 7%
                                  D = greater than 7%
For seed or shine fault: S = 1.1 to 3%
                                  L^1 = 3.1 to 7%
                                  L^2 = greater than 7%

Apart from merino and crossbred fleece wool, the AWC type list also covers categories for:

- carpet wool;
- merino skirtings;
- crossbred skirtings;
- weaners, lambs;
- locks;
- crutchings;
- stains;
- cotted wool;
- black and grey wool;
- overgrown and double fleece;
- plucked and dead wool.

Belly wool is indicated by the prefix 'R'.

Australian Wool Corporation type list, July 1987

MERINO SKIRTINGS

Micron	Best Style Lth.	Good Lth.	Av. Lth	Av. Short Lth.	Short	Very Short	Micron
18.5 up	140P	146PP	156	161P	165P	169P	18.5 up
18.6–19.5	140	146P	157	161	165	169	18.6–19.5
19.6–20.5	141	146	158	162	166	170	19.6–20.5
20.6–21.5	142	147	159	163	167	171	20.6–21.5
21.6–22.5	143	148	159A	163A	167A	172	21.6–22.5
22.6–23.5	144	149	160	164	168	172A	22.6–23.5
23.6–24.5	145	150	160A	164A	168A	173	23.6–24.5
Micron	———— FNF–B–C–D ————						Micron

MERINO FRENCH COMBING FLEECE

Grade	Best Style Lth.	Good/Av. Lth. Style	Short	Very Short	Grade
80s up	118P	121PP	124PP	127PP	80s up
18.0–18.5	118	121P	124P	127P	18.0–18.5
18.6–19.5	118A	121	124	127	18.6–19.5
19.6–20.5	119	122	125	128	19.6–20.5
20.6–21.5	119A	122A	125A	128A	20.6–21.5

MERINO FRENCH COMBING FLEECE

Grade	Best Style Lth.	Good/Av. Lth. Style	Short	Very Short	Grade
21.6–22.5	120P	123P	126P	129P	21.6–22.5
22.6–23.5	120	123	126	129	22.6–23.5
23.6–24.5	120A	123A	126A	129A	23.6–24.5
Micron	F–B–C	——— FNF–B–C–D ———			Micron

CROSSBRED SKIRTINGS

Micron	Best Style Lth.	Good Lth.	Av. Lth.	Short	Very Short	Micron
19.5 up	467PP	476PP	484PP	491PP	498PP	19.5 up
19.6–20.5	467P	476P	484P	491P	498P	19.6–20.5
20.6–21.5	467	476	484	491	498	20.6–21.5
21.6–22.5	468	476A	484A	492	499	21.6–22.5
22.6–23.5	469	477	485	492A	499A	22.6–23.5
23.6–24.5	469A	477A	485A	493	500	23.6–24.5
24.6–25.5	470	478	486	493A	500A	24.6–25.5
25.6–26.5	470A	478A	486A	494	501	25.6–26.5
26.6–27.5	471	479	487	494A	501A	26.6–27.5
27.6–28.5	471A	479A	487A	495	502	27.6–28.5
28.6–30.5	472	480	488	495A	502A	28.6–30.5
30.6–34.0	473	481	489	496	503	30.6–34.0
34.1 down	474	482	490	497	504	34.1 down
Micron	——— FNF–B–C–D ———					Micron

DOGGY FLEECE

	Micron	Gd./Av. Lth. Style	Inf. Lth. Style
Mer.	22.5 up	340P	341P
	22.6–24.5	340	341
XBD	28.5 up	342P	343P
	28.6 down	342	343
		F–B–C–D	

OVERGROWN AND DOUBLE FLEECE　　　　SKTS

	Micron	Ogwn	Dbl	Ogwn	Dbl
Mer.	24.5 up	344	345	346	347
XBD	28.5 up	348		—— 350 ——	
	28.6 down	349		—— 351 ——	
		——— F–B–C–D ———			

MERINO COMBING FLEECE

Grade	CHOICE			SUPERIOR			SPINNERS			BEST T/M			GOOD T/M			AVERAGE T/M			INFERIOR T/M			Grade
	A	B	C	A	B	C	A	B	C	A	B	C	A	B	C	A	B	C	A	B	C	Micron
100/90s	1PP	9PP																				
90s up	1P	9P	14P	18P	25P	31P		43PP														
80s up	1	9	14	18	25	31	34	43P	49PP	54PP	60PP	65PP	70PP	76PP	82PP	87PP	92PP	97PP	E87PP	E92PP	E97PP	17.5–17.9
74s up	2	10	15	19	26	32	35	43	49P	54P	60P	65P	70P	76P	82P	87P	92P	97P	E87P	E92P	E97P	18.0–18.5
18.6–19.5	3	11	16	20	27	33	36	44	49	54	60	65	70	76	82	87	92	97	E87	E92	E97	18.6–19.5
19.6–20.5							37	45	50	55	61	66	71	77	83	87A	92A	97A	E87A	E92A	E97A	19.6–20.5
20.6–21.5							39	46	51	56	62	67	72	78	84	88	93	98	E88	E93	E98	20.6–21.5
21.6–22.5							40	47	52	57	63	68	73	79	85	89	94	99	E89	E94	E99	21.6–22.5
22.6–23.5							41	48	53	58	64	69	74	80	86	90	95	100	E90	E95	E100	22.6–23.5
23.6–24.5							42A	48A		59	64A	69A	75	81	86A	91	96	101	E91	E96	E101	23.6–24.5
Micron	FREE			FREE			FNF-B			FNF-B-C			FNF-B-C-D									Micron

CROSSBRED FLEECE

Micron	Choice Style Lth.	Super Style Lth.	Best Style Lth.	Good Style Lth.	Av. Style Lth.	Inf. Style Av. Lth.	Good Style Medium Lth.	Av./Short Lth.	Short	Very Short	Micron
19.5 up	400PP	410PP	420PP	430PP	440PP	E440PP	650PP	660PP	670PP	680PP	19.5 up
19.6–20.5	400P	410P	420P	430P	440P	E440P	650P	660P	670P	680P	19.6–20.5
20.6–21.5	400	410	420	430	440	E440	650	660	670	680	20.6–21.5
21.6–22.5	401	411	421	431	441	E441	651	661	671	681	21.6–22.5
22.6–23.5	402	412	422	432	442	E442	652	662	672	682	22.6–23.5
23.6–24.5	402A	412A	422A	432A	442A	E442A	652A	662A	672A	682A	23.6–24.5
24.6–25.5	403	413	423	433	443	E443	653	663	673	683	24.6–25.5
25.6–26.5	403A	413A	423A	433A	443A	E443A	653A	663A	673A	683A	25.6–26.5
26.6–27.5	404	414	424	434	444	E444	654	664	674	684	26.6–27.5
27.6–28.5	404A	414A	424A	434A	444A	E444A	654A	664A	674A	684A	27.6–28.5
28.6–30.5	405	415	425	435	445	E445	655	665	675	685	28.6–30.5
30.6–32.5	406	416	426	436	446	E446	656	666	676	686	30.6–32.5
32.6–34.5	407	417	427	437	447	E447	657	667	677	687	32.6–34.5
34.6–36.5	408	418	428	438	448	E448	658	668	678	688	34.6–36.5
36.6 down	409	419	429	439	449	E449	659	669	679	689	36.6 down
Micron	FNF		F–B–C				FNF–B–C–D				Micron

DOWNS	Micron	Good	Average	PCS/BLS Lms etc.
FINE	26.5 up	392	395	398
MEDIUM	26.6–30.5	393	396	399
COARSE	30.6 down	394	397	399A
		F–B–C–D	F–B–C–D–K	F–B–C–D–Y–K

COTTED WOOL
FLEECE   SKTS

	Micron	Soft	Med.	Hard	Md./Hd.
Mer.	22.5 up	329	330	331	335
	22.6–24.5	329A	330A	331A	335A
XBD	28.5 up	332	333	334	336
	28.6 down	332A	333A	334A	336A
	——— F–B–C–D–K ———				

XBD WEANERS AND LAMBS
COMBING   CARDING

Micron	Best Style Lth.	Gd./Av Style Lth.	Short	Best	Av.	Short	Very Short	Micron
19.5 up	515PP	521PP	527PP	547PP	551PP	555PP	559PP	19.5 up
19.6–20.5	515P	521P	527P	547P	551P	555P	559P	19.6–20.5
20.6–21.5	515	521	527	547	551	555	559	20.6–21.5
21.6–23.5	516	522	528	547A	551A	555A	559A	21.6–23.5
23.6–25.5	517	523	529	548	552	556	560	23.6–25.5
25.6–27.5	518	524	530	549	553	557	561	25.6–27.5
27.6–28.5	518A	524A	530A	549A	553A	557A	561A	27.6–28.5
28.6 down	519	525	531	550	554	558	562	28.6 down
Micron	F	FNF–B–C–D		FNF–Y	FNF–Y–C–K			Micron

XBD CARBO FLEECE
P & B

G. Lth. Colour	Short Fair Colour	Micron
533	539	22.6–23.5
534	540	23.6–25.5
535	541	25.6–28.5
536	542	28.6–29.5
538	544	29.6 down
——— K ———		Micron

MERINO WEANERS AND LAMBS

	COMBING					CARDING						
Grade	Choice	Best Lth. Style	Good Lth. Style	Av Lth.	Short	Choice	Super	Good	Av.	Short	Very Short	Grade
80s up	206	211PP	213PP	215P								80s up
18.0–18.5	207	211P	213P	215	219	243	247	251	255			18.0–18.5
18.6–19.5	208	211	213	216	220	244	248	252	256	259	265	18.6–19.5
19.6–20.5	209	212P	214P	217	221	245	249	253	257	260	265A	19.6–20.5
20.6–21.5	210	212	214	218	222	246	250	254	258	261	266	20.6–21.5
21.6–23.5	210A	212A	214A	218A	222A	246A	250A	254A	258A	261A	266A	21.6–23.5
Micron	FNF	F–B–C	FNF–B–C–D			FNF	FNF–Y	FNF–Y–C–K				Micron

MERINO CARBO

Flc.	Gd. Length P.B	Short P.B.	Micron
198 199	202 203	204 205	22.5 up 22.6–24.5
—— K ——			Micron

WIGGINGS AND SHANKS

Merino	Crossbred
24.5 up	
309	604
——— F–Y–C–K ———	

BRANDS

Merino	Crossbred	
24.5 up	28.5 up	28.6 down
326	327	328
——— F–B–C–D–K ———		

PLUCKED AND DEAD

	Micron	Good	Av.	Carbo
Mer.	24.5 up	311	313	316
XBD	28.5 up 28.6 down	611 612	613 614	615
		— F–B–C–D —		K

CROSSBRED LOCKS   CROSSBRED CRT

Micron	Good B & C	Good B Fair Col.	Average B & C	Inferior B & C	Good B & C	Good B Fair Col.	Av B Fair Col.	Micron
25.0 up	563	567	571	575	583	587	591	25.0 up
25.1–27.5	564	568	572	576	584	588	592	25.1–27.5
27.6–31.0 31.1 down	565 566	569 570	573 574	577 578	585 586	589 590	593 594	27.6–31.0 31.1 down
Micron	——— FNF–Y–C–K ———				——— FNF–Y–C–K ———			Micron

MERINO LOCKS

Micron	Good Bulk & Colour	Good Bulk Fair Colour	Av. Bulk Fair Colour	Inf. Bulk Colour	Dags
18.5 up 18.6–21.0	279P 279	282P 282	285	288	
21.1–23.0	280	283	286	289	
23.1–24.5	281	284	287	289A	290
Micron				——— FNF–Y–C–K ———	

MERINO CRT

Micron	Good Bulk & Colour	Good Bulk Fair Colour	Av. Bulk Fair Colour	Inf. Bulk Colour
18.5 up 18.6–21.0	296	299	303	
21.1–23.0	297	300	304	306
23.1–24.5	298	301	305	307
Micron			——— FNF–Y–C–K ———	

MERINO COMBING STAIN

Micron	Gd. Lth.	Av. Lth.
22.5 up 22.6–24.5	194 195	196 197
Micron	— FNF–B–C–D —	

MERINO CARDING STAIN

Micron	Gd. Lth. Med. Stain	Gd. Lth. Heavy Stain	Av. Lth. Med. Stain	Av. Lth. Heavy Stain	Short Med. Stain	Short Heavy Stain
22.5 up 22.6–24.5	267 267A	268 268A	270 271	273 274	275 276	277 278
Micron	——— FNF–Y–C–K ———					

CARPET WOOL

	STYLE	LENGTH	V.M.	TYPE	GRADE
FLEECE	Best	Best	FNF–Y	700	Av. 38
	Gd./av.	Gd./av.	FNF–Y	701	Av. 36
	Gd./av.	Prem	FNF–Y	702	Av. 36
	Inf.	Av.	FNF–Y–C–K	703	Av. 36
	Inf.	Short	FNF–Y–C–K	704	Av. 36
	Cotts	Av.	FNF–Y–C–K	705	Av. 36
LMS	Gd./av.	Gd.	FNF–Y	730	Av. 36
	Gd./av.	Short	FNF–Y	731	Av. 36
	Inf.	Av./sht	FNF–Y–C–K	732	Av. 36
SKTS	Gd.	Gd.	FNF–Y	760	Av. 36
	Av.	Av.	FNF–Y	761	Av. 36
	Inf.	Av.	FNF–Y–C–K	762	Av. 36
	Inf.	Short	FNF–Y–C–K	763	Av. 36
	Colour	Bulk	V.M.		
LKS/CRT	Fair	Gd	FNF–Y–C–K	780	Av. 36
	Av./inf.	Av./inf.	FNF–Y–C–K	781	Av. 36
STAIN		Av.	FNF–Y–C–K	790	Av. 36 Micron

XBD COMBING
STAIN

XBD CARDING
STAIN

Micron	Good Lth.	Av. Lth.	Micron	Medium	Heavy	Dags
23.6–25.5	505	510	23.6–25.5	595	597	
25.6–28.5	506	511	25.6–28.5	595A	597A	
28.6–29.5	507	512	28.6–29.5	596	598	
29.6 down	508	513	29.6 down	596A	598A	600
Micron	—— F–B–C–D ——		Micron	—— F–Y–C–K ——		

BLACK AND GREY

MERINO	FLEECE	PCS/BLS	CROSSBRED FLEECE			PCS/BLS
Good	Average	Lms etc	26.5 up	26.6–30.5	30.6 down	Lms etc.
24.5 up	24.5 up	24.5 up				
320	321	322	606	607	608	609
—— F–B–C–D ——		B–C–D–Y–K	—— F–B–C–D ——			B–C–D–Y–K

Greasy mm Lengths for Particular Australian Combing Wool Types Offered with Additional Measurement

MERINO WEANERS AND LAMBS

Micron	Choice	Best Lth./ Style	Good Lth./ Style	Av. Length	Short	Micron
80s up	206 M 66	211PP M 66	213PP M 61	215P M 55		80s up
18.0–18.5	207 M 66	211P M 66	213P M 61	215 M 56	219 55–46	18.0–18.5
18.6–19.5	208 M 67	211 M 67	213 M 62	216 M 57	220 56–46	18.6–19.5
19.6–20.5	209 M 68	212P M 68	214P M 63	217 M 58	221 57–48	19.6–20.5
20.6–21.5	210 M 69	212 M 69	214 M 64	218 M 59	222 58–48	20.6–21.5
21.6–23.5	210A M 70	212A M 70	214A M 65	218A M 61	222A 60–50	21.6–23.5

MERINO FRENCH COMBING FLEECE

Micron	Best Lth./Style	Good/Av. Lth./Style	Short	Very Short
80s up	118P 67–62	121PP 65–57	124PP 56–47	127PP Max 46
18.0–18.5	118 67–62	121P 65–57	124P 56–47	127P Max 46
18.6–19.5	118A 67–62	121 65–57	124 56–47	127 Max 46
19.6–20.5	119 68–62	122 66–57	125 56–47	128 Max 46
20.6–21.5	119A 69–63	122A 67–58	125A 57–48	128A Max 47
21.6–22.5	120P 70–64	123P 68–59	126P 58–49	129P Max 48
22.6–23.5	120 71–65	123 69–60	126 59–50	129 Max 49
23.6–24.5	120A 72–66	123A 70–61	126A 60–51	129A Max 50

MERINO SKIRTINGS

Micron	Best	Good Lth.	Av. Length	Av./Sht. Lth.	Short	Very Short	Micron
18.5 up	140P M 84	146PP M 81	156 80–74	161P 73–66	165P 65–56	169P Max 55	18.5 up
18.6–19.5	140 M 85	146P M 82	157 81–74	161 73–66	165 65–56	169 Max 55	18.6–19.5
19.6–20.5	141 M 86	146 M 83	158 82–75	162 74–68	166 67–58	170 Max 57	19.6–20.5
20.6–21.5	142 M 87	147 M 84	159 83–75	163 74–68	167 67–58	171 Max 57	20.6–21.5
21.6–22.5	143 M 88	148 M 85	159A 84–76	163A 75–69	167A 68–60	172 Max 59	21.6–22.5
22.6–23.5	144 M 89	149 M 86	160 85–76	164 75–69	168 68–60	172A Max 59	22.6–23.5
23.6–24.5	145 M 90	150 M 87	160A 86–77	164A 76–70	168A 69–61	173 Max 60	23.6–24.5

MERINO CARBO

Micron	Flc.	Good Length P & B	Short P & B
22.5 up	198	202 M 76	204 Max 75
22.6–24.5	199	203 M 76	205 Max 75

Greasy mm Lengths for Particular Australian Combing Wool Types Offered with Additional Measurement (for AWC Reserve Price Application Purposes), AWC, July 1988

Key: M = Minimum
      Max = Maximum

MERINO COMBING FLEECE

Micron	Spinners			Best Topmaking			Good Topmaking			Average Topmaking			Inferior Topmaking			Micron
17.0–17.4	34 M 85	43PP 84–76	49PP 75–67													17.0–17.4
17.5–17.9	35 M 86	43P 85–77	49P 76–68	54PP M 85	60PP 84–76	65PP 75–68	70PP M 83	76PP 82–74	82PP 73–66	87PP M 81	92PP 80–72	97PP 71–66	E87PP M 79	E92PP 78–69	E97PP 68–57	17.5–17.9
18.0–18.5	36 M 87	43P 86–78	49 77–68	54P M 86	60P 85–77	65P 76–68	70P M 84	76P 83–75	82P 74–66	87P M 82	92P 81–73	97P 72–66	E87P 79–70	E92P M 80	E97P 69–57	18.0–18.5
18.6–19.5	37 M 89	44 88–79	50 78–68	54 M 88	60 87–78	65 77–68	70 M 86	76 85–76	82 75–66	87 M 84	92 83–74	97 73–66	E87 M 82	E92 81–71	E97 70–57	18.6–19.5
19.6–20.5	39 M 91	45 90–81	51 80–69	55 M 90	61 89–80	66 79–69	71 M 88	77 87–78	83 77–67	87A M 86	92A 85–76	97A 75–67	E87A M 84	E92A 83–73	E97A 72–57	19.6–20.5
20.6–21.5	40 M 93	46 92–82	52 81–70	56 M 92	62 91–81	67 80–70	72 M 90	78 89–79	84 78–68	88 M 88	93 87–77	98 76–68	E88 M 85	E93 84–74	E98 73–58	20.6–21.5
21.6–22.5	41 M 94	47 93–83	53 82–71	57 M 93	63 92–82	68 81–71	73 M 91	79 90–80	85 79–69	89 M 89	94 88–78	99 77–69	E89 M 86	E94 85–75	E99 74–59	21.6–22.5
22.6–23.5	42 M 95	48 94–84		58 M 94	64 93–83	69 82–72	74 M 92	80 91–81	86 80–70	90 M 90	95 89–79	100 78–70	E90 M 87	E95 86–76	E100 75–60	22.6–23.5
23.6–24.5	42A M 96	48A 95–85		59 M 95	64A 94–84	69A 83–73	75 M 93	81 92–82	86A 81–71	91 M 91	96 90–80	101 79–71	E91 M 88	E96 87–77	E101 76–61	23.6–24.5

MERINO DOGGY FLEECE

Micron	Good/Av. Lth./Style	Inf. Lth. Style
22.5 up	340P M 80	341P Max 79
22.6–24.5	340 M 81	341 Max 80

CROSSBRED SKIRTINGS

Micron	Best	Good Lth.	Av. Lth.	Short	Very Short	Micron
19.5 up	467PP M 88	476PP M 84	484PP 83–74	491PP 73–66	498PP Max 65	19.5 up
19.6–20.5	467P M 89	476P M 85	484P 84–75	491P 74–68	498P Max 67	19.6–20.5
20.6–21.5	467 M 90	476 M 86	484 85–75	491 74–68	498 Max 67	20.6–21.5
21.6–22.5	468 M 91	476A M 87	484A 86–76	492 75–69	499 Max 68	21.6–22.5
22.6–23.5	469 M 92	477 M 88	485 87–77	492A 76–69	499A Max 68	22.6–23.5
23.6–24.5	469A M 94	477A M 90	485A 89–78	493 77–70	500 Max 69	23.6–24.5
24.6–25.5	470 M 97	478 M 93	486 92–79	493A 78–70	500A Max 69	24.6–25.5
25.6–26.5	470A M 99	478A M 95	486A 94–80	494 79–71	501 Max 70	25.6–26.5
26.6–27.5	471 M 100	479 M 96	487 95–81	494A 80–72	501A Max 71	26.6–27.5
27.6–28.5	471A M 101	479A M 97	487A 96–82	495 81–73	502 Max 72	27.6–28.5
28.6–30.5	472 M 102	480 M 98	488 97–83	495A 82–74	502A Max 73	28.6–30.5
30.6–34.0	473 M 103	481 M 99	489 98–84	496 83–75	503 Max 74	30.6–34.0
34.1 down	474 M 104	482 M 100	490 99–85	497 84–76	504 Max 75	34.1 down

MERINO COMBING STAIN

Micron	Good Length	Av. Length
22.5 up	194 M 76	196 Max 75
22.6–24.5	195 M 76	197 Max 75

XBD WEANERS AND LAMBS

Micron	Best Style/Lth.	Good/Av. Style/Lth.	Short
19.5 up	515PP M 67	521PP M 57	527PP 56–50
19.6–20.5	515P M 69	521P M 59	527P 58–50
20.6–21.5	515 M 70	521 M 60	527 59–51
21.6–23.5	516 M 73	522 M 63	528 62–56
23.6–25.5	517 M 75	523 M 65	529 64–56
25.6–27.5	518 M 78	524 M 68	530 67–56
27.6–28.5	518A M 80	524A M 70	530A 69–61
28.6 down	519 M 83	525 M 73	531 72–61

XBD DOGGY FLEECE

Micron	Good/Av. Lth./Style	Inf. Lth./Style
28.5 up	342P M 95	343P Max 94
28.6 down	342 M 97	343 Max 96

CROSSBRED FLEECE

Micron	Choice	Super	Best	Good	Av.	Inf.	Micron	Gd. Style Medium Lth.	Av./Short Lth.	Short	Very Short	Micron
19.5 up	400PP M97	410PP M97	420PP M92	430PP M87	440PP M80	E440PP M80	19.5 up	650PP 86–72	660PP 79–61	670PP 60–48	680PP Max 47	19.5 up
19.6–20.5	400P M99	410P M99	420P M94	430P M89	440P M82	E440P M82	19.6–20.5	650P 88–74	660P 81–63	670P 62–48	680P Max 47	19.6–20.5
20.6–21.5	400 M100	410 M100	420 M95	430 M90	440 M83	E440 M83	20.6–21.5	650 89–75	660 82–64	670 63–49	680 Max 48	20.6–21.5
21.6–22.5	401 M101	411 M101	421 M96	431 M91	441 M84	E441 M84	21.6–22.5	651 90–76	661 83–65	671 64–50	681 Max 49	21.6–22.5
22.6–23.5	402 M102	412 M102	422 M97	432 M92	442 M85	E442 M85	22.6–23.5	652 91–77	662 84–66	672 65–51	682 Max 50	22.6–23.5
23.6–24.5	402A M105	412A M103	422A M98	432A M93	442A M86	E442A M86	23.6–24.5	652A 92–78	662A 85–67	672A 66–52	682A Max 51	23.6–24.5
24.6–25.5	403 M108	413 M104	423 M99	433 M94	443 M87	E443 M87	24.6–25.5	653 93–79	663 86–68	673 67–53	683 Max 52	24.6–25.5

continued

CROSSBRED FLEECE

Micron	Choice	Super	Best	Good	Av.	Inf.	Micron	Gd. Style Medium Lth.	Av./Short Lth.	Short	Very Short	Micro
25.6–26.5	403A M110	413A M105	423A M100	433A M95	443A M88	E443A M88	25.6–26.5	653A 94–80	663A 87–69	673A 68–54	683A Max 53	25.6–26.5
26.6–27.5	404 M112	414 M107	424 M102	434 M97	444 M90	E444 M90	26.6–27.5	654 96–82	664 89–71	674 70–56	684 Max 55	26.6–27.5
27.6–28.5	404A M115	414A M110	424A M105	434A M100	444A M93	E444A M93	27.6–28.5	654A 99–84	664A 92–72	674A 71–57	684A Max 56	27.6–28.5
28.6–30.5	405 M117	415 M112	425 M107	435 M102	445 M95	E445 M95	28.6–30.5	655 101–86	665 94–73	675 72–58	685 Max 57	28.6–30.5
30.6–32.5	406 M120	416 M115	426 M110	436 M105	446 M98	E446 M98	30.6–32.5	656 104–88	666 97–74	676 73–59	686 Max 58	30.6–32.5
32.6–34.5	407 M122	417 M117	427 M112	437 M107	447 M99	E447 M99	32.6–34.5	657 106–90	667 98–75	677 74–60	687 Max 59	32.6–34.5
34.6–36.5	408 M125	418 M120	428 M115	438 M110	448 M100	E448 M100	34.6–36.5	658 109–90	668 99–75	678 74–60	688 Max 59	34.6–36.5
36.6 down	409 M130	419 M125	429 M120	439 M115	449 M100	E449 M100	36.6 down	659 114–90	669 99–75	679 74–60	689 Max 59	36.6 down

XBD COMBING STAIN

Micron	Good Length	Av. Length
23.6–25.5	505 M 79	510 Max 78
25.6–28.6	506 M 82	511 Max 81
28.6–29.5	507 M 83	512 Max 82
29.6 down	508 M 84	513 Max 83

XBD CARBO

Micron	Good Length Colour	Short Fair Colour
22.6–23.5	533 M 77	539 Max 76
23.6–25.5	534 M 79	540 Max 78
25.6–28.5	535 M 82	541 Max 81
28.6–29.5	536 M 83	542 Max 82
29.6 down	538 M 84	544 Max 83

Note: This list is to be used in conjunction with the complete July 1987 AWC type list.

Yield

Yield is the quantity of clean scoured wool, recovered from a given weight of greasy wool, after processing for the removal of all impurities. It is expressed as a percentage of the original greasy weight.

In the broad sense, yield may be classed under three headings, namely:

1. *Washing*. This is the percentage of clean wool, including burrs and seed — if any — left after scouring a line of greasy wool.
2. *Top and noil*. This percentage is calculated from the actual weight of top and noil, plus recognised regains, obtained after scouring, carding and combing.
3. *Carbonising*. This is the percentage of absolutely clean wool resulting from a line of burry greasy wool after it has been scoured and carbonised. In addition to the loss

of vegetable matter, this wool loses a small percentage of fibre substance in the carbonising process.

When calculating yield percentage, allowance is made for moisture regain. Standard regains for semi-processed wools are as shown in Table 15.1.

Table 15.1   Standard regains for semi-processed wools.

	International Wool Textile Organisation
Clean scoured	17%
Tops in oil	19%
Tops dry combed	18¼%
Lister and Noble noils	14%
Schlumberger noils	16%

Note: Wool can absorb up to 33⅓% of its weight in moisture before becoming saturated.

Regain is the weight of water in a sample expressed as a percentage of the weight of the moisture-free fibre in the sample.

$$\text{Regain} = \frac{\text{Weight of water}}{\text{Oven dry weight}} \times \frac{100}{1}$$

Moisture content is the weight of water in a sample expressed as a percentage of the air dry weight of the wool sample.

$$\text{Moisture content} = \frac{\text{Weight of water}}{\text{Air dry weight}} \times \frac{100}{1}$$

To determine oven dry weight of wool, the IWTO conditioning regulations require that test samples must be dried in an oven at a constant temperature of $105° \pm 2°C$. The drying of samples is considered as complete when two consecutive weighings made at an interval of fifteen minutes show a difference equal to less than 0.05 per cent of the weight of the sample.

Assessing Yield

The following characteristics are closely associated with yield of greasy wool and should be taken into account as a basis for visual estimation.

Length

A long wool usually yields better than a short wool, due to the lesser proportion of tip to the overall length of staple. Most impurities in wool, such as burr and dust, are carried in the tip of the staple.

Quality Number

Generally speaking, the coarser the fibre, the higher the yield.

Grease

The amount of grease in wool may be detected by the eye and by feel. Heavy condition wools usually have a creamy yellow colour and feel greasy. A black greasy staple tip also indicates a low yield. Rams' wool may be detected by its strong odour and by the fact that, although usually bright in colour, it tends to be heavy in condition and low-yielding.

Vegetable Matter

The quantity of vegetable matter in wool and the variety has a direct relationship to the yield. Noogoora and Bathurst burrs, for example, are very heavy, and when found in wool greatly reduce the yield.

The following allowances should be made for the Australian Wool Corporation burry wool classification: Free — nil, F/NF up to one per cent, B — one to three per cent, C — three to seven per cent, D — greater than seven per cent.

Dust
Amount visible and type, for example, whether light powdery dust or heavy sand.

Density
Generally a well-grown wool with a bulky staple yields better than a thinly-grown wasty wool.

Approximate Washing Yields for Free Wools
Table 15.2 gives approximate yields for wool types of good average quality. To obtain a more accurate estimate, the student should adjust these figures by a few percentage points either way to make due allowance for variation in the factors which influence the yield of greasy wool.

Table 15.2   Approximate washing yields for free wools.

Merino			Crossbred		
Free FLC	80	67%	Free FLC	60 CBK	73%
	74	68%		58 CBK	74%
	70	69%		56 XD	75%
	64	70%		50 XB	76%
	60	71%		46/44 XB	77%
BST top		67%		40/36 Lincoln	70%
top		62%		CBK/LMS	73%
NKS		68%		CBK PCS	65%
Dusty backs		46%		CBK BLS	60%
LMS		65%		XB LMS	75%
PCS		62%		XB PCS	66%
BLS		59%		XB BLS	61%
STD PCS		40%		XB STD PCS	42%
Locks		46%		XB LKS	48%
Crutchings		44–48%		XB Crutchings	50%

Note: In the case of burry wool, estimate the percentage of vegetable fault and reduce the yield accordingly.

Wool Valuation
The value of greasy wool is controlled by the yield and one must be able to 'type and yield' wool before this value can be ascertained. Although impurities contained in the greasy wool have some commercial value, for valuing purposes they are usually ignored. Much of the greasy weight consists of these impurities, and the proportion must be known so that the price per kilogram in the grease, that will correspond with the clean-scoured cost, can be fixed. This price then forms the 'limit' that the buyer is prepared to bid for a line of wool.

The greasy value is affected by the amount of extraneous matter present in the wool. The higher the yield, or the greater the percentage of pure wool remaining from every 100 parts treated, the higher the greasy value will be, all things being equal.

As an example, say the clean-scoured value of a certain type of wool is 260 cents per

Conditioning oven and balance   *CSIRO rapid dryer and direct reading regain tester*

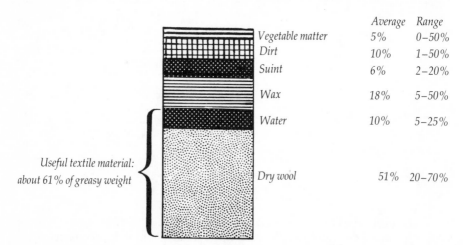

	Average	Range
Vegetable matter	5%	0–50%
Dirt	10%	1–50%
Suint	6%	2–20%
Wax	18%	5–50%
Water	10%	5–25%
Dry wool	51%	20–70%

Useful textile material: about 61% of greasy weight

Figure 15.1   Composition of greasy merino wool
The amount of actual dry wool contained in greasy merino wool can vary from 20 per cent to 70 per cent according to the amount of other material present. On the average, 1 kg (2 lb 4 oz) of greasy wool contains 0.6 kg (1 lb 5 oz) of useful textile material, which includes both bone-dry wool and the water which it absorbs from the atmosphere.

kilogram. The estimated yield is 60 per cent. The equivalent greasy value will be found by the following method:

$$\frac{\text{Clean-scoured value} \times \text{Yield}}{100} \quad \text{or} \quad \frac{260 \times 60}{100} = 156 \text{ cents per kilogram}$$

From this illustration it will be seen that the proportion of foreign matter present in wool must be carefully considered when appraising its value.

Measuring Yield in the Laboratory from a Hand Sample

Sampling
Before the sampling table is used, make sure that it is clean. The sample to be tested is spread evenly over the table with the tip uppermost. The sub-sample is then drawn by taking small plucks from the sample; these plucks should be similar in size and evenly spaced for distance. Commencing at one end of the table the plucks are taken working backwards and forwards across the sample and at the same time down the whole length of it. In this manner the whole sample is covered, thus giving a truly representative sub-sample.

Weighing
Before the sub-sample is weighed, the level of the scale should be checked by making sure that the bubble is central and the weighing tray cleaned. If a mesh bag is to be used, turn the bag inside out to ensure that no wool has been left in it. Select a number tag, record its number on the scour sheet, and place the tag in the mesh bag. Proceed to weigh the sample, record the correct weight on the scour sheet then place the sub-sample in the mesh bag, tying the throat of the bag with a rubber band. Check the weight off the scales.

Scouring
The sub-sample is scoured in the four-bowl test scour; it is gently agitated and passed from bowl to bowl through the squeeze rollers. The time each sample is in the bowls varies and is largely a matter of experience but the approximate times given below can be used as a guide:

Bowl	
No. 1	12 minutes
2	10 minutes
3	8 minutes
4	6 minutes

The mixture for each bowl varies according to individual preference and experience. The following mixture is for bowls of 90-litre capacity.

Bowl		Temperature
1	100 cc Teepol (detergent) 140 g Soda Ash	58°C
2	125 cc Teepol 120 g Soda Ash	55°C
3	75 cc Teepol 70 g Soda Ash	52°C
4	Water rinse bowl only	44°C

After the sub-sample has passed through the scour it is placed in the Hydro Extractor. When the spin drying is completed the sample is removed from its bag and the sub-sample 'opened up' to assist drying. The sub-sample is then placed in an oven and dried to a constant weight.

Yield Calculations

$$\text{Washing yield} = \frac{\text{Scoured oven dry weight}}{\text{Greasy air dry weight}} \times \frac{100 + \text{regain}}{1}$$

Example

$$\text{Greasy air dry weight} = 125 \text{ grams}$$
$$\text{Scoured oven dry weight} = 75 \text{ grams}$$
$$\text{Regain} = 17\%$$

$$\text{Washing yield} = \frac{75}{125} \times \frac{100 + 17}{1}$$

$$= \frac{75}{125} \times \frac{117}{1}$$

$$= \frac{3}{5} \times \frac{117}{1} = \frac{351}{5}$$

$$= 70.2\%$$

$$\text{Wool base} = \frac{\text{Scoured oven dry weight}}{\text{Greasy air dry weight}} \times \frac{(100 - A - E - T)}{1}$$

A = Ash content, as a percentage of the oven-dry scoured test specimen.

E = Alcohol extractable matter, as percentage of the oven-dry scoured test specimen.

T = The total ovendry, ash-free, extractives-free, alkali-insoluble impurities present in, and expressed as a percentage of, the oven-dry test specimen. See test method IWTO–19–76 (E).

Australian Carbonising Yield 17% Regain is proving popular in Australia, Japan, Korea and Belgium as the basis for trade in carbonising and carding types (locks, crutchings, lambs etc). The yield calculation from wool base and vegetable matter (VM) base allows for expected processing losses during carbonising.

Australian Carbonising Yield

ACY% = IWTO Clean Wool Content Yield % + .1616 VM% − 5.12

Note: IWTO Clean Wool Content = Wool Base × 1.1972

Japanese Clean Scoured Yield is the basis for most trade with Japan. Standard allowances for residual grease and dirt, and 16% moisture regain is added to wool base. Although this yield has the vegetable matter (VM) deducted, no allowance is made for fibre loss which would occur during processing.

Japanese Clean Scoured Yield

Japanese Clean Scoured Yield = Wool Base × 1.1777.

IWTO TEST CERTIFICATE

2-562920-P5   IWTO-19.29

20 BALES   CNE/AUROA   AAAM

REF.   S02S 5955   CNE/AUROA   AAAM   BALE NO. GROSS TARE
                                                77258   4   201   2
                                                                    5   190   2
TOTAL BALE WEIGHTS: 7   197   2
                                                                    8   164   2
TEST HOUSE GROSS   3 574 KG   TARE   40 KG   NETT   3 534 KG   10   185   2
                                                                   11   188   2
                                                                   12   141   2
TEST RESULTS: 14   165   2
                                                                   15   174   2
1.   WOOL BASE - 2 SUBSAMPLES   56.85 %   18   176   2
                                                                   19   179   2
2.   MEAN FIBRE DIAMETER - 2 SPECIMENS   20.9 MICRONS   22   180   2
                                                                   24   169   2
3.   VEGETABLE MATTER BASE   2.8 %   26   189   2
       INCLUDING **** % HARD HEADS-TWIGS   29   199   2
                                                                   30   177   2
                                                                   34   181   2
4.   IWTO SCHLUMBERGER DRY TOP & NOIL YIELD   64.7 % 2 286 KG   35   190   2
                                                                   37   193   2
5.   IWTO SCOURED YIELD, 17% REGAIN   71.4 % 2 523 KG   39   136   2

6.   JAPANESE CLEAN SCOURED YIELD   67.0 % 2 368 KG   GROSS 3574 KG TARE 40 KG NETT 3534 KG

7.   AUSTRALIAN CARBONISING YIELD, 17% REGAIN 63.4 % 2 241 KG

ADDITIONAL INFORMATION FOR APPRAISAL PURPOSES ONLY:

8.   VEGETABLE MATTER COMPOSITION
       BURR-MEDIC 50 % SEED-SHIVE   50 % HARD HEADS-TWIGS   0 %

CHARGES:

     CERTIFICATION FEE $27.30

     TOTAL $27.30

                                                        BALES WERE SAMPLED ON   7.8.85
                                                        AWTA LTD SERVICE TIME   3 DAYS

25.0094 6759   13 386   9 812 273 474   0.6759   96070

For and on behalf of Australian Wool Testing Authority Ltd — Incorporated in Victoria

D J WARD B Sc M.A.I.A.S   H.W. HOPKINS   1.10.85   3
MANAGING DIRECTOR

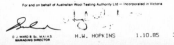

TEST CERTIFICATE

2-596222-L6   AUSTRALIAN STANDARD   AS2810

10 BALES   JMA/KINCORA   SUPAAA

REF.   S05S 5536   JMA/KINCORA   SUPAAA   BALE NO. GROSS TARE
                                                85393   16   191   2
                                                                   17   188   2
TOTAL BALE WEIGHTS: 19   193   2
                                                                   20   180   2
TEST HOUSE GROSS   1 841 KG   TARE   20 KG   NETT   1 821 KG   23   187   2
                                                                   24   190   2
                                                                   29   178   2
TEST RESULTS: 30   199   2
                                                                   31   190   2
1.   MEAN STAPLE LENGTH - 63 STAPLES   110 MM   36   145   2

2.   COEFFICIENT OF VARIATION OF STAPLE LENGTH   15 %   GROSS 1841 KG TARE 20 KG NETT 1821 KG

3.   MEAN STAPLE STRENGTH - 63 STAPLES   26 NEWTONS/KTEX

4.   DISTRIBUTION OF POSITION OF BREAK:

       BROKE IN THE TIP REGION   5 %

       BROKE IN THE MIDDLE REGION   90 %

       BROKE IN THE BASE REGION   5 %

CHARGES:

     CERTIFICATION FEE $22.00

     TOTAL $22.00

25.0094 3548   10 629   8 788 266 028   0.3548   165600

For and on behalf of Australian Wool Testing Authority Ltd — Incorporated in Victoria

D J WARD B Sc M.A.I.A.S   H.W. HOPKINS   1.10.85   5
MANAGING DIRECTOR

Figure 15.3   Australia Wool Testing Authority certificates

16

WOOL TESTING

All wool sold by sample must be core tested for yield, vegetable matter base, and mean fibre diameter. On receival at the wool broker's store and prior to the auction, all bales are core sampled and weighed independently by officers of the Australian Wool Testing Authority, or the sampling and weighing operation is conducted by wool store staff under the supervision of the AWTA.

Test certificates are issued according to the regulations of the International Wool Textile Organisation (IWTO). They are usually available in the selling centre within ten working days of sampling. The certified test results provided are as follows: 1. Wool base; 2. Mean fibre diameter; 3. Vegetable matter base; 4. IWTO Schlumberger top and noil yield; 5. IWTO scoured yield 17% regain; 6. Japanese clean scoured yield; 7. Australian carbonising yield 17%; and 8. Vegetable matter composition.

Additional measurements for staple length, staple strength and average yellowness of wool are also now commercially available.

The IWTO test procedures may be found in the following appendices to this chapter:
- I. IWTO core test regulations;
- II. Determination of wool base, vegetable matter base, IWTO clean wool content and IWTO scoured yield in raw wool (IWTO-19-76-(E));
- III. Method of determining wool fibre diameter by projection microscope (IWTO-8-61(E));
- IV. Determination by the airflow method of the mean fibre diameter of core samples of raw wool (IWTO-28-82(E)).

Appendix I

IWTO Core Test Regulations

Regulations relating to Core Testing for the determination of:

1. IWTO Clean Wool Content in greasy or scoured wool.
2. Estimation of a Commercial Top and Noil Yield by the application of agreed Processing Allowances to the Theoretical Top and Noil Yield. This estimation applies to greasy wool only.
3. Estimation of a Commercial Card Sliver Yield by the application of agreed Processing Allowances to the Theoretical Card Sliver Yield. This estimation applies to greasy wool only.
4. IWTO Scoured Yield in greasy or scoured wool.

These Regulations should be read in conjunction with the current versions of:

IWTO Method for the Determination of Wool Base, Vegetable Matter Base, IWTO Clean Wool Content and IWTO Scoured Yield in Raw Wool. IWTO International Trade Agreement Applicable to Transactions in Raw Wool.

1. Contracting parties must come to prior agreement on the testing house to issue the Core Test Certificate. Such testing house having undertaken to carry out yield tests in accordance with the IWTO Core Test Regulations.

    When the Core Test is carried out in the country of origin of the merchandise, the Core Test Certificate should be sent to the buyer as soon as possible.

    When the Core Test is carried out in the country of destination, the Core Test Certificate shall be sent to the seller without delay, and in any case, not later than 90 days after the arrival of the wool at its port of final destination, unless otherwise agreed. Intention to retest is to be declared within 15 days of receipt of the original Core Test Certificate, and the retest is to be completed within 120 days of the arrival of the wool.

    When a Core Test Certificate is issued by a testing house, which has been agreed between buyer and seller, the result shall be final and binding on both parties to the contract. However, a retest may be made in the case of obvious error or if both parties agree that a retest should be made. If a retest is necessary as above, only one retest shall be admitted and it shall be made on samples obtained by recoring the consignment if the wool is still available. If the wool is not available, the remaining core samples from the original test shall be used. Any certificate resulting from a retest shall bear the word 'retest' and all copies of the certificate originally issued shall be returned to the testing house for destruction.

    Note: This clause will be subject to review when an international scheme for the control of core testing laboratories, as approved by IWTO, has been set up.

2. All the bales in the test lot shall be undamaged and of similar dimensions and weight within commercial limits, and shall be uniformly and randomly packed, the wool being from one

Hand coring

country of origin and of the same general type and yield. In addition, all the scoured wool must have received similar treatment.

3. The bales must be core sampled in accordance with the procedure laid down in Appendix A.

4. *Samples drawn for appraisal.* If appraisal samples are drawn from the consignment after core sampling, their weight shall be recorded by a party who is independent of buyer and seller (the testing house staff or accredited representatives of the testing house may be recognised for this purpose) and the record of this weight shall be sent to the testing house who will record it on the Core Test Certificate, which shall be recalculated to take account of the wool removed. If the weight of such samples exceeds 1% of the weight of the consignment at core sampling the Core Test Certificate will become invalid.

5. The calculation of:
   (a) IWTO Clean Wool Content;
   (b) Theoretical Top and Noil Yield;
   (c) Estimated Commercial Top and Noil Yield;
   (d) Theoretical Card Sliver Yield; and
   (e) Estimated Commercial Card Sliver Yield is given in Appendix B4 and B5.

6. The calculation of IWTO Scoured Yield is given in Appendix C.

7. **Certification**
   There are two IWTO Test Certificates that can be issued, namely,
   (a) The IWTO Clean Wool Content Certificate (greasy or scoured wool).
   (b) The IWTO Scoured Yield Certificate (greasy or scoured wool).
   Both certificates must show the following information:
   - number of bales in the test lot;
   - whether the wool is greasy or scoured;
   - clear identification of the test lot;
   - gross weight at time of coring;

Commercial core and grab sampling line

- declared tare;
- weight of samples (if any) removed for appraisal;
- number of subsamples tested.

In addition to the above, data appropriate to each particular type of certificate, as indicated below, must be given. Percentages (shown to 2 decimal places) are based on the weight of the core sample.

(a) For the IWTO CLEAN WOOL CONTENT CERTIFICATE (greasy or scoured wool):
- The title: 'IWTO Clean Wool Content'
- Vegetable Matter base (including . . . % hard heads and twigs)   %
- Wool Base   %
- IWTO Clean Wool Content   % and weight.

(Here the following statement, which qualifies the additional *optional* information that can be included, must be placed.)

"The following values are calculated from the tested and certified values given above using standard Processing Allowances incorporated in the IWTO Core Test Regulations. These values are *not certified*."

Optional additional information to obtain an Estimated Commercial Top and Noil Yield (greasy wool only):
- Theoretical Top and Noil Yield (without allowance for processing loss)   %
- Processing Allowance   %
- Estimated Commercial Top and Noil Yield   % and weight.

Optional additional information to obtain an Estimated Commercial Card Sliver Yield (greasy wool only):
- Theoretical Card Sliver Yield (without allowance for processing loss)   %
- Processing Allowance   %
- Estimated Commercial Card Sliver Yield   % and weight.

(Here will follow the declaration and signature of the testing house).

(b) For the IWTO SCOURED YIELD CERTIFICATE (greasy or scoured wool):
- The title: 'IWTO Scoured Yield'   %
- Vegetable Matter Base   %
- Wool Base   %
- IWTO Scoured Yield at R% regain (representing Maximum Scoured Yield at that regain without allowance for processing loss*)   % and weight

(Here will follow the declaration and signature of the testing house.)

Note: Since this yield is without allowance for processing loss in commercial scouring and incorporates the standard ash and alcohol extractives (see Appendix C), it may differ from 'clean scoured' yield defined in Article 7 (a) of the IWTO International Trade Agreement Applicable to Transactions in Raw Wool.

8. **IWTO Combined Test Certificate**

Where delivery comprises individual parts, all of which have been tested in accordance with appropriate IWTO Test Methods and the IWTO Core Test Regulations, a combined certificate may be calculated by the testing house nominated in the contract for delivery against the contract. The Certificate shall be known as an IWTO Combined Test Certificate, as defined in IWTO(E)-4-73 (E).

9. **Test House Guarantees**

Test Houses are financially responsible for ensuring that testing is correctly carried out and that test results are within the precision limits of the methods used (Article 16 of the IWTO International Trade Agreement Applicable to Transactions in Raw Wool).

IWTO Combined Test Certificates shall be fully guaranteed financially by the Test House issuing the certificate (Article 18 of the IWTO International Trade Agreement Applicable to Transactions in Raw Wool).

Appendix A   Core Sampling

A.1. Principle

1. Every bale in the lot (as defined in Section 2 of the Regulations) shall be core sampled at the time of weighing. Weighing and sampling should be carried out at the same time so as to ensure that no change in bale weight occurs between the operations of weighing and sampling.

2. a) When a test is to be conducted post-sale (i.e. after auction, tender, etc.) and/or when a pre-sale test is to be conducted on samples drawn manually, weighing and sampling shall normally be carried out by the testing house staff. Alternatively, accredited representatives of the testing house may be used on terms approved by National Committees of IWTO of the countries concerned.

   b) When a test is to be conducted pre-sale, on samples to be drawn by machine, the core sampling and weighing functions may be performed by accredited representatives of the testing house provided full-time supervision by Test House staff is operative.

      In all cases where accredited representatives are used, pre-sale or post-sale, it is essential that:
      - the sampling schedule is laid down by the test house in terms of IWTO Regulations;
      - the accredited representatives are not principals (or employees of principals) in the sale transaction;
      - the accredited representatives are fully trained, sworn weighers and samplers whose accreditation is reviewed at least annually, and lodged with the National Committee.
      The testing house shall record:
      - signatures of those responsible for weighing and sampling, together with an indication as to whether they are testing house staff or accredited representatives;
      - the date and location of the weighing and sampling operations;
      - if specifically required, these details must be available to interested parties through the preparation of a sampling report signed by the testing house.

3. The weight of cores taken shall be sufficient to provide five sub-samples, each preferably 200 grams and not less than 150 grams.

4. Sampling equipment and the number of cores taken per lot shall be such as to produce a sampling precision of ± 1% IWTO Clean Wool content at a probability level of 0.95*.

A.2. Sampling Procedure

(a) Cores may be taken by:
   - manually operated pressure coring equipment;
   - power driven rotary coring equipment;
   - automated power pressure coring machines.

(b) The number of cores taken per bale and the size of tube must be such as to satisfy the requirements given in (3) and (4) above.

(c) The tube must be inserted in the direction of compression of the bale.

(d) The point of penetration must be randomly positioned over the surface to be cored but must not be nearer than 3 inches to the bale edge.

(e) • When manual pressure coring equipment or power driven rotary coring equipment is used the following procedure shall apply:
   — Where one core per bale is taken, the core shall be taken alternately through the cap and base of succeeding bales. (The terms 'cap' and 'base' refer to the bale surfaces upon which the baling pressure is applied).
   — Where two cores per bale are taken, a core shall be taken through the cap and base of each bale.
   — Where four, six, eight or ten cores are taken, then two, three, four or five shall be taken through cap and base respectively.

---

* The minimum number of cores to be taken per bale to produce a sample having the required precision may be estimated from the following formula:

$$K = \frac{4\sigma^2_w}{N}$$

where K = minimum number of cores to be taken from each bale in the lot.

   N = number of bales in the lot.

   σ_w = standard deviation of Clean Wool Content of cores within bales. Values of σ_w for a wide variety of wool types are shown in American Society for Testing Materials Method D 1060-65 (1971), Table A1.

Note: For scoured wool, σ_w should preferably be obtained by testing large numbers of cores from consignments of similar type. As an interim guide σ_w may be taken as 1.0 for bales which have been packed at least two days previously (or after equivalent conditioning by binning or spreading before packing). For wools packed immediately from the drier in the scouring works σ_w should be increased by up to 2.0 due to non-uniform moisture content.

Where K is not a whole number its value shall be rounded off to the next highest whole number.

— Where three cores per bale are taken, alternate bales shall be cored 2 cap-1 base and 1 cap-2 base.
— Where five, seven or nine cores are taken, the same principle shall apply — with the corresponding increases in numbers.

• When automated power operated pressure coring machines are used which take a core from 95 per cent or more of the length of the bales, the core shall be taken through the base of the bale where contamination with bagging is more readily avoided.

(f) Before the tube is inserted sufficient bagging must be removed from the penetration area to avoid contamination of the sample with bagging.

(g) On withdrawal of the tube the core shall be extruded into a sample container without loss of material and without undue exposure to the atmosphere. The sample container shall be so constructed that the sample stored therein shall not show a material change in its moisture content during storage, before weighing.

Appendix B   Calculations

B.1. IWTO Clean Wool Content

This is calculated from 'wool base'. Wool base, V.M. base and IWTO Clean Wool content are defined as follows:

Wool Base: The oven-dry weight of wool fibre free from all impurities, i.e. ash-free, alcohol extractives-free and free from all vegetable matter and other alkali-insoluble impurities, expressed as a percentage of the weight of the core sample.

Vegetable Matter Base (V.M. Base): The oven-dry weight of ash-free, alcohol extractives-free burrs (including hardheads), twigs, seeds, leaves and grasses present, expressed as a percentage of the weight of the core sample.

Note 1: 'Hard heads' are Noogoora burrs (*Xanthium pungens*), Bathurst burrs (*Xanthium spinosum*) and similar burrs of a bean-like character covered in readily removable spines. They do not contribute to a loss of wool during processing.

Note 2: 'Twigs' are small pieces of stick, woody leaf stalks and similar woody material. They do not contribute to a loss of wool during processing.

IWTO Clean Wool Content is the amount of wool base adjusted to a standard ash plus alcohol extractives content of 2.27%* (i.e. expressed as a percentage of wool base plus the standard ash and alcohol extractives) and brought finally to a regain of 17%.

That is:

$$\text{IWTO clean wool content} = \text{Wool base} \times \frac{100}{97.73} \times \frac{117}{100}$$
$$= \text{Wool base} \times 1.1972$$

B.2. Theoretical Top and Noil Yield (without allowance for processing loss)

This is defined as 'the maximum yield which theoretically could be obtained if all the wool fibre in a lot could be converted to top and noil yield without processing loss'.

A uniform tear of 8:1 is assumed and the Theoretical Top and Noil Yield is calculated from 'Wool Base' using appropriate conversion factors as follows:

Noble Combed in oil (4.6% T.F.M.)	= Wool Base × 1.255
Noble Combed dry (1.0% T.F.M.)	= Wool Base × 1.205
Schlumberger Combed in oil (4.6% T.F.M.)	= Wool Base × 1.257
Schlumberger Combed dry (1.0% T.F.M.)	= Wool Base × 1.207

Note: The above calculations of Theoretical Top and Noil Yield are based on the following standard % values for regain and total fatty matter (T.F.M.) content (dichloromethane soxhlet extract, dry fat-free basis):

IWTO standard description	Standard regains		Total fatty matter content of top
	Top (R%)	Noil (r%)	(% T.F.M.)
Noble Oil (4.6% T.F.M.)	19	14	4.6
Noble Dry (1.0% T.F.M.)	18¼	14	1.0
Schlumberger Oil (4.6% T.F.M.)	19	16	4.6
Schlumberger Dry (1.0% T.F.M.)	18¼	16	1.0

If contracting parties agree to use non-standard oil contents, the factors for conversion of wool base into a theoretical top and noil yield for a range of "Total Fatty Matter contents" at commonly used top and noil re-

---

* Note: This value of 2.27% is equivalent to 2 parts of ash plus alcohol extractives per 86 parts of wool base.

gains are given in Table 1 (†). It is essential that the contracting parties specify:
- Total Fatty Matter Content   (% T.F.M.)
- Top regain   (R%)
- Noil regain   (r%)

† For details see WIRA publication *Core Test Conversion Factors*, R. Bownass, 13.8.73.

Table 1   Conversion Factors for Non-Standard Total Fatty Matter Contents at
Commonly Used Top and Noil Regains

Total Fatty Matter (% T.F.M.)	R = 19% r = 14%	R = 18¼% r = 14%	R = 18¼% r = 16%	R = 19% r = 16%
0.5	—	1.199	1.201	—
0.75	—	1.202	1.204	—
1.0	—	1.205	1.207	—
1.25	—	1.208	1.210	—
1.5	1.218	1.211	1.213	1.220
1.75	1.221	1.214	1.216	1.223
2.0	1.224	1.217	1.219	1.226
2.25	1.227	1.220	1.222	1.229
2.5	1.230	1.223	1.225	1.232
3.0	1.236	—	—	1.238
3.5	1.242	—	—	1.244
4.0	1.248	—	—	1.250

Note 1: Theoretical Top and Noil Yield = Wool Base × Conversion Factor.

Note 2: Top and Noil Yields at non-standard conditions are not used in the IWTO International Trade Agreement Applicable to Transactions in "Raw Wool", and are supplied for information purposes only.

B.3. Estimated Commercial Top and Noil Yield

This is defined as 'the Theoretical Top and Noil Yield less an appropriate allowance for loss in processing of the wool to the top stage'.

The conversion from Theoretical Top and Noil Yield (above) to Estimated Commercial Top and Noil Yield is to be made in accordance with the following formulae:

a) *Noble Combed Wools*

Estimated Commercial Top and Noil Yield = Theoretical Top and Noil Yield (2 − VA).

b) *Schlumberger Combed Wools*

Estimated Commercial Top and Noil Yield = Theoretical Top and Noil Yield −(2.5 − VA).

Where VA is an allowance for wool removed with vegetable matter during processing and is based on VM′, the actual content of vegetable matter base excluding hard heads and twigs, calculated according to the following formula:

$$VA = 5.20 - \frac{40.60}{7.80 + VM'}$$

Table 2 gives values of VA, calculated to two decimal places using the above formula, for values of VM′.

Note on processing allowances

The above formulae have been found to be in close relation to combing yields on a large proportion of combing length wools. There are, however, certain cases, e.g. very short combing wools, heavy burry wools, lambs wools, certain Montevideo wools, and dead wools (i.e. wools removed from the carcasses of sheep which have died other than by slaughter), where the fibre loss may be different in commercial combing from the above allowances.

B.4. Theoretical Card Sliver Yield

This is defined as 'the maximum yield of card sliver which could theoretically be obtained if all the wool fibre in a lot could be converted into card sliver without processing loss'. This is derived from wool base adjusted to a standard ash plus alcohol extractives content of 2.27% and brought finally to a regain of 18.25%.

That is:

$$\text{Theoretical Card Sliver Yield} = \text{Wool Base} \times \frac{100}{97.73} \times \frac{118.25}{100}$$

$$= \text{Wool base} \times 1.210$$

B.5. Estimated Commercial Card Sliver Yield

This is defined as 'the Theoretical Card Sliver Yield less an appropriate allowance for loss in processing of the wool to the card sliver stage'.

The conversion from Theoretical Card Sliver Yield above to Estimated Commercial Card Sliver Yield is to be made in accordance with the following formula:

Estimated Commercial Card Sliver Yield = Theoretical Card Sliver Yield -(2-VA)

where VA is an allowance for wool removed with vegetable matter during processing and is as defined in Appendix B3 above and in Table 2.

Appendix C   Calculation of IWTO Scoured Yield at R% Regain

IWTO Scoured Yield at R% Regain is derived from wool base and vegetable matter base as defined in Appendix B1 according to the following formula:

$$\text{IWTO Scoured Yield at R\% Regain} = (\text{Wool Base} + \text{V.M. Base}) \times \frac{100}{97.73} \times \frac{100 + R}{100}$$

---

[1] For details see "Conversion factors and processing allowances used in IWTO core testing for yield". R. Bownass, WIRA Report 246, March, 1975.

Table 2

VM'	VA	VM'	VA	VM'	VA	VM'	VA
0.0	.00	2.8	1.37	5.6	2.17	8.4	2.69
0.1	.06	2.9	1.41	5.7	2.19	8.5	2.71
0.2	.13	3.0	1.44	5.8	2.21	8.6	2.72
0.3	.19	3.1	1.48	5.9	2.24	8.7	2.74
0.4	.25	3.2	1.51	6.0	2.26	8.8	2.75
0.5	.31	3.3	1.54	6.1	2.28	8.9	2.77
0.6	.37	3.4	1.58	6.2	2.30	9.0	2.78
0.7	.42	3.5	1.61	6.3	2.32	9.1	2.80
0.8	.48	3.6	1.64	6.4	2.34	9.2	2.81
0.9	.53	3.7	1.67	6.5	2.36	9.3	2.83
1.0	.59	3.8	1.70	6.6	2.38	9.4	2.84
1.1	.64	3.9	1.73	6.7	2.40	9.5	2.85
1.2	.69	4.0	1.76	6.8	2.42	9.6	2.87
1.3	.74	4.1	1.79	6.9	2.44	9.7	2.88
1.4	.79	4.2	1.82	7.0	2.46	9.8	2.89
1.5	.83	4.3	1.84	7.1	2.48	9.9	2.91
1.6	.88	4.4	1.87	7.2	2.49	10.0	2.92
1.7	.93	4.5	1.90	7.3	2.51	10.1	2.93
1.8	.97	4.6	1.93	7.4	2.53	10.2	2.94
1.9	1.01	4.7	1.95	7.5	2.55	10.3	2.96
2.0	1.06	4.8	1.98	7.6	2.56	10.4	2.97
2.1	1.10	4.9	2.00	7.7	2.58	10.5	2.98
2.2	1.14	5.0	2.03	7.8	2.60	10.6	2.99
2.3	1.18	5.1	2.05	7.9	2.61	10.7	3.01
2.4	1.22	5.2	2.08	8.0	2.63	10.8	3.02
2.5	1.26	5.3	2.10	8.1	2.65	10.9	3.03
2.6	1.30	5.4	2.12	8.2	2.66	11.0	3.04
2.7	1.33	5.5	2.15	8.3	2.68	11.1	3.05

Table 2 (cont.)

VM'	VA	VM'	VA	VM'	VA	VM'	VA
11.2	3.06	13.4	3.28	15.6	3.46	17.8	3.61
11.3	3.07	13.5	3.29	15.7	3.47	17.9	3.62
11.4	3.09	13.6	3.30	15.8	3.48	18.0	3.63
11.5	3.10	13.7	3.31	15.9	3.49	18.1	3.63
11.6	3.11	13.8	3.32	16.0	3.49	18.2	3.64
11.7	3.12	13.9	3.33	16.1	3.50	18.3	3.64
11.8	3.13	14.0	3.34	16.2	3.51	18.4	3.65
11.9	3.14	14.1	3.35	16.3	3.52	18.5	3.66
12.0	3.15	14.2	3.35	16.4	3.52	18.6	3.66
12.1	3.16	14.3	3.36	16.5	3.53	18.7	3.67
12.2	3.17	14.4	3.37	16.6	3.54	18.8	3.67
12.3	3.18	14.5	3.38	16.7	3.54	18.9	3.68
12.4	3.19	14.6	3.39	16.8	3.55	19.0	3.69
12.5	3.20	14.7	3.40	16.9	3.56	19.1	3.69
12.6	3.21	14.8	3.40	17.0	3.56	19.2	3.70
12.7	3.22	14.9	3.41	17.1	3.57	19.3	3.70
12.8	3.23	15.0	3.42	17.2	3.58	19.4	3.71
12.9	3.24	15.1	3.43	17.3	3.58	19.5	3.71
13.0	3.25	15.2	3.43	17.4	3.59	19.6	3.72
13.1	3.26	15.3	3.44	17.5	3.60	19.7	3.72
13.2	3.27	15.4	3.45	17.6	3.60	19.8	3.73
13.3	3.28	15.5	3.46	17.7	3.61	19.9	3.73
						20.0	3.74

Note VM' = 'Vegetable Matter Base' excluding hard heads and twigs.

Scouring the greasy sample after weighing

The centrifuge or spin-dryer

CSIRO rapid dryer

Weighing the dry scoured sample

Vegetable matter is recovered for measurement by dissolving the wool in a 10 per cent solution of sodium hydroxide.

Drying the recovered vegetable matter

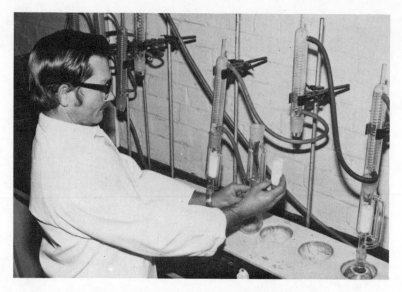

The Soxhlet apparatus for removing residual grease from the scoured sub-sample

The furnace is used to reduce a sub-sample to ash so that the residual mineral matter may be measured

Appendix II

Method for the Determination of Wool Base, Vegetable Matter Base, 'IWTO Clean Wool Content' and 'IWTO Scoured Yield' in Raw Wool (Original Version)

1. Foreword

This method has been prepared by the members of the IWTO Raw Wool Certification Sub Committee Technical Section. It is based on the experience of laboratories which have carried out a series of cooperative trials over several years.

The method includes the determination of clean wool content in samples of greasy cores, which enables an estimate of a theoretical top and noil yield for a particular lot of greasy wool to be made.

The method is associated directly with the regulations laid down in the IWTO Core Test Regulations concerning the method of sampling a consignment of bales so as to obtain a determination of scoured yield of raw wool and an estimate of top and noil yield for combing wools by use of appropriate allowances.

2. Scope

The method is applicable to core samples only and covers all forms of raw wool as defined below. It is associated with the IWTO Core Test Regulations which lay down the method of obtaining the cores from the bales in a test parcel.

3. Principle

Weighed subsamples of the cores are scoured (if in the greasy state), dried and weighed. Test specimens of the scoured subsamples are then taken for the separate determination of vegetable matter (which can be expressed as 'vegetable matter base'), ash and alcohol extractives. The weight of these non-wool constituents is subtracted from the weight of the dry scoured wool subsamples to give the dry weight of wool fibres free from all impurities ('wool base').

The 'wool base' is converted to 'IWTO Clean Wool Content' by the addition of standard amounts of ash, alcohol extractives and moisture.

The 'wool base' plus 'vegetable matter base' is converted to 'IWTO Scoured Yield' by the addition of standard amounts of ash, alcohol extractives and moisture.

4. Definitions

For this specification, the following definitions shall apply:

Raw Wool: greasy wool; wool which has been commercially scoured, carbonised, washed or solvent-degreased; scoured skin wools; and slipe wools. It consists of wool fibre together with variable amounts of vegetable matter and extraneous alkali insoluble impurities, mineral matter, wool waxes, suint and moisture.

Greasy Wool: Wool from the sheep's back or sheepskins which has not been commercially scoured, solvent degreased or carbonised.

Sample: The wool drawn by coring methods from, and representative of, a lot, bulk or consignment for testing.

Subsample: The randomly drawn portion, representative of the sample, for determination of wool base.

Test Specimen: The randomly drawn portion, representative of the scoured and dried subsample, to be used for determining a specific impurity in that subsample.

Corrected Oven Dry Weight: The weight obtained by oven drying the scoured subsamples in accordance with Appendix C of this method and calculating as instructed in clauses 7.4 and 7.5.

Oven Dry Scoured Content: The Corrected Oven Dry Weight, as defined above, expressed as a percentage of the sample.

Wool Base: The oven-dry weight of wool fibre free from all impurities, i.e. ash-free, alcohol extractives-free and free from all vegetable matter and other alkali-insoluble impurities, expressed as a percentage of the weight of the sample.

Vegetable Matter Base (V.M. Base): The oven-dry weight of ash-free, alcohol extractives-free burrs (including hard heads), twigs, seeds, leaves and grasses present, expressed as a percentage of the weight of the sample.

Note 1: 'Hard heads' are Noogoora burrs (*Xanthium pungens*), Bathurst burrs (*Xanthium spinosum*) and similar burrs of a bean-like character covered in readily removable spines. They do not contribute to a loss of wool during processing.

Note 2: 'Twigs' are small pieces of stick, woody leaf stalks and similar woody material. They do not contribute to a loss of wool during processing.

Total Alkali-insoluble Impurities: The vegetable matter as defined above and all other alkali-insoluble substances such as skin, dags, dung locks, strings, paper and lime that may be present after scouring. It is the oven-dry weight of the total ash-free, alcohol extractives-free alkali-insoluble impurities expressed as a percentage of the weight of the oven-dry scoured cores.

IWTO Clean Wool Content: The wool base as defined above adjusted to a standard ash plus alcohol-extractives content of 2.27%* (i.e. expressed as a percentage of wool base plus the standard ash and alcohol extractives) and brought finally to a regain of 17%.

That is:

$$\text{IWTO clean wool content} = \text{Wool base} \times \frac{100}{97.73} \times \frac{117}{100}$$
$$= \text{Wool base} \times .1972$$

IWTO Scoured Yield at R% Regain is derived from wool base and vegetable matter base as defined above according to the following formula:

$$\text{IWTO scoured yield at R\% regain} = (\text{Wool Base} + \text{V.M. Base}) \times \frac{100}{97.73} \times \frac{100 + R}{100}$$

5. Sampling

The methods of sampling a bulk consignment of bales are set out, as mentioned above, in the IWTO Core Test Regulations and for the purpose of this Specification it is only necessary to refer to the sample of cores taken. The weight of these cores is relevant to the methods used in preparing the subsample for the scouring and subsequent analysis of the scoured subsample.

In the sampling regulations it is prescribed that the weight of cores taken shall be sufficient to provide five subsamples (preferably of 200 grams each, but not less than 150 grams) with the proviso that the sampling methods shall be such as to produce a sampling precision of ± 1% of IWTO Clean Wool Content at a probability level of 0.95.

6. Procedure

Determine the wool base and V.M. base of the sample by the following procedure:

6.1 *Subsampling*

Determine the weight of the sample to the nearest 0.1 g as the difference between the gross weight and the weight of the container.

Draw and weigh to the nearest 0.1 g five representative subsamples, each preferably of 200 g, and not less than 150 g, by the method described in Appendix A.

Weigh the remainder left after the subsamples are taken to the nearest 0.1 g.

Test three of the subsamples separately in accordance with 6.2, 6.3 and 6.4.

Keep the remaining subsamples and other sample material, if any, in an air-tight container for check testing if required as detailed in 7.3.

6.2 *Scouring of Subsample*

If the sample comprised scoured wool, the scouring step is omitted. Otherwise proceed as follows:

Scour the greasy subsample in appropriate vessels so as to remove the extraneous matter normally removed in commercial scouring, care being taken to ensure that no loss of short fibre or vegetable matter occurs. (See Appendix B for detailed procedure.)

6.3 *Drying of Scoured Subsample*

Determine the oven-dry weight of the scoured subsample (or subsample from scoured wool) by the method given in Appendix C.

6.4 *Determination of Extraneous Materials in the Scoured Subsample (or subsample from scoured wool).*

Determine the ash, alcohol extractives and vegetable matter in the scoured subsample by the methods given in Appendices D, E and F.

7. Calculations and Expression of Results

7.1 *Notation*

The reported data shall be denoted as follows:

---

* This value of 2.27% is equivalent to 2 parts of ash plus alcohol extractives per 86 parts of wool base.

W = weight of sample as received.

W_B = sum of weights of all subsamples plus remainder of the sample.

W_i = the weight of the ith subsample, drawn from the blended core sample.

P_i = the oven-dry weight of the scoured ith subsample.

E_i = alcohol extractable matter, as a percentage of the oven-dry scoured test specimen drawn from the ith subsample.

A_i = ash content, as a percentage of the oven-dry scoured test specimen drawn from the ith subsample.

V_i = oven-dry, ash-free, extractives-free hard heads and twigs, as a percentage of the oven-dry scoured test specimen drawn from the ith subsample.

H_i = oven-dry, ash-free, extractives-free hard heads and twigs, as a percentage of the oven-dry scoured test specimen drawn from the ith subsample.

T_i = the total oven-dry, ash-free, extractives-free alkali-insoluble impurities present in, and expressed as a percentage of, the oven-dry test specimen drawn from the ith subsample.

B_i = the wool base (per cent) of the ith subsample.

B = the wool base (per cent) of the sample.

V.M.B. = the oven-dry weight of ash-free, extractives-free vegetable matter expressed as a percentage of the weight of the sample.

P = the oven-dry scoured content (per cent) present in the sample.

7.2 *Wool Base in a single subsample*

Calculate the wool base (B_i) present in the ith subsample from equation 1.

$$B_i P_i \frac{(100 - E_i - A_i - T_i)}{W_i} \tag{1}$$

7.3 *Check Tests*

If the range of wool base values for the three subsamples exceeds 1.00 in the case of a greasy sample or 0.75 in the case of a scoured sample, then repeat the tests on the two remaining subsamples. Include all five results in equations 3, 4 and 5 unless there is conclusive evidence that one or more of the results is in error. In the latter case the number of valid results included must not be less than three.

7.4 *Oven-Dry Scoured Content, Wool Base and Vegetable Matter Base in the Sample*

Calculate the oven-dry scoured content (per cent), P, the wool base (per cent) B and the vegetable matter base (per cent) V.M.B. present in the sample from equations 2, 3 and 4 respectively:

$$P = \frac{W_B \, \Sigma \, P_i \times 100}{W \, \Sigma \, W_i} \tag{2}$$

$$B = \frac{W_B \, \Sigma \, [P_i \, (100 - E_i - A_i - T_i)]}{W \, \Sigma \, W_i} \tag{3}$$

$$V.M.B. = \frac{W_B \, \Sigma \, P_i \, V_i}{W \, \Sigma \, W_i} \tag{4}$$

where i = 1, 2, 3, normally and i = 1, 2, 3, 4, 5, when additional check tests are required.

7.5 *Percentage of hard heads and twigs in the sample (greasy wool only)*

Calculate the hard heads and twigs (per cent) H present in the greasy sample from equation 5.

$$H = \frac{W_B \, \Sigma \, P_i \, H_i}{W \, \Sigma \, W_i} \tag{5}$$

Include the value H in the expression of the vegetable matter base content (V.M.B.) as follows: 'Vegetable Matter Base (including H% hard heads and twigs) . . . per cent.'

7.6 *IWTO Clean Wool Content*

Calculate the IWTO Clean Wool Content (per cent) of the sample from equation 6.

$$\text{IWTO clean wool content} = \text{Wool base } (B) \times 1.1972 \tag{6}$$

7.7 *IWTO Scoured Yield*

Calculate the IWTO Scoured Yield (per cent) of the sample to the required regain (R%) from Equation 7.

$$\text{IWTO scoured yield at } R\% \text{ regain} = (B + \text{V.M.B.}) \times \frac{100}{97.73} \times \frac{100 + R}{100} \qquad (7)$$

8. Report
Report as follows:

8.1 *For an IWTO Clean Wool Content test*

Oven Dry Scoured Content	=	%
Wool Base	=	%
Vegetable Matter Base (including . . . % hard heads and twigs)	=	%
IWTO Clean Wool Content	=	%

8.2 *For an IWTO Scoured Yield test*:

Oven Dry Scoured Content	=	%
Wool Base	=	%
Vegetable Matter Base (including . . . % hard heads and twigs)	=	%
IWTO Scoured Yield at R% regain	=	%

Appendix A Method for Drawing Representative Subsamples

A.1. Introduction
A number of methods are applicable for preparing and selecting the subsamples from the cores. The methods are detailed in A.2; other methods may be used but in all cases *it is essential to ensure that the subsamples are representative of the whole of the cores*. All weighings shall be made in conformance with Section 6.1.

A.2. Blending and Selection of Subsamples
A.2.1 *Hand-Selection Method*
After weighing (see Section 6.1), spread out the sample on a suitable table and blend into a homogeneous state by hand to give a layer of fibre 3 cm to 6 cm thick. Divide the layer into squares by halving, quartering, etc. Draw small handfuls at random from the squares; a representative subsample shall contain at least 16 such handfuls.

This method may be modified by the use of a tray over which the cores from one bag are *thoroughly mixed* and spread evenly. A number of dividers are placed on the tray equal in number to the subsamples to be drawn and the cores within a divider are removed successively into appropriate subsample containers. The process is repeated for each bag of cores, and the subsamples are collected in air-tight boxes.

Weigh to the nearest 0.1 g the portion of the sample remaining after the subsamples have been taken.

Add the weight of the sample remainder to the sum of the weights of the subsamples for determining the factor W_B/W

A.2.2 *Mechanical Blending Method*
After weighing (see Section 6.1), the sample shall be thoroughly blended mechanically without any loss of wool fibre and vegetable matter.

Reweigh the blended sample to determine the factor W_B/W immediately prior to drawing the subsamples.

Note A.2: Care must be taken to ensure that a proportionate amount of any material shaken out in blending is included in the subsamples.

A.3. Number and Weighing of Subsamples
Take and weigh to the nearest 0.1 g five representative subsamples, each preferably of 200 g and not less than 150 g.

Appendix B Method of Scouring the Subsample

B.1. Principle
Each subsample is scoured (and dried, see Appendix C) to remove, as in commercial scouring, those extraneous constituents present in greasy wool, but without loss of the wool fibre and vegetable matter present in the subsample prior to scouring.

Note B.1: If the subsample consists originally of scoured wool then it shall be rescoured if it contains more than 5% of alcohol extractable matter.

B.2. Essential Requirements of the Method
B.2.1 The oven-dried scoured produce must contain all the wool fibre and vegetable matter present in the subsample before scouring.

B.2.2 If the scouring solution contains sodium carbonate, the scouring temperature shall be maintained at $52 \pm 3°C$.

B.2.3 If synthetic detergents are used (e.g. as may be desirable for wools containing lime), these shall be non-ionic and may be used with builders at temperatures up to 70° C under conditions appropriate to the specific detergent. Care should be taken that such solutions do not damage the wool.

Note B.2.3: If the subsamples are also to be used for the determination of airflow fineness by the method

IWTO-28-75 (E)* and synthetic detergents are used in the scouring of the subsample, then the conditions for such scouring as detailed in IWTO-28-75(E) Appendix 4 must be employed.

B.3. Scouring Methods

As an example, suitable apparatus and a procedure (which is based on the method given in ASTM Designation D584-72) for scouring the subsample are described in B.4. to B.6. *Other forms of apparatus and procedures may be used providing they meet the essential requirements detailed in B.2.*

B.4. Apparatus

B.4.1 *Scouring Bowl or Bowls*

A rectangular or cylindrical vessel of 30 to 50 litres capacity, with an attached drain board. The lower portion of the bowl is in the shape of an inverted pyramid or cone that is connected to a sliding-disc valve and a short length of drain pipe. At the bottom of the bowl, above the valve and drain pipe, is a close fitting, removable perforated plate (B.4.1.1). The drain pipe is centred over a 74 micron (200 mesh per inch) sieve, 12 to 20 cm in diameter, supported in a catch-basin.

B.4.1.1 *Two plates*, one with 1 to 2 mm openings, the other a similar plate covered on its upper surface with 149 micron (100 mesh per inch) woven wire cloth.

B.4.1.2 *Thermostatic Device*, capable of delivering water to the scouring bowl at a desired temperature with a tolerance of ±3°C.

B.4.1.3 *Paddle or other Stirring Device.*

B.4.1.4 *Spray or Shower Head*, with a flexible connection, for use in rinsing.

B.4.2 *Flotation Jar.* A glass or transparent plastic vessel of 1 or 2 litre capacity, approximately 20 cm tall, for separating by flotation the short wool fibres retained by the 74 micron sieve from associated sand and other heavy impurities.

B.4.3 *Wringer or Basket Centrifuge.* For the removal of excess water from the scoured sample before drying in an oven.

B.4.4 *Net Bag.* Having openings of 250 micron (60 mesh per inch) or finer. Bags are used with a squeeze roll type of wringer or with a centrifuge.

B.4.5 *Metal Can.* With bottom formed from 149 micron (100 mesh per inch) wire screen supported by a perforated metal plate, may be used with basket centrifuges. The dimensions of the can must be such that the can is capable of containing the scoured sample and fitting into the centrifuge.

B.5. Reagents

B.5.1 *Scouring Solution A.* A solution containing approximately 0.3 per cent of Na_2CO_3 and 0.1 per cent of soap having a temperature of not over 25°C. Addition to the solution of approximately 0.3 per cent of lime-sequestering agent of the polyphosphate type is recommended.

B.5.2 *Scouring Solution B.* A solution containing approximately 0.15 per cent of Na_2CO_3 and 0.05 per cent of soap having a temperature of not over 25°C. Addition to the solution of approximately 0.3 per cent of a lime-sequestering agent of the polyphosphate type is recommended.

B.6. Procedure

If the apparatus used is as defined in B.4, proceed as follows:

B.6.1 With the coarse perforated plate in place in the scouring bowl, immerse the subsample in Scouring Solution A at a temperature of 52 ±3°C (not less than one litre for each 15 g of wool) and stir for 3 min. Drain the solution through the 74 micron sieve. Spray the wool with a strong stream of warm water (35° to 45°C) so as to flush out as much as possible of sand and other soil. Dung and dags should be crushed and flushed away with the sand and soil. Remove the wool from the bowl and place it on the drain board. Raise the plate, and remove and discard any impurities *other than vegetable matter* lodged thereon.

B.6.2 Spray the material on the 74 micron sieve with warm water, then transfer to the flotation jar. Fill the jar with warm water, using spray to cause agitation and separation. After the sediment has settled, decant the floating wool and vegetable matter into the bowl. Refill the jar with the spray, allow to settle, and again decant. If the sediment still contains wool or vegetable matter estimated to exceed 0.05 per cent of the specimen weight, repeat once more before discarding the sediment.

B.6.3 With the coarse perforated plate in place in the scouring bowl and using Scouring Solution B, repeat the stirring, draining, water spraying and flotation operations described in B.6.1 and B.6.2.

B.6.4 With the 149 micron screen in place in the scouring bowl, repeat the preceding operations twice more but with warm water (35° to 45°C) instead of scouring solution. During the rinsing remove by hand and discard as much as possible of the remaining strings, skin pieces, and other extraneous material *free from wool fibre or vegetable matter.*

B.6.5 Thoroughly mix the wool fibres and vegetable matter recovered by flotation after the last rinse with the main body of scoured wool, place in the net bag or other container (B.4.4, B.4.5) and centrifuge or pass through the wringer to remove excess water, prior to drying to constant weight (see Appendix C).

---

* Previously Test Method Under Examination IWTO (E)-1-71(E).

Appendix C   Method for Determining the Dry Scoured Weight

C.1. Apparatus

(i) Drying apparatus capable of drying subsamples at 105 ± 2°C. Ventilated ovens with a mechanically in-duced air current or apparatus designed for forced air drying are suitable.

(ii) Suitable instruments for measuring the temperature and humidity of the air entering the drying oven.

(iii) Balance for determining the oven-dry scoured weight to the nearest 0.05 g (see Note C.1).

Note C.1: For 5 min. weighing intervals the limit drying rate of 0.01% of the weight of the scoured subsample per min. gives a 0.05% weight change between successive weighings. For the minimum oven-dry weight of 65 g which is normally required for analysis, this requires weighing to 0.03 g in order to determine the drying rate. Thus a balance capable of detecting relative changes on successive weigh-ings of 0.02 g is needed.

C.2. Procedure

Each scoured subsample shall be dried at a temperature of 105 ± 2°C in a ventilated oven or apparatus de-signed for forced air drying.

Determine the final oven-dry weight to the nearest 0.05 g as follows:

(i) When drying either within ventilated ovens or in forced draught driers, subsamples shall be weighed at the end of the estimated drying time. Further weighings at intervals equal to 20 per cent of the drying time (minimum interval 5 mins.) shall be made until the drying rate is less than 0.01 per cent minute of the weight of the subsample (see Note C.1)

(ii) Subsamples shall be weighed

(a) hot, while inside the drying oven (see Note C.2)

or

(b) hot, while inside the drying can but on an external balance (see Note C.3)

(iii) If the drying air is not drawn from a room maintained at the standard atmosphere conditions of relative humidity 65 ± 2% and temperature 20 ± 2°C, correct the oven-dry weight by multiplying by a correction factor for the moisture content of the drying air. The correction factors* are determined directly from the following formula:

$$\text{Correction factor} = 1 + 0.00053 \left(9.470 - \frac{622.e.r.}{760 - e.r.} \right)$$

Where e = saturation vapour pressure in mm of mercury

r = relative humidity per cent/100

The values of e and r may be obtained from wet and dry bulb thermometer readings of the ambient air by the use of hygrometric tables.

Note C.2: Accurate hot weighing requires considerable care to avoid errors due to convection currents, draughts and buoyancy effects. When weighing inside the oven, both inlet and outlet air ports must be closed to avoid draughts or convection currents.

Note C.3: Accurate hot weighing on a balance external to the drier or heating oven is only possible if great care is taken to minimise undesirable effects such as buoyancy and draughts. In order to be sure of at-taining the required precision of 0.05 g it is necessary to carry out the following tests:

(a) *Draughts*. The balance should be enclosed by an open-fronted box. Test for freedom from draughts by at least three successive weighings of a 100 g scoured wool sample which has been dried for twice the normal drying time. Differences between weighings must not exceed 0.025 g.

(b) The buoyancy correction must be determined for the apparatus and cans in use. Steel wool cleaned in petroleum ether is packed into the cold can to a density of 60 g per litre and the weight of the loaded can determined at room temperature. The can is then heated on the drier for 15 minutes and weighed hot. The difference between two weighings is the sum of the buoyancy and convection effects and is the correction to be added to hot weighings on the particular apparatus. This correction must be determined for each 10°C step in the range of drying room air temperatures.

(c) When determining the rate of drying by taking intermediate weighings, the total time that cans are removed from the heater or drier must not exceed 30 seconds. In addition, the time taken to complete a hot weighing must not exceed 20 seconds.

(d) *Cans*. The cans which hold the wool subsample during drying and weighing should be as small

---

* A table, "Correction Factors for Moisture Content of Drying Air", which has been prepared by the New Zealand Wool Testing Authority, enables the correction factors to be determined directly from the wet and dry bulb readings of an aspirated or whirling psychrometer. Copies of this table may be obtained direct from them at the following address: New Zealand Wool Testing Authority, P.O. Box 1379, Wellington, 1, New Zealand.

and light as practicable. The packing density of the wool should be 60 g or more per litre. A can which will hold 65–200 g of clean wool at this packing density is satisfactory. The shape of the can should be such as to avoid dead air space, i.e. air space without wool in it. Dead air spaces can cause variations in buoyancy effects. The cans must be fitted with mesh ends not coarser than 149 microns (100 mesh per inch) to retain the wool under the full force of the drying air and prevent loss of short fibre.

To maintain a uniform temperature of 105 ± 2°C throughout the wool in a can on a forced air drier, it is necessary to shroud the cans on the drier by a shield which ensures a flow of the exhaust air around the cans inside the shield.

C.3. Report

Report the oven-dry scoured weight of the ith subsample as P_i where $i = 1, 2, 3 \ldots$

Appendix D   Method for Determining Ash Content

D.1. Principle

From each scoured subsample a representative test specimen of between 9.5 g and 10.5 g is drawn. This is re-dried if necessary before being weighed. It is then charred to remove as much volatile matter as possible before ashing in a furnace.

D.2. Apparatus

 (i) Tared crucible, of accurately known weight after firing at 700–800°C.
 (ii) Gas burners, if used for charring the test specimen before ashing.
(iii) Ventilated ashing furnace, to be capable of maintaining a temperature of 750 ± 50°C.
(iv) Desiccator, for cooling the crucible and residue.
 (v) Balance, for determining the weight of ash to the nearest 0.001 g.

D.3. Test Procedure

Draw a representative test specimen of between 9.5 g and 10.5 g from each dry scoured subsample in such a manner as to avoid any change in mineral content, and test this as follows:

D.3.1 Determine the dry weight of the test specimen either by re-drying as set out in C.2 or by rapidly sampling and weighing while the dried scoured subsample is still hot. Weigh each specimen to the nearest 0.01 g.

D.3.2 Place the test specimen in a crucible which has previously been tared after firing at 750 ± 50°C, followed by cooling in a desiccator. Note that a new tare figure must be obtained prior to the ashing of each test specimen. Char the wool over gas burners to remove as much volatile matter as possible (see Note D.1).

Note D.1: When heat is applied, wool will swell and froth, and care must be taken to avoid losses during this period. Frothing may be reduced by damping the wool with distilled water.

D.3.3 Place the crucible in a ventilated ashing furnace maintained at 750 ± 50°C until all carbonaceous matter has been oxidised.

D.3.4 Remove the crucible and residue and cool it in a desiccator; determine the weight of the mineral ash to the nearest 0.001 g.

D.4. Report

Report for each test specimen the dry weight of mineral ash, expressed as a percentage of the dry scoured weight. For the ith subsample denote this as a A_i where $i = 1, 2, 3 \ldots$

Appendix E   Method for Determining Alcohol Extractable Matter

E.1. Principle

From each scoured subsample, a representative test specimen of between 9.5 g and 10.5 g is drawn. This is re-dried if necessary and then weighed. It is then extracted with alcohol to determine the amount of extractable matter which remains in the wool after scouring.

E.2. Apparatus and Reagent

  (i) Apparatus for re-drying the test specimen (if needed).
 (ii) Balance for weighing the test specimen with an accuracy of 0.01 g.
(iii) Soxhlet extractor assembled with ground glass joints. The barrel shall have a capacity of not less than 100 ml.
(iv) Device for heating the Soxhlet flask.
 (v) Thimble of filter paper or other material capable of retaining all fine solids and of sufficient length to extend above the top of the siphon. If coarser material than filter paper is used for the Soxhlet thimble, further filtration may be needed after extraction.
(vi) Drying oven controlled to 105 ± 4°C.
(vii) Desiccator (if used).
(viii) Analytical balance accurate to 0.0005 g.
(ix) Reagent containing at least 94% (anhydrous) ethyl alcohol by volume. A 'blank' test using the normal quantity of solvent in the Soxhlet must not leave more than 1 mg of residue.

E.3. Procedure

Draw a representative test specimen of between 9.5 g and 10.5 g from each dry scoured subsample.

E.3.1 Determine the dry weight of the test specimen either by re-drying as set out in C.2 or by rapidly sampling and weighing while the dried scoured subsample is still hot. Weigh each specimen to the nearest 0.01 g.

E.3.2 Transfer the test specimen to a filter thimble and place it in the Soxhlet barrel ensuring that absorption of moisture prior to extraction does not exceed 5%.

E.3.3 Fit a 250 ml flask (round or conical) to the Soxhlet barrel and add alcohol in sufficient amount for siphoning to continue throughout the extraction without risk of the flask becoming dry. Ensure that siphoning continues correctly throughout the extraction period and that no part of the wool is higher than the top of the siphon.

Extract for a minimum of 20 siphonings.

E.3.4 Remove the extraction flask, preferably at a time when the Soxhlet barrel is almost full, filter the contents if necessary and transfer to a tared flask or beaker of not more than 100 ml capacity. Evaporate off the alcohol, using a hot plate or preferably a steam bath, and remove the last traces of alcohol by heating in a drying oven at 105°C ± 4°C for 30 minutes. Transfer to a desiccator and weigh. Alternatively, stopper the flask on removal from the oven, cool, release stopper and weigh.

E.4. Report

Report for each test specimen the dry weight of alcohol extractable matter expressed as a percentage of the dry scoured weight. For the ith subsample denote this as E_i where $i = 1, 2, 3 \ldots$

Appendix F   Method for Determining Vegetable Matter Content and Total Alkali-Insoluble Impurities

F.1. Principle

The wool in a test specimen of at least 40 g representative of the scoured subsample is dissolved in sodium hydroxide solution and the alkali-insoluble residue collected on a sieve, washed free from alkali and dried. The fractional weights of the various types of vegetable and non-vegetable matter in the residue are determined or estimated. The weight of ash in the total residue is then determined or calculated. Corrections are applied to the fractional weights of the vegetable and non-vegetable components of the residue for partial solubility in the caustic soda solution and for alcohol extractives. From these corrected weights the percentage of oven-dry, ash-free, alcohol extractives-free vegetable matter and total alkali-insoluble impurities in the test specimen are determined.

F.2. Apparatus and Reagents

  (i) Alkali-resistant container of not less than 2 litres capacity.

  (ii) Hot plate or other suitable heating apparatus for heating the alkali to boiling point.

  (iii) Alkali-resistant 380 micron (40 mesh per inch) sieve for collecting the alkali-insoluble matter. A suitable tray (alkali-resistant) in which to stand the sieve while washing insoluble matter so that:

    (a) at least 2 cm of liquid constantly remains in the sieve during spraying;

    (b) the sieve mesh is clear of the floor of the tray by at least 0.7 cm to allow the rinse liquid to leave the sieve through its mesh base and finally to over-flow the sides of the tray. A spray nozzle should be adjusted so that the fine spray jets cover the surface of the liquid inside the sieve evenly without agitating the vegetable matter. It is an advantage to have one or two holes (diameter 0.7 cm) in the floor of the tray in order to drain away caustic during filtering and to facilitate complete drainage after the spray is turned off.

  (iv) Tared crucible of accurately known weight after firing at 750° ± 50°C.

  (v) Dryer capable of drying the insolubles in air at 110° ± 2°C without any loss of material.

  (vi) Desiccator, for cooling the crucible and alkali-insoluble material under moisture-free conditions.

  (vii) Balance capable of determining the weight of alkali-insoluble matter and mineral ash to the nearest 0.001 g.

  (viii) Sodium hydroxide solution — 10% (w/v) solution of sodium hydroxide in water. Laboratories must ensure that the sodium hydroxide concentration is 10% at the point of use.

F.3. Test Procedure

Draw a representative test specimen of a least 40 g from each dry scoured subsample and test as follows:

Determine the dry weight of the test specimen as set out in C.2. Immerse the test specimen in 600 ml (per 40 g of test specimen) of boiling 10% NaOH solution and without further heating stir continuously for 180 ± 10 seconds. Then add 1 litre of cold water, re-stir and allow the undissolved residue to settle. Decant the solution through the 380 micron sieve and wash the residue on to the sieve with cold water. By spraying, wash the residue in the sieve for at least 3 minutes. During the entire period of spraying, the sieve shall be mounted in the tray described so as to ensure that a constant level of at least 2 cm of liquid is retained within it. Determine to the nearest 0.001 g the dry weight of the total residue from the sieve and, after hand separation, the dry weight fractions of the types of vegetable matter and other alkali-insoluble impurities (as designated in the first column of Table F1) present in the residue in accordance with C.2 except that a drying temperature of 110°

± 2°C should be used. Where the total alkali-insoluble impurities do not exceed 5 per cent of the test specimen, the weight fractions of the various types of material may be estimated by careful visual examination.

Transfer the whole of the residue to the tared crucible and determine the ash content of the residue by the procedure described in D.3.2 to D.3.4. To facilitate ashing, large burrs may be cut into small sections.

Alternatively, an estimate of the ash content of the residue can be made as described in Section F.5.

F.4. Calculation of Results

F.4.1 *Vegetable Matter*

Using Equation F1 and the appropriate values in Table F1, calculate F_v, the factor for converting the weight of recovered oven-dry, ash-free, alkali-insoluble material to the original weight of oven-dry, ash-free, extractives-free vegetable matter present in the test specimen. Calculate V_i the percentage of such vegetable matter in the test specimen according to Equation F2.

$$F_v = f_1 v_1 \, f_2 v_2 + \ldots \tag{F1}$$

$$V_i = \frac{100 \, F_v (W_1 - A_T)}{W_2} \tag{F2}$$

where

F_v = conversion factor for vegetable matter only.
V_i = percentage oven-dry, ash-free, extractives-free vegetable matter present in the test specimen.
$f_1, f_2,$ = weight fractions of the individual types of vegetable matter present in the recovered alkali-insoluble impurities.
v_1, v_2 = correction factors for the individual types of vegetable matter present (from Table F1).
W_1 = weight of oven-dry recovered alkali-insoluble impurities.
W_2 = oven-dry weight of test specimen.
A_T = weight of ash from recovered alkali-insoluble impurities.

F.4.2 *Total Alkali-insoluble Impurities*

Using Equation F3, and the appropriate values in Table F1, calculate F_T, the factor for converting the weight of recovered oven-dry, ash-free, alkali-insoluble material to the original weight of total oven-dry, ash-free, extractives-free alkali-insoluble impurities present in the specimen. Calculate T_i, the percentage of such total impurities present in the test specimen according to Equation F4.

$$F_T = F_v + f_5 \, p_5 + f_6 \, p_6 \tag{F3}$$

$$T_i = \frac{100 \, F_T \, (W_1 - A_T)}{W_2} \tag{F4}$$

where F_v, W_1, W_2 and A_T are defined in F.4.1, and

F_T = average correction factor for total alkali-insoluble impurities.
T_i = percentage oven-dry, ash-free, extractives-free total alkali-insoluble impurities present in the test specimen.
f_5 = weight fraction of skin pieces present in the recovered alkali-insoluble impurities.
f_6 = weight fraction of the impurities other than vegetable matter and skin that is present in the recovered alkali-insoluble impurities.
p_5 = correction factor for skin pieces (from Table F1).
p_6 = correction factor for impurities other than vegetable matter and skin that is present in the recovered alkali insoluble impurities (from Table F1).

Table F1   Correction Factor for Alkali-Insoluble Impurities

Average values of the ratio of scoured, oven-dry, ash-free, extractives-free weight of individual types of vegetable matter and other alkali-insoluble impurities to the corresponding weight of oven-dry, ash-free material recovered by this method.

Type of alkali-insoluble impurity	Correction factor
Seed and Shive	1.40
Spiral burr	1.20
Hard heads and twigs	1.03
Bidi-bidi (New Zealand)	1.56
Skin pieces	2.00
Other alkali-insoluble impurities (mainly paper and string)	1.05

F.4.3 *Hard heads and twigs*

Using Equation F5 and the value for hard heads and twigs given in Table F1 calculate H_i, the percentage of the oven-dry, ash-free, extractives-free hard heads and twigs present in the test specimen from:

$$H_i = \frac{100\,f_{hvh}\,(W_1 - A_T)}{W_2}$$ (F5)

where

f_h = weight fraction of hard heads and twigs present in the recovered alkali-insoluble impurities.

v_h = correction factor for hard heads and twigs (from Table F1).

W_1, W_2, and A_T are defined in F.4.1.

F.5. Alternative Calculations of Results

Analysis of data from 16 laboratories has indicated a suitable method of estimation of the ash of total alkali insolubles as an alternative to carrying out the determination.

In the alternative method of calculation, Equations F2, F4 and F5 can be substituted by Equations F6, F7 and F8 when ash of total alkali insolubles is not determined.

$$V_i = \frac{93.\,F_v.\,W_1}{W_2}$$ (F6)

$$T_i = \frac{93.\,F_T.\,W_1}{W_2}$$ (F7)

$$H_i = \frac{93.\,f_h.\,v_h.\,W_1}{W_2}$$ (F8)

Where V_i, T_i, H_i, F_v, F_T, f_h, v_h, W_1 and W_2 are defined in F.4.1, F.4.2, and F.4.3.

Note: Where samples contain significant proportions of dag material, determination of the ash content of the total alkali insolubles rather than estimation is recommended.

F.6. Report

Report for each test specimen:

(i) the oven-dry weight of ash-free, extractives-free vegetable matter expressed as a percentage of the dry scoured weight to the nearest 0.1 per cent.

(ii) the oven-dry weight of the total ash-free, extractives-free, alkali-insoluble impurities expressed as a percentage of the dry scoured weight to the nearest 0.1 per cent.

(iii) the oven-dry weight of ash-free, extractives-free hard heads and twigs expressed as a percentage of the dry scoured weight to the nearest 0.1 per cent.

For the *i*th subsample, denote (i), (ii) and (iii) by V_i, T_i and H_i respectively, where $i = 1, 2, 3\ldots$

An Introduction To Statistical Method For Wool Measurement

The following introduction to statistical method for wool measurement is given to provide a better understanding of the method used to describe a 'population'. This example refers to the method used for calculating the mean fibre diameter of a population of wool fibres measured by using a projection microscope (see Appendix III).

Population: This refers to the entire entity. In the case of wool it would be a complete parcel or consignment, e.g. wool buyer's order of 100 bales.

Sample: Is a subset of the population.

Note: Wool is not uniform in nature and while the population has one true measurement for yield, sample yields are variable.

Describing The Population

The Array: This is the arrangement of data in order of magnitude. It helps to clarify the range covered and general distribution of the data.

Frequency Tables: As a further step towards organisation, data may be arranged for presentation in groups to cover a range of values. Such groups are termed frequency tables. For example, micron groups for fibre fineness:

18.6–19.5

19.6–20.5

20.6–21.5

Median: When data is arranged in order of magnitude, the median is the central value. For example, 1, 2, 4, 7, 9, 10, 11, 14, 14, 15, 17. The median is 10.

Mode of fashion: This refers to the item occurring most frequently. In the above example, the only value occurring more than once is 14. Technically this can hardly be called the mode, as the mode only occurs when there is a very large number of cases.

The mean or arithmetic mean ($\bar{x}$):

$$\text{Mean or average value} = \frac{\text{Total}}{\text{Number of observations}}$$

$$\text{i.e.} \quad \bar{x} = \frac{\Sigma x}{n}$$

where $\bar{x}$ = Mean
  Σ (sigma) = Sum of
  x   = Observations
  n   = Number of observations

Variability: Is the distribution about the mean.

(a) Range — is stated as the two extremes or as the differences between the extreme values.

(b) Variance (s^2) $= \dfrac{\Sigma(x - \bar{x})^2}{n}$

i.e. variance is the mean of the squared deviations from the mean.

(c) Standard deviation $= \sqrt{\text{Variance}}$
  e.g. Variance = 9 cm
  Standard deviation = 3 cm

(d) Coefficient of variation $= \dfrac{\text{Standard deviation}}{\text{Mean}} \times \dfrac{100}{1}$

(Resulted expressed as a percentage)
Example

Sample measurements x	Deviation from mean $(x - \bar{x})$	Deviations squared $(x - \bar{x})^2$
4	−1	1
2	−3	9
6	1	1
10	5	25
3	−2	4
25	0	40

Mean $\dfrac{25}{5} = 5.0$

Variance $= \dfrac{\Sigma(x - \bar{x})^2}{n}$

$= \dfrac{40}{5} = 8$

Standard deviation $= \sqrt{\text{Variance}} = \sqrt{8} = 2.8$ approx.

Coefficient of variation $= \dfrac{\text{Standard deviation}}{\text{Mean}} \times \dfrac{100}{1}$

$$= \frac{2.8}{5} \times \frac{100}{1}$$

$$= \frac{280}{5} = 56\%$$

Distribution:

Normal Distribution

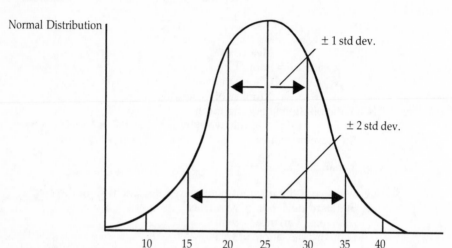

The mean and standard deviation completely describe the normal distribution. In every normal population 68 per cent of the population lies between $\bar{x} - 1$ SD and $\bar{x} + 1$ SD.

Ninety-five per cent of the population will have values between range of $\bar{x} - 2$ SD and $\bar{x} + 2$ SD.

Sampling

Four criteria for evaluating the sampling method are:

1. It must be repeatable.
2. The method must be correlated to the characters that are measured, (i.e. truly representative).
3. The method needs to be efficient.
4. The method must be unbiased.

Random Sample

There must be no selection by which some observations are shown, and some are deliberately rejected.

Variability of Population

Will affect the accuracy of sampling.

Sample Size

As the number of measurements increase so will the accuracy of the method.

Standard error:

$$S.E. = \frac{S.D.}{\sqrt{N}}$$

S.E. = Standard error
S.D. = Standard deviation

N   = Number of measurements

Note: Accuracy of sampling increases in relation to $\sqrt{N}$ where N is the number of samples. Consider the following examples. A constant standard deviation is used for the purpose of illustration.

No. of samples	Standard deviation	Standard error
4	4	$\dfrac{4}{\sqrt{4}} = \dfrac{4}{2} = 2$
9	4	$\dfrac{4}{\sqrt{9}} = \dfrac{4}{3} = 1.33$
16	4	$\dfrac{4}{\sqrt{16}} = \dfrac{4}{4} = 1$
25	4	$\dfrac{4}{\sqrt{25}} = \dfrac{4}{5} = .8$
36	4	$\dfrac{4}{\sqrt{36}} = \dfrac{4}{6} = .66$

Keep in mind that as the number of samples are increased, the sample mean will also be getting closer to the true mean of the population. The standard error is also being reduced, but not in direct proportion to the number of samples. To reduce the standard error by half, the samples are increased from 4 to 16. After this stage, the rate of improvement is reduced, so it would seem that in this case 16 is the optimum number of samples.

Confidence Level

When stating the mean for samples of the population it is important to give the confidence level as an indication of the relationship between the sample mean and the true mean. It is quite permissible to work to any confidence level, although in practice a confidence level of two standard errors is regarded as fairly acceptable. It gives a small degree of error to the estimation of the mean and a probability of 95 per cent that it is correct. For example, if the sample mean for yield of a line of wool was 71 per cent and the standard error was .5 per cent, the result could be expressed as follows:

(a) At a confidence level of one standard error or 68 per cent 71 ± .5%
(b) At a confidence level of two standard errors or 95 per cent 71 ± 1%
(c) At a confidence level of three standard errors or 97 per cent 71 ± 1.5%

Appendix III

Method of Determining Wool Fibre Diameter by the Projection Microscope (IWTO-8 61 (E))

1. Foreword

This specification concerns the determination of the diameter of wool fibres by the projection microscope and conforms in all essential details to the procedure recommended by the International Standards Organisation (I.S.O. Recommendation No. 181).

2. Scope

The method is applicable to woollen and worsted products at all stages from raw material to yarns.

3. Outline of the Method

The principle of the method is the projection on a screen of magnified images of the profiles of short pieces of fibre and the measurement and recording of the widths of these images by a graduated scale. At each stage a special technique is used to avoid operator errors. The arithmetic mean diameter is calculated.

4. Testing Equipment

4.1 *Projection Microscope*

The microscope proper must comprise a light source, a light condenser, a stage which supports the mounted specimen of fibres, an objective, an ocular and a circular screen.

The stage must be moveable in two directions at right angles by means of sliding mechanisms capable of successive displacements at 0.5 mm steps.

The objective and ocular must be capable of providing a magnification of 500×.

The circular screen must be capable of rotation in its own plane about its centre. If this screen is not transparent it should carry a moveable scale 5 cm wide graduated in millimetres. This scale must be capable of movement diametrically across the screen between guides. Transparent screens may carry a scale graduated in millimetres along a diameter. A moveable scale, as shown in Figure 3, is generally preferred. The circular screen must contain a marked central circle whose diameter is equal to one-quarter of the optical distance between the ocular and the centre of the screen. All measurements are made in this circle.

4.2 *Microtome*

For cutting the fibres to a predetermined length the fibre holder and pushers described below must be provided. Alternatively a conventional microtome may be used if it is capable of fulfilling the requirements of section 6.12 regarding the cutting of the fibre pieces.

Fibre Holder and Pushers. These are are shown in Figures 1 and 2. The holder is a short piece of smooth steel G about 3 mm thick with a 1.5 mm slot into which slides the tongue of part H. The tongue of part H is fixed by a screw and may thus be adjusted to project different distances into the slot of G. The pushers J consist of three steel stems with short plates near their ends; all the stems have the same width as the slot, namely 1.5 mm. The stem of one pusher extends 0.8 mm beyond the stop plate, that of the second 0.6 mm and that of the third 0.4 mm.

4.3 *Mounting Media*

Provide a mounting medium with the following properties:

 (i) a refractive index between 1.43 and 1.53,
 (ii) suitable viscosity,
(iii) does not absorb water.

Cedar wood oil and liquid paraffin are examples of suitable media.

4.4 *Miscellaneous*

Glass microscope slide approximately 75 mm × 40 mm, and square or rectangular cover glass No. 1 thickness (i.e. 0.13 mm − 0.7 mm). Suitable dimensions for the cover slip are 50 mm × 35 mm.

5. Standard Atmosphere for Conditions and Testing

Cedar wood oil and liquid paraffin preserve the moisture content of the fibres at the value just prior to immersion in the medium. The sample must therefore be conditioned to equilibrium and the slide prepared in the I.S.O. standard atmosphere for testing, 65 ± 2% R.H. and 20 ± 2°C.

6. Selection and Preparation of Test Specimens

6.1.1 *Raw Wool*

The material is first sampled in the following way. The total amount is divided into 40 zones and a handful of fibres taken from each zone. Each handful is divided into two (taking care to avoid fibre breakage) and half is rejected, the choice of which half to reject being made at random. (If the fibres are naturally grouped in locks where the fibres are parallel the division into two halves is made lengthwise, i.e. in a direction which avoids selection by fibre ends.) The retained half is again divided into two and one half rejected at random. This progressive subdivision is continued until about 25 fibres remain from each zone. The composite sub-sample of about 1,000 fibres is then given a washing treatment consisting of two extractions in benzene, petrol, ether or similar organic solvent. The sample is then conditioned to equilibrium in standard atmosphere. All the fibres are then cut up with scissors or any other suitable means on a glass plate into pieces between about 0.5 and 1 mm in length. *Note*: To give a length biased sample similar

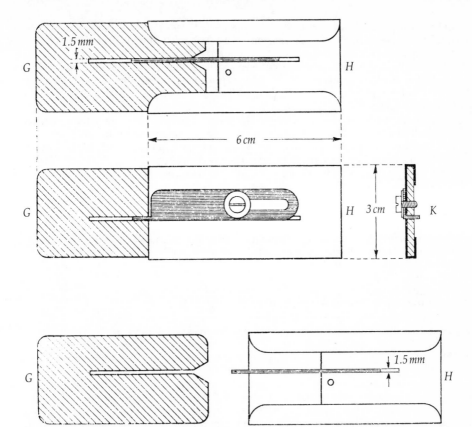

Figure 1   Fibre microtome in which the wool sample is cut into pieces of pre-determined length.

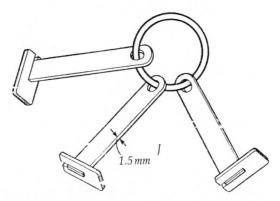

Figure 2   Pushers by which lengths of 0.4, 0.6 and 0.8 mm of fibre can be pressed out.

to that obtained in section 6.2 it is absolutely essential that *all* parts of *all* fibres should be cut up. The cut pieces are divided into 16 zones, from each of which a small quantity is taken and mounted as described in 7.

Note: The above procedure is suitable for dealing with a laboratory sample of about 1 kg of wool but is not suitable for sampling from a bulk containing several hundred kg or more of wool, which sampling is to be done following the general rules (Specification IWTO-13-64.)

6.1.2 *Sliver or Roving*

Place a single thickness of sliver or several thicknesses of roving in the slot of the fibre holder G (as specified in 4.2) and insert the part H so that the tongue compresses the fibres firmly in the slot. To ensure satisfactory retention of the fibres the length of the tongue should be suitably adjusted, then locked by means of the screw. Then using a sharp razor blade cut off the projecting tuft of fibres flush on both sides of the holder. Insert the appropriate pusher (as specified in Table 1) in the slot and slide it backwards and forwards so as to cause a fringe of fibres to project from the opposite side of the holder; with a razor blade cut off this fringe of fibres flush with the surface of the holder and mount as described in 7.

6.1.3 *Yarn*

A representative sample of yarn is wound or folded into a hank of such dimensions that it is firmly held in the slot of the holder G. Then follow the same procedure as detailed for sliver and roving.

6.2 *Choice of Pushers*

Table 1   Choice of Pushers

| | Fibres of average diameter | |
	Above 27 microns	Below 27 microns
Slivers and rovings	0.8 mm	0.4 mm
Yarns	0.6 mm	0.4 mm

Note: If it is not certain whether the average diameter is likely to be less than or greater than 27 microns the longest pusher specified should be used.

7. Mounting of Specimen

The cut fibres are placed in a few drops of mounting medium on a glass slide (as specified in 4.4). The fibre pieces are then stirred well into the mounting medium, using a dissecting needle and employing a circular motion so that a uniform distribution on the slide is obtained. Sufficient of the mixture is then wiped cleanly away with a soft cotton cloth to ensure that no mixture is subsequently squeezed from under the cover glass. This avoids preferential removal of thin fibres. A cover glass (as specified in 4.4) is lowered on to the mixture by placing one edge in contact with the slide and gently lowering the opposite edge.

8. Test Procedure

8.1 *Calibration*

The projector must be calibrated from time to time with a certified micrometer scale and the magnification adjusted to 500 ×. A convenient size is 1 mm divided into 0.01 mm. The scale is

Figure 3   Centre graduated transparent scale which slides between guides as shown in Figure 1.

mounted on the stage and the magnification adjusted so that 0.1 mm of the micrometer scale gives 50 mm on the screen *as measured by the 5 cm wide scale.*

8.2 *Measurement Technique*

8.2.1 *Sampling on the Slide*

Note: Each part of this technique is designed to ensure that a random sample of the fibre pieces present on the slide is chosen for measurement, that the chance of measuring the same fibre piece twice is negligibly small and that the operator has no free choice of the fibres to be measured. Without following such a technique the danger of bias has been found to be very real.

Place the slide on the microscope stage, cover glass towards the objective. After the fibres have settled, the slide is examined in different fields. The distance between the centres of the fields must be theoretically greater than the length of the cut fibres, otherwise the same cut fibre could be measured twice. However, if the centres are 0.5 mm apart the probability of measuring the same cut fibre twice is slight.

Begin the examination by moving the slide until a corner of the cover slip is focused. Then traverse the slide 0.5 mm (to B) then 0.5 mm in the transverse direction, thus bringing the first field into view on the screen. Measure the widths of every fibre image lying within the field of view except the following, which are not measured:

(a) images with more than half their width outside the central circle;
(b) images that end within the width of the 5 cm graduated scale;
(c) images that cross another image at the point of measurement.

Measurement of an image is made as follows. The graduated scale or screen is moved with its length at right angles to the image until a centimetre division coincides with one side of the image. The width of the fibre image is then read off in millimetres. The place of measurement must be where the longitudinal line of the scale crosses the image. The width is taken as the distance between the extremities of the fibre image at this line even if the line happens to coincide with a fibre scale or some other irregularity of the fibre. The stage should remain stationary during all measurements in a given field. It may happen that in a field there will be no measurable fibres or only one or two. Traverse the slide in 0.5 mm steps, using the sliding mechanism described in 4.1, and measure the fibres in each field as before. Continue traversing until the edge of the cover glass C is reached. From the number of fibres measured in the first traverses and the total number of measurements desired, calculate the required number of traverses and thus the amount of each cross traverses (C D) required. If this turns out less than 0.5 mm, 0.5 mm traverses must be made and the extra measurements obtained from a second slide. Cross-traverse the slide this distance and continue with a second traverse and then a third, etc., following the path A B C D E F G . . . until the whole slide has been covered.

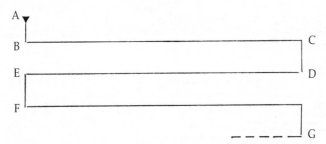

8.2.2 *Focusing*

When the microscope objective is too near the slide the edge of the image shows a white border. When the microscope objective is too far from the slide the edge of the image shows a black border. (These borders are called Becke lines).

When in focus the edge of the image shows a fine line with no border. However, in general both edges of the image will not be in focus together, since wool fibres are in general non-circular

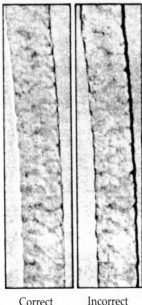

Correct   Incorrect

Figure 4   Correctly focused fibre with one sharp edge and white Becke line and incorrectly focused fibre with black Becke line.

in cross-section. When measuring an image whose edges are not in focus together, adjust the focusing so that one edge is in focus and the other shows a white line. Measure the width from the edge that is in focus to the *inside* of the white line. Figure 4 shows an image correctly and incorrectly focused.

8.2.3 *Recording of Measurements*

In general the second side of a focused image falls between two millimetre divisions of the scale and is entered under the lower whole number of millimetres. In the subsequent calculation all images recorded under N are assigned a diameter equal to N + 0.5 mm. However, sometimes the second side of the image lies exactly on a millimetre division on the scale.

To avoid bias these images must be assigned alternately to this group and to the lower millimetre group. This avoids assigning half an image to each group.

There are several methods of recording the results. A good method in frequent use is given in the Appendix overleaf.

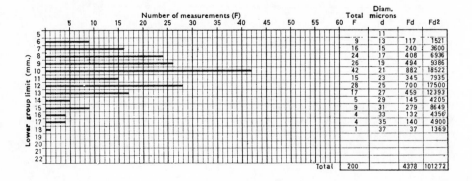

Lower group limit (mm.)	Number of measurements (F)	Total F	Diam. microns d	Fd	Fd2
5			11		
6		9	13	117	1521
7		16	15	240	3600
8		24	17	408	6936
9		26	19	494	9386
10		42	21	882	18522
11		15	23	345	7935
12		28	25	700	17500
13		17	27	459	12393
14		5	29	145	4205
15		9	31	279	8649
16		4	33	132	4356
17		4	35	140	4900
18		1	37	37	1369
19					
20					
21					
22					
Total		200		4378	101272

9. Number of Tests

A small proportion only of the fibres present in the bulk can generally be measured and the average fibre diameter is thus subject to the usual errors of random sampling. The coefficient of variation of fibre diameter for unblended wool tops and worsted yarns lies between 20 and 28 per cent. The confidence limits when testing n fibres are given by $\pm$ $ts/\sqrt{n}$ where s is the standard deviation and t can be taken as equal to 1.96 for a 95% degree of probability.

Table 2 gives an idea of the 95% confidence limits expressed as a percentage of the arithmetic mean for different numbers of measurements. These limits may be taken as an approximate estimate of the extreme deviations of the arithmetic mean of the sample from the true arithmetic mean fibre diameter of the bulk.

Table 2

95% confidence limits	Number of measurements
± 1%	2,500
± 2%	600
± 3%	300
± 5%	100

For blends of lots of wool of different grades, the coefficient of variation is higher and different confidence limits must be applied.

10. Calculation and Expression of Results

The measurements in mm (lower limit + 0.5 mm) multiplied by 2 give for each class the diameter in microns. Calculate the arithmetic mean and express the average fibre diameter in microns.
    Calculate the coefficient of variation.

11. Use of Reference Wools

To enable the deviations observed between different laboratories using this method to be reduced, the Secretariat of the International Wool Textile Organisation (IWTO) supplies fineness type samples.

Merino standard   : 21.7 microns
Average crossbred standard: 28.3 microns

In this way a laboratory can check that its tests agree perfectly with the measurements fixed, or note the divergence. In the latter case, the divergence observed should be systematically deducted from subsequent measurements or added thereto if it is statistically significant.

Appendix   Example of Calculation

Arithmetic mean (microns):

$$\bar{d} = \Sigma Fd/\Sigma F$$
$$= 4378/200$$
$$= 21.89$$

Standard deviation (microns):

$$s = \frac{\sqrt{\Sigma Fd^2 - [(\Sigma Fd)^2/\Sigma F]}}{\Sigma F - 1}$$
$$= \frac{\sqrt{101272 - 4378^2/200}}{199}$$
$$= 5.23$$

Coefficient of variation %:

$$C \text{ of } V\% = 100s/\bar{d} = 100 \times 5.23/21.89 = 23.89\%$$

The lanameter or projection microscope for measuring fibre diameter.

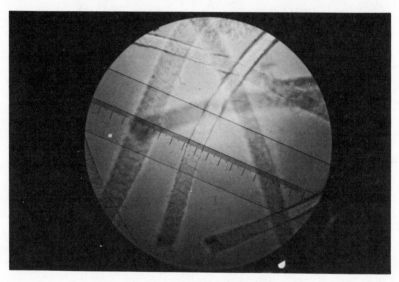

An image of wool fibres as seen with the projection microscope.

Appendix IV

Determination by the Airflow Method of the Mean Fibre Diameter of Core Samples of Raw Wool (IWTO-28-32 (E))

0. Brief History

During the 1960s the core test method for the determination of the yield of parcels of greasy wool gained general acceptance in the wool trade. In Europe, two types of test procedure were used: the mechanical method of the British Wool Federation (BWF); and the chemical methods of the American Society for Testing and Materials (ASTM) and later of the IWTO.

The BWF method produced carded wool, free from vegetable matter, in a form ideal for the measurement of Mean Fibre Diameter by the Airflow Method, which has been used successfully on tops for many years. Initially, Airflow Fineness Meters calibrated for measurements on tops were used for this carded loose wool, but it was soon realised that the results obtained were from 0.5 to 1.0 micrometre too high.

A systematic comparison was therefore made in several laboratories. Samples of top of known

Mean Fibre Diameter were cut into lengths of 20 to 30 mm and passed through the Shirley Analyser. This work indicated that for Mean Fibre Diameters in the region of 20 micrometres, the readings on the cut and carded top were about 0.3 micrometre too high; in the region of 34 micrometres, the results were about 1 micrometre too high. This meant that the over-estimate on cut and carded top was very similar to that on carded loose wool. Hence it was agreed that the Airflow Fineness Meter for measuring the diameter of cleaned carded raw wool should be calibrated with standard tops that had been cut into short lengths and passed through the Shirley Analyser.

This method was jointly developed by the Raw Wool Fineness Working Group of the Technical Committee, and the Raw Wool Certification Subcommittee. It was first issued as a Test Method under Examination in May 1971, and was converted to a full Test Specification in June 1975.

The current text was adopted by the IWTO Technical Committee in London, June 1979 and includes a revised Appendix 9, based on data obtained from a round trial carried out in Australia, 1980.

1. Foreword

When a current of air is passed through a mass of fibres packed in a chamber with perforated ends the ratio of airflow to differential pressure is primarily determined by the total surface area of the fibres. This was predicted from the hydrodynamic equations of Kozeny and others.

For fibres of circular or near-circular cross-section and constant density, such as non-medullated wool, the surface area of a given mass of fibres is inversely proportional to the Mean Fibre Diameter. This principle can be utilised to construct a simple apparatus for rapid estimation of Mean Fibre Diameter; the Airflow Method. Since the method is indirect, the apparatus must first be calibrated, using fibres of known mean diameter.

It has been shown that the estimate of Mean Fibre Diameter given by the Airflow Method is $d(1 + c^2)$, where d is the Mean Fibre Diameter (length biased), given by the projection microscope and c the fractional coefficient of variation. Since c normally lies within comparatively small limits for unblended slivers it is usual to calibrate the apparatus directly in terms of d.

2. Scope

This method sets out a procedure for the determination of the Mean Fibre Diameter of wool fibres sampled from bales of greasy or scoured wool by coring, using cutters of nominal diameter from 12 to 25 mm (½ to 1 in). It describes the procedure to be used to remove extraneous matter from the wool samples prior to measuring the Mean Fibre Diameter by the Airflow Method. Medullated wool or lamb's wool may give misleading results (see Appendix 10).

3. Principle

A sample of raw wool is scoured and carded by the methods described herein and the Mean Fibre Diameter of specimens of the prepared wool is measured by the Airflow Method.

If the sample originates from commercially scoured wool with an ethyl alcohol extractable matter content of less than 1.5%, the subsamples drawn from bales by coring are not subjected to the preparation, i.e. scouring and drying, described in 5.2.2.

4. Definitions

4.1 *Raw Wool*

Greasy wool; wool which has been scoured, carbonised, washed or solvent-degreased; scoured skin wools; and slipe wools. It consists of wool fibres together with variable amounts of vegetable matter and extraneous alkali-insoluble impurities, mineral matter, wool waxes, suint and moisture.

4.2 *Greasy Wool*

Wool from the sheep's back or sheepskins which has not been scoured, solvent-degreased or carbonised.

4.3 *Total Sample*

A representative sample of wool obtained by coring techniques. (When issuing an IWTO Certificate, sampling must comply with the IWTO Core Test Regulations.)

4.4 *Subsample*

A subsample drawn from and representative of the total sample, prepared, scoured and dried by the method described in Appendix 7.

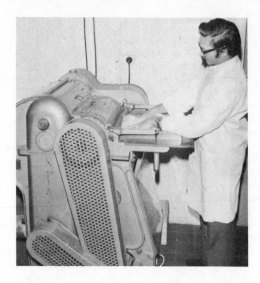

Preparing the sample in the Shirley Analyser prior to measurement for mean fibre diameter.

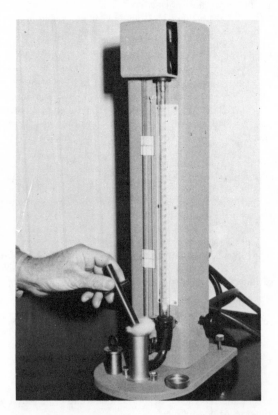

Weighing 2.5 g (1/10 oz) sample for measurement in the WIRA air-flow machine.

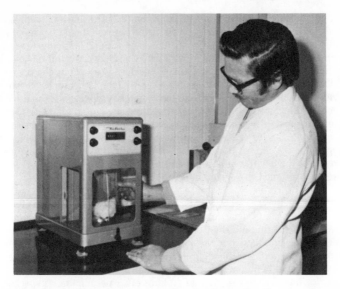

WIRA air-flow machine for measurement of mean fibre diameter.

4.5 *Laboratory Sample*
The randomly drawn sample representative of the scoured subsamples, prepared for measurement in the Airflow Fineness Meter.

4.6 Test Specimen
The conditioned, randomised representative of the laboratory sample, weighing 2.500 ± 0.004 g (Constant Pressure) or 1.500 ± 0.002 g (Constant Flow), for insertion in the Airflow Fineness Meter.

5. Test Method

5.1 *Apparatus*

5.1.1 *Airflow Fineness Meter*
Two alternative forms of apparatus are described: 'Constant Flow' and 'Constant Pressure'. Both forms of apparatus have the same arrangement of parts as illustrated in Figure 1.

In the Constant Flow apparatus, the flow is adjusted to a fixed value and the Mean Fibre Diameter derived from the manometer reading. The relationship between Mean Fibre Diameter and manometer reading is not linear, the distance between successive micrometre intervals decreasing with increasing diameter.

In the Constant Pressure apparatus, the presure is adjusted to a fixed value and the Mean Fibre Diameter derived from the flowmeter reading. The Constant Pressure apparatus gives a nearly linear relationship between fibre diameter and flowmeter reading. Detailed procedures for calibrating both types of apparatus are given in Appendix 3.

The apparatus consists of the following components (see Figure 1):
   (i) Air Valve B, giving sufficiently fine control of the air supply, such that the level of the flowmeter or manometer may be quickly adjusted to the working value.
   (ii) Suction Pump of a type providing a smooth output of at least 30 1/min at 200 mm H_2O with minimum fluctuation of the float of the flowmeter. A filter may be inserted between the pump and the air valve B to trap any loose fibres (see Appendix 4).
   (iii) Chamber A, of brass or steel which consists of the following parts:
      • the cylindrical base into which the test specimen is packed;
      • the plunger which slides into the base and compresses the test specimen into a fixed volume; and
      • the cap which secures the plunger to the base.

The dimensions of chamber A are given in Figure 2. The finish of the inside walls of the cylindrical base and of the outside walls of the plunger should be smooth and polished so that the plunger slides easily into the base without trapping fibres from the test specimen. If the plunger is held in the base by a screw cap a thin plate of brass or steel should be fixed to the top of the plunger which can be held by the operator to prevent the plunger from turning whilst the cap is screwed on. Other types of clamp can be used provided that they hold the flange of the plunger tightly against the lip of the cylinder.

Each Airflow Fineness Meter is calibrated initially and at intervals so that small variations in the dimensions of chamber A are not critical. However, to ensure that the packing density of the test specimen is similar on all Airflow Fineness Meters the length of plunger, the distance of the perforated plate from the lip of the cylinder, the internal diameter of the cylinder and the diameter of the plunger should be within 25 micrometres of the dimensions given in Figure 2.

(iv) A Manometer Reservoir D, with fluid as specified in Table 1, mounted at sufficient height to give a clear working distance ZH (Figure 1) of 350 mm in the glass limb of the manometer.

(v) A Manometer, made of glass tubing with internal diameter at least 5 mm to reduce surface tension effects. In both the Constant Flow and Constant Pressure apparatus a small amount of dye may be added to the manometer fluid to make it more readily visible. When distilled water is used, a trace of chromic acid may be added to give a clear meniscus.

(vi) A Scale, graduated in millimetres, approximately 350 mm in length.

(vii) A Glass Flowmeter, of working range, in litres of air per minute, as specified in Table 1.

(viii) • Rubber or Plastic Tube connecting the manometer reservoir D to the chamber A, which should be pressure tubing of small internal diameter to avoid constriction at the bends.
• Rubber or Plastic Tube from A to the Flowmeter F, of internal diameter not less than 6 mm. It should be as short as possible and should not be twisted or kinked between calibration of the apparatus and its subsequent use.

5.1.2 Packing Rod (Optional)

The packing rod is illustrated in Figure 3 and is made from a non-metallic substance, e.g. polythene, to minimise wear on the Constant Volume chamber. The packing rod is helpful when packing the test specimen into the Constant Volume chamber. The rod is held by the long end and the short end used to press the fibres into the Constant Volume chamber.

The cross-piece ensures that the rod only penetrates as far into the chamber as the plunger and so prevents uneven and overtight packing of the test specimen into the chamber.

5.1.3 Balance

A balance capable of weighing the test specimen to an accuracy of ±1 mg.

5.1.4 Mechanical Sample Card

A mechanical sample card is required, capable of homogenising and fully opening the scoured laboratory sample, without excessive loss of wool, while eliminating dust and most of the vegetable matter. The card may be the Shirley Analyser (Wool Model) (see Appendix 1), or any other mechanical card meeting the necessary requirements (see Appendix 2).

5.1.5 Standard Atmosphere

Means of producing and maintaining the ISO standard laboratory atmosphere of 20 ± 2°C and 65 ± 2% relative humidity is required.

5.2 Preparation of Wool for Measurement

5.2.1 Sampling

The weight of raw wool cores taken as the total sample shall be sufficient to provide a minimum of two subsamples of between 150 g and 200 g.

If an IWTO Certificate is to be issued, the packaged wool must be core sampled and weighed in accordance with the current IWTO Core Test Regulations.

5.2.2 Preparation of Subsamples

The representative subsamples are prepared, i.e. scoured and dried, as described in Appendix 7.

5.2.3 Preparation of the Laboratory Sample

The laboratory sample is prepared from the scoured subsamples as follows:

From each of a minimum of two scoured subsamples draw at random approximately equal

Table 1   Manometer and Flowmeter Details

	Constant flow	Constant pressure
Minimum diameter of resevoir	150 mm	60 mm
Manometer fluid	n-propyl alcohol	Distilled water
Working range of flowmeter	10–20 litres/min	4–24 litres/min

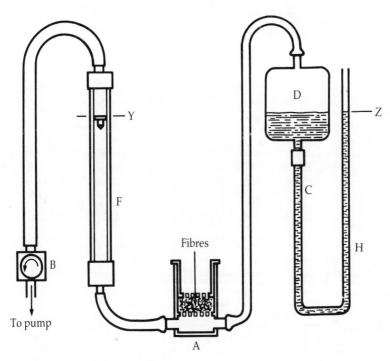

Figure 1   General arrangement of apparatus

quantities of wool which when combined will provide about 30 g to constitute the laboratory sample.

Note 1:   A representative sample of each scoured subsample may be drawn by spreading out the scoured subsample on a board, dividing it into 8 equal parts and removing pinches from each of the 8 parts until sufficient wool has been accumulated (15 g and 10 g respectively when 2 and 3 subsamples are used).

Note 2:   If the ethyl alcohol extract of the scoured subsamples exceeds 1.5%, as determined using the procedure in the current version of IWTO-19, re-wash or rinse the subsamples so that it meets this requirement.

5.2.4 *Carding of Laboratory Sample*

Card the laboratory sample on the mechanical card. If the Shirley Analyser (Wool Model) is used, refer to Appendix 1. If another form of mechanical card is used, refer to Appendix 2.

5.2.5 *Conditioning of Laboratory Sample*
Ensure that the carded laboratory sample has a regain not exceeding 10% before conditioning it to equilibrium in the standard atmosphere (20° ± 2°C, 65 ± 2% r.h.) and testing in that atmosphere.
Note 1: With wool dried after laboratory scouring as described in IWTO-19, a regain of 10% can be obtained by placing the wool in a ventilated oven at about 50°C for 30 minutes. Rapid driers may also be used.

If the wool has been passed through a Shirley Analyser in a room atmosphere of 65% r.h. or if it has been dampened to facilitate passage through a Shirley Analyser dry for 30 minutes at 70°C prior to conditioning in a standard atmosphere.

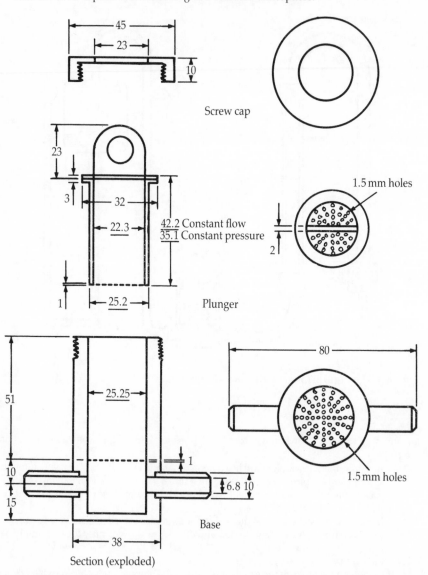

Screw cap

Plunger

1.5 mm holes

42.2 Constant flow
35.1 Constant pressure

Base

1.5 mm holes

Section (exploded)

Figure 2   Suggested dimensions of constant Volume Chamber A. All dimensions are in millimetres. Important dimensions are underlined.

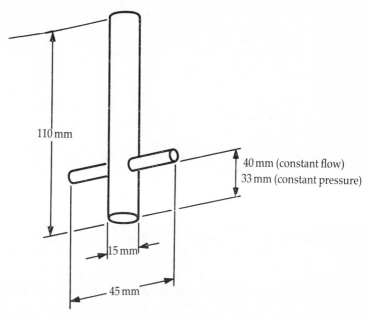

Figure 3   Packing rod

Note 2: If the test specimen does not reach equilibrium with the atmosphere because of insufficient conditioning time, or if the relative humidity of the atmosphere fluctuates, the result may be incorrect. A minimum exposure of four hours is required for a well-opened scoured sample to attain equilibrium with the atmosphere in a well-ventilated room maintained at a constant temperature and relative humidity. Much longer may be required for samples held in mesh bags or other containers in which case overnight conditioning (16 hours) is recommended.

5.2.6 *Preparation of Test Specimens*

Passage through a Shirley Analyser or a laboratory card blends most wools satisfactorily. If difficulties are experienced in obtaining good agreement between test specimens the following procedure is recommended:

Spread out the carded laboratory sample and divide it into 8 equal parts. Remove pinches from each of the 8 parts until just over 5 g (Constant Pressure) or 3 g (Constant Flow) of wool fibre have been accumulated. Using forceps, remove any vegetable matter and large accumulations of fibres (slubs, neps, etc.) that may remain in the specimen after mechanical carding. This is particularly necessary when not using the Shirley Analyser (Wool Model) which removes the greater part of the vegetable matter during carding.

Accurately weigh each test specimen for measurement in the Airflow Fineness Meter (a minimum of two specimens are required).

For a Constant Pressure apparatus each test specimen must weigh 2.5 ± 0.004 g.

For a Constant Flow apparatus each test specimen must weigh 1.5 ± 0.002 g.

Note: All wool for the test specimen must be handled as little as possible as the moisture from hands may change the regain of parts of the specimen. Excessive handling, particularly squeezing wool between fingers or rolling wool in the palm of the hand, must be avoided. The use of forceps wherever possible is recommended.

5.3 *Measurement of Mean Fibre Diameter*

In some wool testing laboratories two or more Airflow Fineness Meters are available for measuring Mean Fibre Diameter. The measurement of one or more test specimens from the same laboratory sample on two Airflow Fineness Meters has the following advantages:

- Sudden malfunctions of an Airflow Fineness Meter, such as the development of an air leak in the meter, will immediately show up.
- Table 1 of Appendix 9 shows that the between laboratories variance is much bigger than the other components of variance, and since between Airflow Fineness Meters variance contributes to the between laboratories variance, the use of two Airflow Fineness Meters in one laboratory will reduce between laboratories variance and should improve the precision of the results.

Two procedures, using either one or two Airflow Fineness Meters, are given below. Where possible the use of two Airflow Fineness Meters is recommended.

5.3.1 *Testing of Specimens*

(i) The wool is tested in the standard atmosphere using the Airflow Fineness Meter as described in 5.1.1. Corrections for atmospheric pressure shall be made as described in Appendix 5.

(ii) Ensure that the instrument is calibrated as described in Appendix 3.

(iii) Ensure that the instrument is level and that the float and liquid level are set to zero.

(iv) Ensure that the flow channels of the Airflow Fineness Meter are free from wool fibre residue and dust.

Note: Particles of broken wool fibres may pass into the flowmeter after its continued use with cored loose wool samples. The correction plug as described in Appendix 5 may detect errors due to this cause providing that the correction due to atmospheric pressure does not exceed 0.3 micrometre. It is also possible to incorporate a dust filter such as described in Appendix 4 of this method.

Regular checking of the equipment using standard samples or orifice plates as described in Appendix 6 is also recommended.

(v) Pack the specimen evenly into the constant volume chamber, a small amount at a time, using forceps, not fingers, to handle the specimen, so as not to contaminate or change the moisture regain of the specimen. Push the fibres into the constant volume chamber, preferably using the short end of the packing rod and taking particular care to ensure that the fibres are uniformly packed and that the walls and bottom of the chamber are not scratched or marked.

(vi) Insert and push down the perforated plunger into the chamber. Secure the retaining cap without rotation of the perforated plunger, ensuring no fibres are trapped between the perforated plunger and the chamber.

(vii) Switch on the air supply, adjust the float or the liquid level, depending on the instrument, to the prescribed settings and read the indicated Mean Fibre Diameter to the nearest 0.1 micrometre.

(viii) Take the specimen out of the chamber then repack it in the reverse direction without teasing it out. Use forceps, not fingers, to handle the specimen, so as not to contaminate or change the moisture regain of the specimen. Repeat the procedure described in (vi) and (vii).

5.3.2 *Use of One Airflow Fineness Meter*

Test two test specimens. If the range of the four readings is greater than that in the table below, measure a further test specimen. If the range of the six readings so obtained is greater than that shown in the table, repeat the test on three additional test specimens (see Note 4, Section 6).

| | *Range* | |
| | *Number of test specimens* | |
	2	3
Less than 26 μm	0.3 μm	0.4 μm
26 μm or greater	0.4 μm	0.6 μm

5.3.3 *Use of Two Airflow Fineness Meters*

Test two test specimens — one specimen through each Airflow Fineness Meter. If the range of the four readings is greater than that in the table below, test a further two specimens (one specimen in each apparatus). If the range of the eight readings so obtained is greater than that shown in the table, an additional two specimens must be measured (one specimen in each apparatus), (see Note 4, Section 6).

	Range Number of test specimens	
	2	3
Less than 26 μm	0.3 μm	0.4 μm
26 μm or greater	0.4 μm	0.7 μm

6. Expression and Calculation of the Results

The results shall be expressed as follows.
  (i) The Mean Fibre Diameter of each test specimen from the measurements taken.
 (ii) The Mean Fibre Diameter of the total sample calculated as the arithmetic mean of all test specimens' Mean Fibre Diameters, expressed to one decimal place.
(iii) As well, the 95% confidence limits of the Mean Fibre Diameter may be expressed to one decimal place if required (the same estimates are given regardless of the number of Airflow Fineness Meters used.)
  • If only two test specimens have been measured, the estimates of the 95% confidence limits of the Mean Fibre Diameter are:
    ± 0.2 μm (for wools of Mean Fibre Diameter 26 μm or lower)
    ± 0.3 μm (for wools of Mean Fibre Diameter above 26 μm).
  • If more than two test specimens have been measured, the estimate of the 95% confidence limits of the Mean Fibre Diameter depends on whether 6, 8 or 12 readings have been made (that is one, two or four extra test specimens), and are calculated as ±0.415W, ±0.295W or ±0.195W respectively, where W is the range of the readings in micrometres.
Note 1: Appendix 8 gives examples of the calculation of confidence limits for the mean and the expression of results.
Note 2: Appendix 9 gives full details of the precision of the results.
Note 3: The confidence limits quoted are for the test alone and take no account of sampling variability, which may be different for wools of different origins. The probable magnitude of the sampling variability is discussed in Appendix 9.
Note 4: The confidence limits are based on the assumption that a range of 1% of the Mean Fibre Diameter is acceptable if four readings only are made, or if the range exceeds 1% of the Mean Fibre Diameter for the original four readings, the additional number of test specimens to be taken should be sufficient to give confidence limits which are not greater than those obtained when the range of the original four readings does not exceed 1% of the Mean Fibre Diameter.
Note 5: If an IWTO Certificate is required, the results must be reported as laid down in the current IWTO Core Test Regulations.

Appendix 1   Sample Preparation with the Shirley Analyser (Wool Model)

The machine consists of a traditional Shirley Analyser fitted with the short feed plate as used for the separation of lint and trash in cotton, together with some modifications.

These include the fitting of a static eliminator, provision for fan speed changes, variable air intake and fitting of negative rake metallic wire on the taker-in. Undesirable draughts and convection currents must be avoided, and it is preferable to operate the machine in a room having an atmosphere of about 20°C and 60–70% relative humidity.

The following operating conditions must be used:

1. Speeds

Taker in	900 r.p.m.
Feed Roller	0.9 r.p.m.
Cage	80 r.p.m.
Fan	See note.

Note on Fan Speeds:
With a 3 step speed pulley suitable speeds are:

1,500 r.p.m. for Merino wool
2,000 r.p.m. medium Crossbred
2,400 r.p.m. coarse Crossbred
When a variable unit is fitted the most effective speed should be used.
Typical speeds are:

Speed r.p.m.	Mean Fibre Diameter of Wool (μm)
1,200	20.5
1,500	22.5
2,000	27.5
2,400	33.0
2,700	38.0

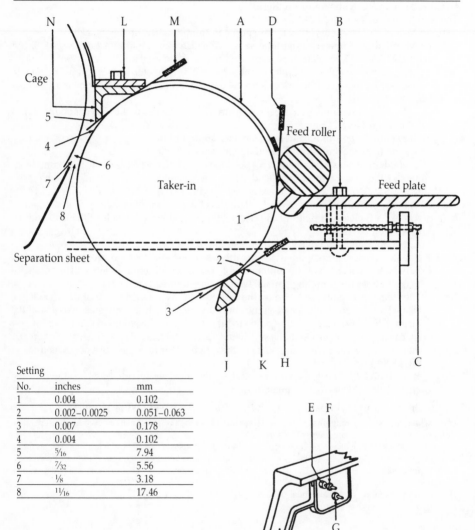

Setting		
No.	inches	mm
1	0.004	0.102
2	0.002–0.0025	0.051–0.063
3	0.007	0.178
4	0.004	0.102
5	$^5/_{16}$	7.94
6	$^7/_{32}$	5.56
7	$^1/_8$	3.18
8	$^{11}/_{16}$	17.46

Figure 4   The Shirley Analyser (Wool Model)

2. Settings

		Inches	mm
1. Feed plate	Taker-in	0.004	0.102
2. Streamer plate (lead-in edge)	Taker-in	0.002–0.0025	0.051–0.063
3. Streamer plate (lead-out edge)	Taker-in	0.007	0.178
4. Stripper knife (bottom edge)	Taker-in	0.004	0.102
5. Stripper knife (bottom edge)	Cage	⁵⁄₁₆	7.94
6. Taker-in	Cage	⁷⁄₃₂	5.56
7. Separation sheet (top edge)	Cage	⅛	3.18
8. Separation sheet (top edge)	Taker-in	¹¹⁄₁₆	17.46
9. Delivery plate	Cage	¹⁄₁₆	1.59

3. Setting Procedure (see Figure 4)

Feed Plate. Remove cover A, slack off nuts B at each end of feed plate. Then, by means of adjusting screw C, move plate up to 'feeler' gauge inserted as at D across full width of machine, whilst revolving taker-in slowly by hand. Tighten nuts B and the locknuts on screw C.

Streamer Plate. Slack off 'brush nuts' E and 'securing nuts' F, G, on each side of machine. Insert 'feeler' gauge as at H, and bring streamer plate J up to the gauge, across full width of machine. Tighten E on both sides of machine, remove gauge and allow streamer plate to swivel to a gauge placed between taker-in and edge K. Tighten F and G on both sides of machine.

Stripper Knife. Remove cover A, slightly slacken nuts L. Insert at M until gauge is 'nipped' by knife N. Tighten screws L, and make quite secure.

4. Condition of Working Parts

The machine will not function efficiently if certain working parts are damaged even to the slightest extent. The main components which must retain their smoothness and freedom from damage are:
1. Striking face of the feed plate.
2. Lead-in and lead-out edges and outer working face of the streamer plate.
3. Lower edge and working face (facing the cage) of the stripper knife.
4. Outer surface of the cage.
5. Inner surface of the celastoid cage cover.
6. Lead-in edge and outside surface of the delivery plate.

The taker-in should be examined occasionally and all bent or damaged teeth repaired.

5. Sample Preparation

The Shirley Analyser (Wool Model) has a carding action which removes vegetable matter and dirt and which opens up the wool fibre, mixing it to form a homogeneous blend. When using the Analyser the following precautions should be observed:
1. Ensure that the Analyser is free from wool and grease and is correctly adjusted. Clean the static eliminator and switch it on.
2. (a) Pass the whole of the laboratory sample through the Analyser by spreading the washed cores evenly across the feed apron and pushing the wool into the input rollers.
   (b) When all the wool has passed through the machine, empty the contents of the reject tray onto the input feed apron and pass through the machine.
3. Repeat 2(b) until it is obvious that no further separation of wool and vegetable matter will take place. Two or three passes of the reject may be needed with light fault, five or six passes with heavy fault or badly entangled wool.
4. Pass the wool from the 'Product' box once through the Analyser. This will mix the mass of fibres thoroughly and remove most of the mineral and vegetable matter remaining after the first pass.

Appendix 2   Sample Preparation on the Laboratory Mechanical Card

The function of the mechanical sample card is to open and homogenise the scoured laboratory sample and simultaneously remove dust and a large proportion of the vegetable matter.

The following precautions are to be taken in using this type of card:
1. In order to avoid any excessive loss of the short core-sampled wool, set the distance between the various rollers and the swift at 0.2–0.3 mm, using a metal feeler gauge.
2. Make sure that the card is clean, particularly that it has been stripped so that the laboratory sample is not contaminated by other wool fibres that may remain in the card clothing.
3. Pass the whole of the laboratory sample through the card and collect the carded web. Run the card until all the wool has passed through.
4. Take the once carded material, place the layers in a direction at right angles to the first passage, and repeat operation 3.

5. Make sure that the loss of material falling under the card and lodging in the card clothing does not exceed approximately 25% of the initial loose wool.

If such a loss occurs, collect the fallen material from below the card, incorporate it in the material obtained after operation 4, and repeat the carding operation a third time.

Note 1: In the case of wools highly entangled after scouring, the precaution should be taken of feeding the samples between the feed rollers in small quantities, previously opened by hand as far as possible.

Note 2: If the card clothing is in good condition, after two passages the web should contain only a small amount of unopened accumulations of fibres (neps or slubs) and the major proportion of vegetable matter should also have been removed without excessive loss of broken wool fibres. The small residual amount of neps, slubs and vegetable matter is to be removed using forceps. If this performance is not achieved, the roller settings should be checked according to operation 1 (above), or the card clothing should be renewed.

Note 3: In the case of highly entangled fine wools two passages on the card may be sufficient to achieve good opening without too many neps or slubs. Operation 3 should then be repeated a third or even a fourth time if necessary.

Appendix 3   Calibration of the Airflow Fineness Meter

1. Principle
Instruments are calibrated using samples of eight dry-combed reference slivers covering the range of Mean Fibre Diameters generally used in wool commerce and industry. The reference slivers to be used are the Interwoollabs Standards.

A set of suitable reference slivers may be obtained from: Interwoollabs Standards, C/o Secretariat Interwoollabs, Boite 13, Rue du Luxembourg 19/21, 1040 Brussels, Belgium.

The reference slivers are supplied in dry-combed form, with a total fatty matter content of less than 1%. The slivers should not be cleaned before use. The Mean Fibre Diameter of each sliver has been measured by many member laboratories of Interwoollabs by Airflow and by Projection Microscope in repeated round-tests organised by Interwoollabs.

Up-to-date information and further details can be obtained from the above address.

2. Method
Cut the reference sliver into lengths of approximately 20 mm. Completely deparallelise the cut fibre pieces necessary for the calibration, using the same technique that will subsequently be used for sample preparation.

The deparallelisation may be effected by carding on the Shirley Analyser (Wool Model) (Appendix 1) or on a mechanical sample card (Appendix 2).

If a dust filter as described in Appendix 4 is used, both calibration and routine measurement must be made with the dust filter in place. Carry out the calibration of the instrument as described below.

3. Leakage Test
After assembling the apparatus as in Figure 1, remove the cap and plunger from the constant volume chamber A and insert a rubber stopper. By means of a Hoffman clip close the rubber tube between A and F after introducing a pressure difference causing the meniscus in the manometer to fall by about 150 mm. Note the position of the meniscus periodically for several minutes. If it changes, the apparatus should be examined for leaks.

4. Graduating the Scale
4.1 *Constant Flow Apparatus*

Make a horizontal mark Y (Figure 1) near the top of the flowmeter scale, avoiding any position giving marked fluctuation of the float. Fix a scale graduated in mm behind the manometer and adjust the zero mark to coincide with the meniscus of the liquid. Condition the reference slivers as described in Section 5.2.5. Then prepare and weigh 5 test specimens (1.5 g ± 2 mg) from each reference sliver, as described in Section 5.2.6.

Test each specimen according to the method described in Section 5.3. Adjust the air valve until the top of the float of the flowmeter coincides with the reference mark Y and note the distance in mm below the zero to which the meniscus falls. Similarly, obtain a total of 10 readings on each reference sliver and calculate the mean reading.

The mean readings of 'h' in mm may be plotted against the known values of the Mean Fibre Diameter 'd' in micrometres, and inspected to ensure that the points lie about a smooth curve.

The results obtained must be fitted to a curve using the Least Square Method given below. The regression equation obtained is the Calibration Equation and must be used to determine the relationship between Airflow readings and Mean Fibre Diameter. This is preferably done by direct calculation, but a Table relating Airflow reading to Mean Fibre Diameter can also be used.

4.1.1 *Fitting a Curve by the Least Squares Method*

The relation between d and h is in the form of hd^c = constant and it is thus necessary to take the logarithms to obtain a linear relation.

For each of the n lots of sliver used for calibration, obtain values for X and Y where:
$$X = \log d \ (d = \text{known Mean Fibre Diameter in } \mu m)$$
$$\text{and } Y = \log h \ (h = \text{manometer reading in mm } H_2O)$$
Then calculate the following values:

$$b = \frac{\Sigma(XY) - n. \bar{X}.\bar{Y}}{\Sigma(X^2) - n.(X)^2}$$

$$a = \bar{Y} - b.\bar{X}$$

Where:

$$\bar{X} = \frac{X_1 + X_2 \ldots X_n}{n}$$

$$\bar{Y} = \frac{Y_1 + Y_2 \ldots Y_n}{n}$$

$$\Sigma(X^2) = X_1^2 + X_2^2 + \ldots X_n^2$$
$$\Sigma(XY) = X_1Y_1 + X_2Y_2 \ldots X_nY_n$$

The regression equation of Y and X which applies to the apparatus is then:
$$Y = a + b(X) \tag{1}$$
which can be transformed to:

$$\log d = \frac{\log h - a}{b} \tag{1a}$$

Finally, prepare a conversion table relating d to h by substituting values of h at 5 mm intervals into equation (1a) and so tabulating d for each value of h. A scale graduated in micrometres may then be fitted behind the manometer.

4.2 *Constant Pressure Apparatus*
Make a horizontal mark 180 mm below the zero mark Z (Figure 1) of the manometer. Fix a scale graduated in mm behind the flowmeter F so that the zero of this scale coincides with a file mark (zero) made near the bottom of the flowmeter. Condition the reference slivers as described in Section 5.2.5. Then prepare and weigh 5 test specimens (2.5 g ± 4 mg) from each reference sliver as described in Section 5.2.6. Test each according to the Constant Pressure method described in Section 5.3.1. Adjust the air valve until the meniscus level of the manometer coincides with the 180 mm reference mark H and note the distance h in mm of the float of the flowmeter from zero. Similarly, obtain a total of 10 readings on each reference sliver and calculate the mean reading.

The mean readings of 'h' in mm may be plotted against the known values of the Mean Fibre Diameter 'd' in micrometres, and inspected to ensure that the points lie on a nearly linear curve.

The results obtained must be fitted to a curve using the Least Square Method given below. The regression equation obtained is the Calibration Equation and must be used to determine the relationship between Airflow readings and Mean Fibre Diameter. This is preferably done by direct calculation, but a Table relating Airflow reading to Mean Fibre Diameter can also be used.

4.2.1 *Fitting a Curve by the Least Squares Method*
Fit a second degree regression line of d on h. This is done by finding the coefficients a, b, c in the equation:
$$h = a + bd + cd^2 \tag{1}$$
The coefficients a, b, and c are calculated as follows:

$$b = \frac{(S_{dh}.S_{zz} - S_{hz}.S_{dz})}{(S_{dd}.S_{zz} - S^2_{dz})}$$

$$c = \frac{(S_{dd}.S_{hz} - S_{dh}.S_{dz})}{(S_{dd}.S_{zz} - S^2_{dz})}$$

$$a = \bar{h} - b.\bar{d} - c.\bar{z}$$

Calculate also the 'Mean Square Error', MSE, from the formula:

$$MSE = \frac{S_{hh} - b.S_{dh} - c.S_{hz}}{(n - 3)}$$

where:
$\bar{d} = \Sigma d/n; \bar{h} = \Sigma h/n; \bar{z} = \Sigma z/n;$ and $z = d^2;$ and Σd denotes the summation e.g. $\Sigma d = d_1 + d_2 + d_3 + \ldots d_n$ and the values $S_{dd}, S_{hh}, S_{zz}, S_{dh}, S_{dz}, S_{hz}$ are defined by the general formula:

$$S_{xy} = \Sigma(x.y) - \frac{\Sigma x \Sigma y}{n}$$

When the values of a, b, and c have been determined, they must be used in Equation (1) to calculate the values of h corresponding to the nominal values of the reference slivers (d_1, d_2 etc.).

If either:

(i) any calculated flowmeter height differs from the observed value by more than 1.00 mm; or
(ii) the Mean Square Error is greater than 1.00 mm² the calibration is unsatisfactory. The results must be discarded and another calibration carried out with fresh specimens of reference sliver.

When satisfactory values have been obtained for a, b, and c, prepare a conversion table relating d to h using the formula:

$$d = \frac{\sqrt{b^2 + 4.c. \, (h - a)} - b}{2.c}$$

A scale in micrometres may then be fitted behind the flowmeter.

Appendix 4   Dust Filter to be Used with Airflow Fineness Meter when Measuring Test Specimens

In the operation of both the 'Constant Flow' and 'Constant Pressure' forms of the Airflow Fineness Meter, air passes through the fibre mass by suction from the exterior ambient air into the interior of the instrument, i.e. into the calibrated tube of the flowmeter, which often contains a hollow float.

When many measurements are made on test specimens originating from raw wool core samples, there is a very rapid build up of dust in the connecting tubes and a dust deposit in the hollow float of the flowmeter.

To avoid this dust deposition, which necessitates the frequent cleaning of the flowmeter and the float, a filter may be used. The filter is positioned between the constant volume chamber and the flowmeter, and is designed to prevent all vegetable matter particles, fine dust and short fibres from entering and being deposited in the calibrated tube or the float.

Description of Filter

The filter consists of a non-oxidizable metal screen woven from wires 0.11 mm diameter with a pore size of 0.16 mm (No. 23 in French Standard AFNOR or sieve No. 100 in the 'Standard Screen Scale' of the ACS Year Book).

It is held in a metal cylinder which is fitted in a cylindrical container provided with entry and exit for the airflow. A recess provided at the end of the filter cylinder allows it to be positioned correctly in relation to a stud fixed in the container. A cover is screwed over the container when the filter is in use.

As an example, sectional diagrams are given of the filter which is used at CTCRS. The type and surface of the screen are the important parameters when ensuring efficient performance of the filter. Also in Figure 5, the filter is shown as mounted on the Airflow Fineness Meter.

It is necessary to clean the filter daily when working with core-sampled wools.

Appendix 5   Corrections for Atmospheric Pressure

The readings of flowmeters are influenced to a small extent by variations in barometric pressure.

1. Correction Formula

This correction has been given by Greuel, Sustmann and Henning, from which the following procedures are deduced:

1.1 Calibrate the apparatus with the reference slivers at 20°C and 65% r.h. at a barometric pressure near to the average for the locality.

1.2 When making routine tests record the barometric pressure if it deviates by more than 10 mm from the calibration pressure.

1.3 The correction to be added to the value obtained from calibration, in micrometres, is (C−H).m.d

where C = atmospheric pressure at calibration (mm Hg)
where H = atmospheric pressure during test (mm Hg)
where d = instrument reading (micrometres)
where m = constant of the apparatus (1/mm Hg)

(For apparatus constructed according to Figures 6 and 7, values of m = 0.00022 have been found. This gives corrections up to 0.2–0.3 μm from pressure deviations of 40 mm Hg within the range 20–35 micrometres.)

1.4 The apparatus constant m can be determined accurately for any apparatus using the equipment shown in Figure 6. By adjusting the valve T the pressure above the constant volume chamber can be reduced by a given amount indicated by the manometer M. The instrument readings given for a range of slivers are determined for depressions of 0, 20, 40, 60, 80 and 100 mm Hg using the procedure of Constant Flow or Constant Pressure as described in section 5. From these results the value of m in the correction equation is derived.

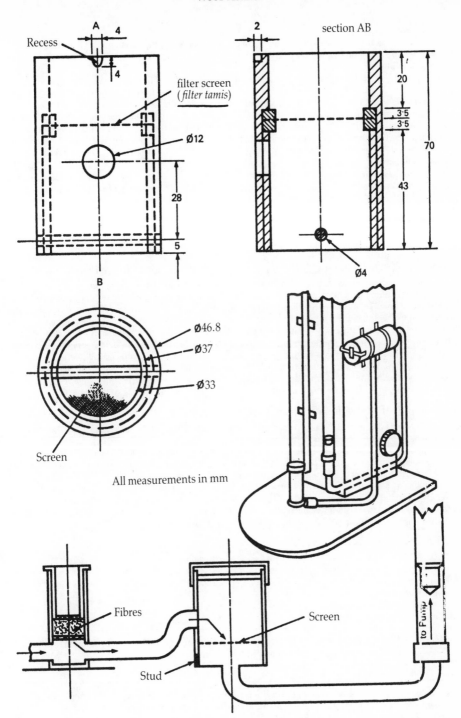

Figure 5   Dust filter

2. Calibration

Alternatively, instead of using the correction formula, a calibration of the apparatus for the value of atmospheric pressure at the time of test may be carried out.

3. Correction Plugs

A correction for barometric pressure may be made with the aid of a plug filled with non-hygroscopic fibre, e.g. polyester. Details of the construction are given in Figure 7. The plug is filled with the correct mass* of non-hygroscopic fibre and the top of the plug sealed on permanently with a resin which sets at a sufficiently low temperature to avoid damaging the fibres (e.g. 100°C). With the chamber empty the plug is clamped** in the Airflow Fineness Meter and the instrument reading taken several times at 20°C when the barometric pressure is within 5 mm Hg of the pressure at calibration. Let the mean reading be C (micrometres). During any series of subsequent tests the plug is inserted and a reading taken. Let this be T (micrometres). Then the results of any test are multiplied by the factor C/T.

Experience has shown this to be a most useful check, and it has explained several interlaboratory discrepancies. However, if used very many times the plug will filter out dust if the atmosphere is dirty, and the reading will change. It is advised therefore that after it has been used for about 400 times (18 months if used once per day) the non-hygroscopic fibre be replaced by clean fibre and a new standard reading for the plug be determined.

Likewise the plug should be kept in a protective container when not in use and should be discarded if it has been in contact with any liquid, powder, etc.

Appendix 6   Checking the Airflow Fineness Meter by Use of Orifice Plates

To make regular daily checks that the apparatus is in good order the use of two orifice plates is recommended. These consist of aluminium discs having the same diameter as the inside of the constant volume chamber, each with a central hole. The discs have a rim which rests on the annular top of the constant volume chamber. The diameter of the central hole in one disc is chosen to give a reading of about one-third of the available scale on the manometer (Constant Flow apparatus) or flowmeter (Constant Pressure apparatus) when clamped and used in the apparatus under working conditions, with no fibre in the chamber. The diameter of the central hole in the second disc is similarly chosen to give a reading of about two-thirds of the available scale.

About once a day the orifice plates are clamped in the apparatus so that air enters only through the central hole and the readings are noted. Variations in the readings should not exceed 2 mm and 4 mm respectively for the two orifice plates. This provides a useful and quick check on the functioning of the apparatus.

Appendix 7   Preparation of Scoured Wool Subsamples

If an IWTO Certificate is to be issued, the subsamples must come from a total sample taken in accordance with the current IWTO Core Test Regulations.

The subsamples may be the same ones as used to determine the Wool Base and Vegetable Matter Base of the raw wool according to the current IWTO Test Method (IWTO-19). If this is the case, then the procedure in that Test Method for taking and scouring the subsamples should be followed.

The following alternative to that procedure may be used. Two or more subsamples each of at least 150 g are taken by the method given in the current IWTO Test Method (IWTO-19). These subsamples are then scoured and dried in accordance with the same Test Method. When non-ionic detergents are used, as provided for in that Test Method, their use shall be limited by the following conditions:

(a) *Hand Washing systems*

	1st washing bowl	*2nd washing bowl*
Maximum concentration of detergent	2 g/litre	1.5 g/litre
Washing temperature	52 ± 3°C	52 ± 3°C
Minimum liquor to wool ratio	80:1	80:1
Maximum time of immersion in bowl	5 mins	5 mins

Rinsing bowls: No limits on liquor to wool ratio, time of immersion and method of rinsing (running water or not) but the temperature must not exceed that of the washing bowls.

---

* The correct mass of non-hygroscopic fibre in grams is given by 1.5p/1.31 for Constant Flow and 2.5p/1.31 for Constant Pressure where p is the density of fibre and the density of wool is assumed to be 1.31. For polyester fibre the correct mass will be 1.58 g for Constant Flow and 2.63 g for Constant Pressure.
** It is important to ensure that there is a good seal between the lip of the chamber and the flange of the plug. A rubber 'O' ring is satisfactory.

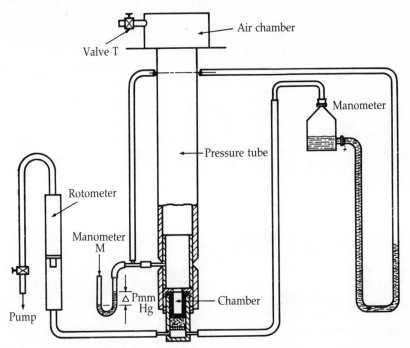

Figure 6   Equipment for determining apparatus constant m

(b) *Semi-Automatic and Automatic Washing Systems*
The same conditions will apply as for hand washing systems except that a temperature up to a maximum of 80°C may be used. In this case, however, the total time of contact between the wool and the detergent liquor at 80°C shall be less than 5 minutes.

Appendix 8   Examples of the Calculation of Results

Using one Airflow Fineness Meter
1st Example
The range, W, of the four readings is greater than the value given in 5.3.2.

1st Test specimen	32.4 μm	32.6 μm ⎱	W = 0.5 ⎱	
2nd Test specimen	32.9 μm	32.8 μm ⎰		W = 0.6
3rd Test specimen	32.7 μm	33.0 μm		

With 6 readings W = 0.6, which is not greater than the range given in 5.3.2.
Calculation of Results
Mean Fibre Diameter of each test specimen: 32.50; 32.85; 32.85 μm
Mean Fibre Diameter of the sample: 32.7 μm

n = 6
W = 0.6
95% confidence limits of the sample = ± 0.415 × 0.6 = ± 0.2 μm

2nd Example
Following 5.3.2. it was necessary to measure six specimens

1st Test specimen	35.0 μm	35.0 μm ⎱	W = 0.5 ⎱		
2nd Test specimen	34.5 μm	34.7 μm ⎰		W = 0.7 ⎱	
3rd Test specimen	34.3 μm	34.4 μm			
4th Test specimen	34.9 μm	35.1 μm			W = 0.9
5th Test specimen	34.8 μm	34.8 μm			
6th Test specimen	35.2 μm	34.9 μm			

Calculation of Results
Mean fibre diameter of each test specimen: 35.00; 34.60; 34.35; 35.00; 34.80; 35.05 μm

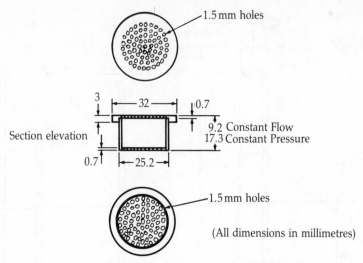

Section elevation

9.2 Constant Flow
17.3 Constant Pressure

1.5 mm holes

1.5 mm holes

(All dimensions in millimetres)

Note: A thin rubber or plastic sealing ring of internal diameter 25.2 mm is placed under the rim of the plug so there is no edge leakage when clamped in position on the constant volume chamber.

Figure 7   Dimensions of correction plug

Mean fibre diameter of the sample: 34.8 μm

n = 12

W = 0.9

95% Confidence limits of the sample = ± 0.195 × 0.9 = ±0.2 μm

Note: The confidence limits given above are within laboratory confidence limits. With more than one laboratory testing the same lot of wool, the confidence limits given in Appendix 9 apply.

Appendix 9   Precision

Table 1 gives the components of variance and confidence limits when two specimens are weighed out and two measurements are made on each specimen. The basic data was taken from an interlaboratory trial carried out in Australia in 1980.

Table 1   Components of Variance for Airflow Measurements and confidence limits for 2 × 2 Determinations

	Single meter method		Two meter method	
	Wools to 26 μm	Wools above 26 μm	Wools to 26 μm	Wools above 26 μm
Between laboratories σ_L^2	0.060	0.105	0.020	0.037
Between meters σ_I^2	—	—	0.021	0.027
Between specimens σ_S^2	0.018	0.058	0.026	0.052
Between measurements σ_M^2	0.006	0.008	0.004	0.005
Variance of 2 × 2 determinations	0.0705	0.136	0.0445	0.0778
Confidence limit of 2 × 2 determinations	0.52	0.72	0.41	0.55

The variance of the 2 × 2 determinations (two measurements on each of two test specimens) was calculated as follows:

One Airflow Fineness Meter $\sigma_T^2 = \sigma_L^2 + \sigma_S^2 + \sigma_M^2$ (1)

Two Airflow Fineness Meters $\sigma_T^2 = \sigma_L^2 + \sigma_I^2 + \sigma_S^2 + \sigma_M^2$ (2)

The 95% confidence limit for the determination is $1.96\,\sigma_T$

The between cores variance has been determined for several types of wool from different origins. This information has been used to calculated the overall variance using Equation (3).

$$\sigma_{OV}^2 = \sigma_T^2 + \sigma_C^2$$ (3)

where σ^2 c is the between cores (within bale) variance and n is the number of cores per lot.

The 95% confidence limit for the determination is $1.96\sigma_{ov}$.

From this data the overall confidence limits were calculated and are given in Tables 2A and 2B overleaf.

Appendix 10   Special Types of Wool

1. Lambswool

Robinet and Franck have tested samples of lambswool and found that the estimates of diameter obtained by the Airflow Method were systematically finer than those given by the projection microscope, the apparatus having been calibrated from slivers of ordinary wool.

The maximum difference they obtained in estimated fibre diameter by two methods was 6.7 per cent.

2. Medullated Wool

The theory of the Airflow Method assumes that the fibres have a constant over-all density, so a fixed weight of fibres of the same average diameter will always give the same amount of fibre surface. Highly medullated fibres may have an appreciably lower fibre density than the accepted value of 1.31 for non-medullated wool fibres. An airflow test carried out on such medullated fibres will give a reading finer than the fibre diameter as measured on a projection microscope. This error may be of significance for fibres coarser than about 35 micrometres. The effect is illustrated by the following data taken from a paper by Richards.

Density (g/cc)	1.31	1.29	1.27	1.25
Apparent Mean Fibre Diameter (μm) (Airflow)	35.00	34.20	33.20	32.20

Additional Measurements

Although introduced only a little over a decade ago, sale by sample and certificate is now recognised as being highly successful and is well established as the perferred method of selling wool in Australia. Through continuing research and the introduction of innovative procedures, rapid progress is being made towards developing scientific equipment capable of objective measurement of all the value-determining characteristics of wool. When this is achieved, it will enable the introduction of sale by description, which will not only help to minimise selling costs for the grower, but will also provide a much more accurate basis for buyers and manufacturers to forecast the processing potential of greasy wool.

In the meantime, sale with additional measurement was recently introduced and will give growers the opportunity to have the following measurements included on their sale lots:

- average staple length;
- variation in staple length;
- average staple strength;
- position of break in the staple;
- clean colour.

The next stage of development might be measurement of resistance to compression and fibre diameter variability.

Commercial Test Procedures

The test procedures are outlined in the following SAA standards: 'Wool — Measurement of mean staple length', and 'Determination of mean staple strength'.

Usually 60 staples sub-sampled from the grab sample are used for length and

Table 2A   One Airflow Fineness Meter
Overall Confidence Limits (95% Level) of the Determination of Mean Fibre Diameter of Raw Wool by the Airflow Method

Country of Origin	Between Cores Variance	Mean Fibre Diameter less than 26 micrometres Cores per lot						Mean Fibre Diameter 26 micrometres and greater Cores per lot					
		10	15	20	30	50	100	10	15	20	30	50	100
America	0.903	0.79	0.71	0.67	0.62	0.58	0.55	0.93	0.87	0.83	0.80	0.77	0.75
Argentina	0.835	0.77	0.70	0.66	0.61	0.58	0.55	0.92	0.86	0.83	0.79	0.77	0.74
Australia	0.083	0.55	0.54	0.54	0.53	0.53	0.52	0.74	0.74	0.73	0.73	0.73	0.73
Brazil	1.034	0.82	0.73	0.69	0.64	0.59	0.56	0.96	0.89	0.85	0.81	0.78	0.75
Canada	0.865	0.78	0.70	0.66	0.62	0.58	0.55	0.92	0.86	0.83	0.80	0.77	0.75
Hungary	0.160	0.58	0.56	0.55	0.54	0.53	0.53	0.76	0.75	0.74	0.74	0.73	0.73
New Zealand	0.523	0.69	0.64	0.61	0.58	0.56	0.54	0.85	0.81	0.79	0.77	0.75	0.74
South Africa	0.152	0.57	0.56	0.55	0.54	0.53	0.53	0.76	0.75	0.74	0.74	0.73	0.73
Uruguay	1.222	0.86	0.76	0.71	0.65	0.60	0.56	1.00	0.91	0.87	0.82	0.79	0.75

Table 2B   Two Airflow Fineness Meter
Overall Confidence Limits (95% Level) of the Determination of Mean Fibre Diameter of Raw Wool by the Airflow Method

Country of Origin	Between Cores Variance	Mean Fibre Diameter less than 26 micrometres Cores per lot						Mean Fibre Diameter 26 micrometres and greater Cores per lot					
		10	15	20	30	50	100	10	15	20	30	50	100
America	0.903	0.72	0.63	0.59	0.54	0.49	0.45	0.80	0.73	0.69	0.64	0.61	0.58
Argentina	0.835	0.70	0.62	0.58	0.53	0.48	0.45	0.79	0.72	0.68	0.64	0.60	0.58
Australia	0.083	0.45	0.44	0.43	0.43	0.42	0.42	0.58	0.57	0.56	0.56	0.55	0.55
Brazil	1.034	0.75	0.66	0.61	0.55	0.50	0.46	0.83	0.75	0.71	0.66	0.62	0.58
Canada	0.865	0.71	0.63	0.58	0.53	0.49	0.45	0.79	0.72	0.68	0.64	0.60	0.58
Hungary	0.160	0.48	0.46	0.45	0.44	0.43	0.42	0.60	0.58	0.57	0.57	0.56	0.55
New Zealand	0.523	0.61	0.55	0.52	0.49	0.46	0.44	0.71	0.66	0.63	0.60	0.58	0.56
South Africa	0.152	0.48	0.46	0.45	0.44	0.43	0.42	0.60	0.58	0.57	0.56	0.56	0.55
Uruguay	1.222	0.80	0.70	0.64	0.57	0.51	0.47	0.88	0.78	0.73	0.67	0.63	0.59

MTS — mechanical tuft sampler, developed by AWTA Ltd

strength measurement. Results are calculated for average staple length, staple length variability, average staple strength and position of break.

Staple length

In a commercial wool testing laboratory staples are individually placed on a conveyor belt which travels at a constant speed between a light source and a detector on opposite sides of the belt. When the light beam is interrupted by a staple, an electronic counter records the duration of time for which the beam is interrupted by the staple. The time is recorded and the staple length is calculated in millimetres.

Staple strength

Staple strength is reported in newtons per kilotex. Newtons measure the force required to break a staple, and kilotex measures the thickness of the staple. Directly after the measurement of staple length, the maximum force to break the staple is measured. Each end of the staple is gripped by jaws which move apart until the staple breaks. The force required is recorded.

Linear density is established by weighing the two broken portions of the greasy staple, and recalculating to clean weight using the yield results for that lot. An alternative is to measure staple thickness (cross-sectional area) which is highly correlated with linear density.

Individual staples range in strength from 0 to more than 80 N/ktex. (1 kilotex = 1 g of clean dry wool per metre of length.)

Values can be related to traditional subjective estimates as follows:

The CSIRO 'Automatic Tester for Length and Strength' (ATLAS)

1. Conditioned staples are loaded onto a moving belt which drops each one individually onto a second belt moving at right angles to the first. 2. This means that each staple then moves lengthwise past a light beam which measures staple length. 3. Each staple then moves into a set of jaws which grip each end of the staple and then one jaw moves horizontally till the staple breaks. 4. The two halves are then collected and weighed to calculate the position of break and staple thickness. Photograph: Courtesy AWC

- sound — above 30 newtons/kilotex;
- part tender — about 20 newtons/kilotex;
- tender — about 15 newtons/kilotex;
- rotten — below 10 newtons/kilotex;

These relationships are only approximate. It is difficult to break staples by hand when their strength exceeds 30 newtons/kilotex, whereas measurement techniques cover the complete range and are far more accurate than subjective procedures.

Position of Break

The broken halves of the staples are weighed to calculate position of break. The position of break in staples from a lot may be recorded in terms of the distance from

Staple length . . . is measured by passing individual staples past a continuous beam of light. Research has shown that even visually uniform wools vary widely in staple length and 60 individual staples need to be measured to obtain an average length with reasonable precision.

Strength . . . After measurement for length staples are then strained to breaking point to determine tensile strength and the position of break. This measurement is a more reliable guide to processing performance than present, subjective assessment.

the tip expressed as a percentage of staple stength. The length/strength instrument has an interfaced computer which calculates the following:

- mean staple length (mm);
- coefficient of variation of staple length (%);
- mean staple strength (N/ktex);
- position of break.

For both length and strength the major component of the precision of the measurements is the sampling precision.

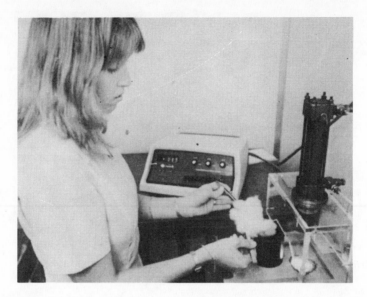

Clean colour . . . Measurements are taken from the scoured core sample. Satisfactory equipment for this test is already commercially available although adoption of SAA standards awaits interlaboratory agreement on scouring and drying procedures.

Clean Colour

When wool is processed, most of the grease, dirt and suint is removed by scouring and the vegetable matter by carding. Therefore, these impurities do not affect the colour of the final product. For this reason, a sample of greasy wool from the lot must be processed to a state of cleanliness comparable to that of the final industry product before relevant colour measurement can be made. The most difficult aspect of the wool colour measurement is in achieving sufficiently consistent preparation procedures.

A colorimeter (such as the HUNTER LAB D25D2M colour difference meter) is used to measure the colour of a 2.500 g sample of clean scoured homogenised wool. This is the same sample used to measure mean fibre diameter employing the airflow apparatus. The colour measuring instrument reduces the appearance of the scoured sample to numbers that are repeatable and relate to visual evaluations. The clean colour of the lot is reported in terms of yellowness (B) and brightness (L). However, this system has not yet been finalised. An alternative method may be the use of Y and Y–Z. Some typical values of these parameters for wool are:

Colour description	L	B
Bright, white	86	10
Creamy	82	12
Dull, yellow	79	21

These figures may be altered, as the colour standard has not yet been finalised.

17

CARBONISING AND SCOURING

Carbonising

Carbonising is the process for removing vegetable matter from raw wool — especially very short, burry wools such as lambs, crutchings and locks — or manufactured fabrics. Carbonising is also resorted to in order to remove vegetable matter from 'union' materials (wool and cotton blends) so that the wool fibres may be recovered for further use.

The first procedure in all cases is the sorting of the greasy wool according to type, length, burr content, etc. Fleece wool is seldom carbonised because the modern carding machines can remove up to 14 per cent of fault with the aid of 'Peralta' rollers which exert a maximum pressure of 3.5 tonnes. With typical 'carbo' wools, however, the length will generally be too short for profitable combing and can be used to better advantage in the manufacturing of woollen materials.

To facilitate subsequent processes after sorting, the wool is passed through an opening machine, as it is necessary to distribute the burr evenly throughout the greasy wool. Opening up is carried out by passing the wool through a type of 'willey' after which it is teased. The wool is then scoured, for which a five-bowl plant is employed, with each bowl differing in temperature. Scouring for carbonising requires slightly better liquors than is necessary for other purposes. This is to ensure the removal of all grease from the wool and vegetable fault, to secure successful penetration of the acid.

After scouring, the wool is subjected to the carbonising process. Two methods are employed, known as the 'wet' and 'dry' methods. In the former the wool is fed wet into a cold acid bath containing 7 per cent sulphuric acid. The wool travels a combined length of 15.5 m (50 feet) to ensure thorough penetration of the acid. Although hot acid gives better penetration, it is very damaging to the bowls. This method is most commonly adopted as it is cheaper and safer than the dry process. In the dry method, the material is exposed to the action of hydrochloric acid gas, specially constructed cylinders or chambers being used. From the aspect of efficiency, this method is excellent, but unfortunately cannot be so conveniently adopted as the wet process.

After acidification, which usually takes about 10–12 minutes, the wool passes into the carboniser, with a temperature of between 96°C and 99°C. Temperatures much in excess of 99°C cause damage to the fibre structure, distortion and brittleness. The wool is then treated in the second drier which is similar to the first, with the exception that the temperature is approximately 93°C.

The action of the acid is to burn or carbonise the vegetable matter which is left in a very brittle state. Most vegetable matter responds very well to treatment, but green trefoil is particularly difficult to remove, necessitating careful handling.

The wool is fed to the crushers, four sets being employed, each set having four fluted rollers. The bottom rollers travel faster than the top so as to produce a slight rubbing effect. The rollers are closely set and pulverise all vegetable matter passing between them. It is necessary for the wool to undergo a second crushing to ensure complete crushing on both sides. After crushing, the wool is passed through two willeying machines which are perforated revolving cages in which the wool is strongly agitated, to shake out all dust.

As any acidity would be harmful to subsequent processes, free acid in wool is neutralized by passing the wool through five bowls containing:

Bowl 1. fresh clean running water; is actually a pre-soak bowl which removes about 33 per cent of the acid;

Bowl 2. contains a mild solution of detergent and soda ash at a temperature of about 46°C;

Bowl 3. similar to Bowl 2, with the exception that the temperature is 43°C;

Bowl 4. has soda ash only, with a temperature of 41°C;

Bowl 5. has cold water only, being using as a rinse.

After the wool has passed through a drier it is allowed to stand for about 48 hours to condition, when it is ready for subsequent manufacturing processes.

Wool Scouring

Wool in its natural, or greasy, state contains certain extraneous substances in the form of yolk and mineral matter than would, if allowed to remain, adversely affect the results likely to be obtained from the many subsequent processes through which it must pass during manufacture. The removal of these substances is effected by the operation known as wool scouring. The objectives of wool-scouring are:

1. The removal of such foreign matter as dirt and yolk from the wool;
2. To deliver the scoured product in good condition, that is, with all the qualities of the wool well preserved;
3. To develop, as fast as possible, some of these properties, notably colour, softness, and fullness of handle;
4. To leave sufficient wash (0.3 to 0.7 per cent) in the wool to assist succeeding manufacturing processes.

Modern wool-scouring is conducted almost entirely by machinery and the present-day types of scouring machines have been evolved from the old method of 'tank-scouring'. An up-to-date scouring set will comprise three to six separate tanks or bowls placed in running order. The number is necessary in order that the impurities may be effectively removed in the minimum amount of time, at the same time preserving the best of the wool. As a rule, merino wools will require a more extensive scouring set than is generally required for the treatment of crossbred wool.

In an ordinary three-bowl set, the first bowl is the longest and has the greatest capacity. It may be from 7.3 m to 9 m (24 to 30 feet) long and capable of containing 6.8 to 8.2 m³ (1500 to 1800 gallons) of water. The bowls are rectangular in shape, and are fitted with false perforated bottoms to allow the sediment to fall through and settle on the true bottom of the bowl, thus keeping the scour cleaner.

The first bowl is used as a 'scourer' and in this tank, scour of the greatest strength is used to remove, or at least to reduce, the foreign matter, especially the wax, to a condition that renders its removal more complete in the following bowls. The scouring strength of the first bowl will be regulated so as to obtain the most perfect product from the class of wool being treated. The 'scouring strength' is made up by adding detergent to the water and bringing the solution to 49°–52°C.

The second bowl is much shorter, 5.5 m (about 18 feet) with a capacity to about 4.5 m³ (1000 gallons) and it is here that the remaining foreign matter that has been loosened in the 'scour' of the first bowl will be removed. The scouring strength of the solution of this bowl must be greatly reduced because the bulk of the yolk has already been removed. The fibres are now in a more or less unprotected condition and consequently the temperature must be reduced to 46°C, but 50 per cent of the amount of detergent is necessary for the purpose of maintaining softness in the wool, besides increasing the cleansing powers of the wash. The last bowl is looked upon as a rinsing tank and no detergent is used, whilst the temperature is about 41°–43°C. This third bowl is about 3.658 m (12 feet) long, with a capacity of about 3.4 m³ (750 gallons). The special function of the rinsing bowl is to remove the more discoloured liquid that still remains from the previous bowl, and also all remaining traces of detergent.

Water is the most important factor in satisfactory wool scouring. The supply must not only be copious, but it must also be soft. If the available water supply is hard, it may be economically softened by the use of a zeolite or ion-exchange water softener. The water is simply passed through one or more softeners, depending upon its hardness. It is then soft enough for use in the scour. The softener is regenerated from time to time by passing brine through in the opposite direction.

Hard water is most unsuitable, because besides wasting a considerable amount of detergent, the detergent and grease will not emulsify with the water, but forms a greasy scum that makes it difficult to pass the wool through the rollers. Moreover, the colour of the product will be considerably depreciated.

The scouring efficiency of the solution is greatly increased by heating. Hot water will more rapidly emulsify wax and consequently hastens the scouring action. Indeed, cold water, even with the addition of scouring agents, would not effectively remove the foreign matter. The temperature, however, must be used with discretion as very hot water has a most harmful effect upon the fibre, causing it to become tender and discoloured.

A commonly used variation of this procedure is the so-called 'desuinting' process which differs only in the use to which the first bowl is put. In this case the first bowl temperature is somewhat lower, 32°–43°C, and no detergent is added. Advantage is taken here of the natural detergency of suint, which is very soluble in water and in solution exerts a strong washing action. The water in this bowl must be changed fairly frequently to maintain the working efficiency. The succeeding bowls contain detergent, and work at the temperatures used in the normal method. Generally speaking, this procedure is more suitable for crossbred wool which contains a higher proportion

CSIRO jet scour.
Photograph: *courtesy CSIRO*

Harrow scour Photograph: courtesy CSIRO

of suint and a lower one of wax, thus making possible the use of a shorter scouring train (fewer bowls).

The wool is fed into the first bowl automatically, and is then taken through the whole set by mechanical propelling devices and carriers. A certain amount of immersion is necessary, especially in the first two bowls, to ensure satisfactory treatment. The length of the bowl has a considerable influence upon the time it will take the wool to pass through it. Agitation, if too great, is likely to result in 'ropey' wool.

The propelling mechanism is of two types, the *swing harrow* and *swing rake*. This mechanism is controlled and driven by overhead gear and is so adjusted as to descend into the suds and, by a forward sweeping motion, propel the wool through the bowl. It is called the 'harrow' type because the projecting prongs are fastened into a frame in much the same way as in the case of the ordinary iron-tooth harrow used for agricultural purposes. It is given four distinct impulses, first down into the suds; secondly, forward towards the delivery end of the machine; thirdly, upwards clear of the scour, and lastly, backwards a similar distance to the first position. The action tends to keep the wool in an open state and is very gentle, which not only facilitates cleansing, but gives the material a more lofty appearance. Vigorous agitation in this respect with merino wools would tend to bring about felting or roping.

The 'swing rake' is so called because the propelling mechanism consists of a number of large forks — somewhat after the style of a large garden fork — which have a reciprocating motion. That is, when the first fork at the receiving end of the bowl reaches the end of its forward traverse, the fork immediately in front of it simultaneously reaches the end of its backward sweep, from which point it commences its forward movement, to be met again by the one immediately in front of it in a similar fashion. By such means the wool is carried from one fork to the next until it eventually reaches the end of the bowl.

This type of machine is quicker in operation, and consequently, will 'turn off' a greater quantity of wool in a given time, but output is not the only consideration, and in the case of fine wool, where there is a greater disposition to felt, the produce will generally be delivered in better condition by the 'swing harrow' type which is more generally used in Australia. The 'swing rake' type of machine is, however, very suitable for the treatment of crossbred wools. These wools do not felt so readily and consequently, if good results can be procured, the increased output is an important factor in economic working.

After passing through each bowl, the wool is fed into a pair of rollers by means of a revolving lattice or apron. The squeezing the wool receives as it passes through these rollers removes the sullied water of the previous bowl, and incidentally quite a lot of dirt. It is then delivered into the succeeding bowl. The squeeze-head, or pair of rollers fitted to the delivery end of the last bowl, is quite an important feature of the plant. Its function is to deliver the wool in a well-wrung condition so that it can be the more readily dried or delivered into the carding room in the right condition for further treatment.

The squeeze-head consists of two heavy rollers mounted in a frame. Pressure is applied by means of screws and strong springs. The lower roller is brass-covered, and the top one is covered with a wrapping of lincoln slubbing or compressed cloth. This is done to protect the fibres from crushing between the two metal surfaces, and assist in carrying the wool through. Considerable pressure is applied to the rollers in order to wring as much surplus moisture as possible from the wool. This lapping of the upper of the two rollers is the usual commercial practice, but in smaller units, as in test scours, the better material, rubber, would be used since the extra cost would be warranted.

If properly treated, the wool should be lofty, with the staple formation well maintained. The colour should be a bright white, with a kind texture and full handle. It should be in good condition (i.e. not overscoured until it becomes lifeless).

When delivered from the scouring machine, the wool contains from 40 per cent to 50 per cent moisture. Drying, therefore, becomes a most important operation in the preparatory treatment of wool for manufacturing and the operation must be performed with three objectives:

1. to bring the material into the required condition, that is, dryness;
2. to do so as speedily as possible so that the drying may keep pace with the scouring;
3. to preserve the quality of the wool.

The modern methods of drying have been evolved from those practised in pre-mechanical days. In the first instance wool was dried 'on the green', the scoured wool being spread out upon a grassy plot adjacent to the works. Obviously such a method is entirely dependent upon the atmospheric conditions, and is consequently rather unreliable. The colour, texture and quality of the product is well preserved and it has a lofty appearance. Several mechanical means to expedite drying have been developed.

1. The *hydro extractor* consists of a round cage with perforated sides, mounted on a spindle within a larger cylinder. The wool is placed in the cage and is revolved at 1,100–1,200 revolutions per minute. The wool is thrown against the sides of the cage and the surplus water is forced through the perforations. This treatment removes only about 75% of the moisture. To complete the drying it is necessary to 'finish off' the wool in a drying machine.
2. In modern *wool-drying machines* the moisture is removed from the wool by heat and continuity of output is assured. There are quite a number of different types, but in all machines the heat is obtained from steam pipes which traverse that portion of the machine where hot air is generated. Moisture is removed by a forced draught of air that is heated by drawing over steam pipes. In passing through the machine the wool is agitated to facilitate drying, but if it is too violent, roping and entangling of fibres ensues.

The modern wool-drying machine is placed in close proximity to the last bowl of the scouring set and often the scoured wool is delivered from the squeeze-head by a revolving lattice, or some other contrivance, on to the feed sheet of the dryer. The wool is dried at the rate of about 272 kg (600 lb) per hour and at this rate the dryer keeps 'in step' with the scour, thus ensuring continuity of processing.

A dryer usually consists of a large rectangular chamber about 6.4 m (21 feet) long, 1.7 m (5 ft 6 in) wide and 3.35 m (11 feet) high, the machine being divided into four or five compartments arranged in tiers. The air, heated by drawing over heated pipes, or through an oven located underneath the machine, traverses these various compartments. The wool is fed into the uppermost compartment and by means of the agitation and draught is carried forward to the opposite end of the machine, where it drops down onto the second tier, and so on, until the wool arrives on the lowest tier, from which it is delivered at the other end of the machine. In falling from one tier to the next, the wool is turned completely over, making drying more thorough.

An alternative and commonly used drying machine is the single-apron dryer. This machine requires a greater floor space for the same through-put of wool, which is sometimes a disadvantage. It consists of a long rectangular box through which a continuous wire apron passes carrying the wool along. Air is drawn through the machine by fans, but in this case it is drawn through in the opposite direction to the movement of the wool. It is again heated by steam pipes. Thus the dried wool passing out of the machine meets the incoming fresh air, heating it up and being itself cooled; also the incoming wet wool meets the wettest outgoing air in the first stage of drying. The temperature is hottest at the wool inlet and falls progressively towards the exit. By these means, the best economy of heat, and therefore of steam is attained and this type of dryer today is considered the most efficient. A further advantage of the single-apron dryer is the ease of cleaning and maintenance. Nevertheless, the saving of space that the other multi-apron type allows, maintains for it a certain popularity.

To remove the moisture speedily, the temperature employed ranges from 60° to 88°C but the actual temperature depends on the type and quantity of wool, the type of machine and the trade requirements of the finished product. In the mill the wool is conveyed from the drying machine direct to the bins in the carding room. If the wool is to be baled up for storage or for sale, it will be delivered from the drying machine on to the floor. The most important considerations in connection with the drying of washed wool are first, the condition of the material and second, economical working.

The condition of the wool is the primary consideration. It is essential that it should be delivered with all its natural properties well preserved, and consequently great care must be taken in the regulation of the temperature and the running speed of the machine. Too high a temperature tends to discolour the wool, and besides, the texture and soundness will be adversely affected, giving it a rather lifeless appearance. If the agitation is too violent the wool is likely to become roped, and fibres unnecessarily entangled. In the interests of economical working it is vitally important that the drying should proceed equivalent of the delivery capacity of the scouring set.

18

TEXTILE MANUFACTURE

Manufacturing — Worsted

In the manufacture of wool into cloth, both the machinery and the methods employed vary according not only to the type of raw wool used, but also to the type of cloth to be made. In the worsted section, where combing types of wools are used and where the aim is to produce a strong, firm and smooth-surfaced fabric, the machinery used is designed to straighten out the fibres in a parallel arrangement. Even in the scouring process, consideration is given to the subsequent use of the wool, in that the swing harrow is employed, as this is less likely to felt or tangle the wool.

Carding

After the wool is scoured and dried, it is fed into the worsted carding machine, which opens up the wool into an even layer, removes as much burr and seed as possible and lays the fibres parallel to each other. The carding machine consists of a number of rollers, 152.4 cm (60 inches) in width and of various diameters, revolving at various speeds. The rollers are covered with card wire, which consists of fine wires standing rigidly out from the surface of the rollers. These wires are close together and about 25 mm (1 inch) in length.

The wool is fed in from a hopper and drops on to the first roller as it revolves; this carries the wool to the next roller where it is picked up by a 'stripper', passed on to the next or 'swift' which passes it on to another set of workers, strippers and swift. Thus its proceeds until it reaches the last roller in a thin gauze-like film. Whilst the wool is passing over the carding machine, it passes through fast-moving burr-removing rollers where the burrs are knocked off.

After leaving the last roller, the film is gathered together, passed through a funnel, and then wound in large balls in cans. The wool is now known as 'carded slivers' and at this stage is frequently backwashed to brighten the wool as well as clean it. After backwashing it is usually 'gilled'.

The Gill Box

The 'gill box' is an excellent method of blending slightly different slivers because numerous slivers are fed together, emerging from the gill box the size of one original sliver. The process it passes through is as follows. The wool is drawn in by fluted rollers to where 'fallers' or long-toothed combs work forward combing the wool. It is then drawn out by another set of rollers which are travelling much faster than the first set. Thus the wool fed in one end is blended evenly, receives a certain amount of combing, and is drawn out as one sliver.

Combing

The wool is then ready for the actual combing process. Worsted combing has three distinct functions:

1. to remove and then separate the short wool fibres less than a predetermined length;
2. to further straighten and to make the fibres parallel;
3. to remove any foreign impurities such as vegetable matter which still remain after carding.

The longer fibres then form the combed slivers. The shorter tippy fibres which are taken away from the longer fibres are known as 'noil' and this is a valuable product used in the manufacture of woollen yarns and piece goods.

Different combs are employed for different types of wool. For instance, the 'Noble' comb is most suitable for medium length fine wool and is used in the Bradford system

Harrow scour Photograph: courtesy CSIRO

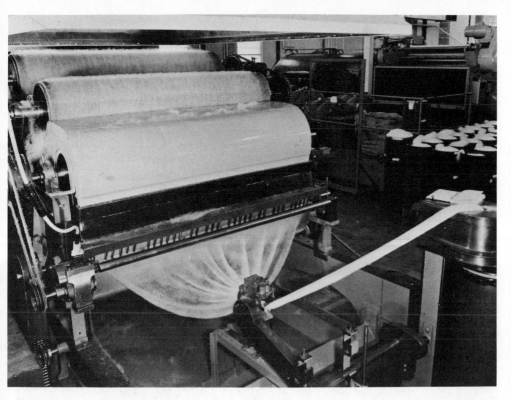

Worsted card output showing the condensing of card web into card sliver

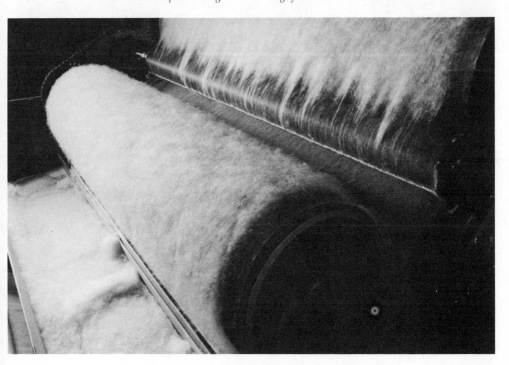

Worsted card showing carding rollers Photographs: courtesy CSIRO

of making worsted yarns. The 'French' or 'Heilmann' comb is very versatile and suitable for medium-length to very short wools. Other types of combs are the 'Nip' or 'Lister' comb which can only be used for long wools and hairs, and the 'square motion' or 'Holden' comb which is used for shorter wools of 76 mm to 152 mm (3 to 6 inches).

Drawing

The next process is to reduce the sliver in size or thickness by 'drawing'. Here the 'top sliver' is passed through machines similar to gill boxes. After the third gilling or drawing, the wool is wound on to bobbins and given its first twist to strengthen the strand. From here the wool is led from one bobbin to another, each new bobbin travelling at a faster speed than the one before it. The thread is thus reduced successively and each time another twist is given.

Spinning

The final stage is the spinning operation. Here the yarn is reduced to the desired thickness and given sufficient twist for the desired strength.

Spinning machines for the worsted system are of four types.

1. The *cap* is used for spinning merinos and fine crossbreds.
2. The *ring* is for medium wools and is the most widely used machine today.
3. The *flyer* is for stronger crossbreds.

In these three methods the action is the same. The thread is drawn off the bobbin

Noble comb

slowly and taken on to another bobbin at a much faster rate, the thread being given the necessary twist at the same time. The only difference is in the method of winding. In the cap method it is by air resistance, with the ring, by a small clip travelling round the bobbin, and in the flyer method by the bobbin itself turning around.

4. The Mule spinner is an entirely different type. This consists of a large frame and carriage, with drafting rollers on one side and the spindles on the carriage. The carriage moves away drawing the thread from the drafting rollers, the thread receiving a twist as it leaves the rollers. When the carriage reaches the full extent of its action, it stops, backs off a little, whilst the full amount of twist is inserted, then travels back on to the frame, winding the thread on to the drafting rollers on the carriage.

After spinning, different yarns are twisted together to make either different ply threads or fancy threads.

Manufacture of Woollen Yarn

In this process the thread produced is quite different to worsted yarn. Although any type of wool can be used, it is the short-fibred lines which are most suitable.

Woollen Carding

After scouring or carbonising, the wool is blended to assure an even batch and it passes to the woollen carding machine. This machine consists of a series of 152.4 cm

French combs (foreground), Gill boxes and auto-levellers (middle ground) and worsted card (background)

Photographs: courtesy CSIRO

(60 inch) wide cylinders, covered with card wire. There are three types of cylinders: the worker, the stripper and the swift, each revolving at different speeds.

The wool is fed on to the first roller in an even layer; it is then taken from this worker, which is a large-diametered cylinder, by the stripper, which is smaller in diameter, then placed on the swift and back again on to another set of workers, strippers and swifts. These rollers are set close together but not touching. The action is to open up the wool into an even layer or film, the main part of this work being done between the swift and the worker. This action is repeated until it reaches the end of the machine when the layer of wool is removed by the doffer and laid transversely across the next machine. In the first machine the fibres are made parallel, but when placed transversely on the next machine, the necessary criss-crossing of fibres occurs. In this machine the cylinders are placed close together and are covered by a finer card wire, and after the wool has passed over this, it is removed to the tape condenser in a thin film. The tape condenser consists of approximately 150 leather tapes, each tape taking its quota of fibres and passing it through leather rubbers which draw the wool in and give it a slight rub but no twist. The sliver is then wound on to spools ready for spinning. In woollen yarn manufacture, no combing or gilling takes place.

Woollen Spinning

There are two main types of machine used for woollen spinning. The older type, the woollen mule, is still common in the United Kingdom, and until quite recently it was regarded as more versatile and capable of making yarns which were not only better but more typically woollen in appearance than the yarns produced on its competitor, the ring-frame. In the early 1960s, however, improved ring-frames were introduced by several makers, and many advantages are claimed for these machines. Consequently there are many more ring-frames than mules used in the woollen industry today.

Sometimes the wool is dyed before carding, but it is usually left until after spinning. In dyeing, *loose wool* and *tops* are placed in perforated containers and the dye solution is forced through to ensure even dyeing; *yarn* is hung on rods and submerged in the dye; *piece goods* have their ends sewn together to give an endless band and are drawn through the liquid; and *hosiery* is put into the liquid and agitated.

Weaving

In worsted cloth the threads that run along the material are known as the 'warp' threads, whilst those running across are called the 'weft'. In the weaving of worsted cloth, the procedure is to wind the warp threads on to a cylinder in whatever colours, pattern, width or length desired. The ends of this warp thread are then tied to a loom beam and wound on to the beam. Each thread must be drawn through the eyelet of the 'heald', the order of drawing determining the pattern of the cloth. The beam is then placed in the loom with the 'healds' tied in the correct position. The ends of the warp threads are then tied to the cloth apron at the front of the loom. The weft threads, wound on to the shuttles, are placed into their respective boxes.

When the loom is started the healds lift the required warp threads up and the shuttle carries the weft thread across the loom in a predetermined manner to produce a pattern. There are many types of materials woven, but the weave is determined by the movement of the healds in relation to that of the shuttles. Double cloths, such as rugs, plain on one side and patterned on the reverse side, are made with two sets of threads.

After weaving, the material is checked for any faults, and is then passed to finish-

Continuous top dyer

Photograph: courtesy CSIRO

ing machines. Milling is the most common finishing method for woollens. The piece is saturated with soap and water and passed through rollers which reduce the cloth in width and length but give thickness. It is then washed and dried under tension.

The material then passes through a cropping machine which removes protruding fibres according to the desired finish. In some materials such as blankets, a woolly finish is obtained by teasing the surface; with velours a series of raisings and croppings is carried out to give the desired finish and suitings may be brushed and steamed, giving a 'set' to the cloth.

Knitting

There are two methods of knitting: flat bed knitting and tubular knitting. In flat bed knitting the needles operate from slots which are arranged along a flat bed, but in the tubular method, fabric is knitted on a tubular cylinder with needles in grooves at the base. The fineness of the finished product is determined by the number of needles to the centimetre.

Felting

Felt is built up by interlocking fibres in a suitable combination of mechanical work, chemical action, moisture and heat without spinning, weaving or knitting. The felting process is distinct and separate from other textile construction methods and is done on specially designed machines. The first and last stages of felting are, however, the same as those used in woollen weaving. In both felting and weaving, the raw fibres must be selected, graded, scoured, mixed and carded. The essence of the felting process is the hardening and fulling or milling process. Hardening is peculiar to felting alone; fulling is one of the finishing operations in other woollen manufacture, but in felting it is a basic operation.

Wool Textile Research

Si-Ro-Mark

Unscourable tar brands on the fleece have been a heavy burden on the textile industry for hundreds of years. In 1954, for instance, tar brands in the Australian clip cost

CSIRO/Repco self-twist spinning machine

Photographs: courtesy CSIRO

almost \$4 million to remove. Development of the scourable branding fluid Si-Ro-Mark circumvents this cost and so helps wool in its competition with artificial fibres. Australian firms under licence from the CSIRO made enough of the fluid last year to brand more than 112 million sheep. Eventually it is hoped all sheep throughout the world will be branded with Si-Ro-Mark.

Solvent-Degreasing

This is a new process for getting the wax, suint and dirt out of raw wool. It results in a greater quantity of combed top than the traditional method of scouring with soda-soap. The CSIRO process uses a petroleum solvent directed through jets on to the wool, as it is carried along on a conveyor. Most of the solvent is recovered, and along

Ring spinner

with it almost all of the lanolin — a valuable by-product which is largely lost in the old soap scours. Solvent-degreased wool is relatively untangled and in subsequent processing can be carded and combed much more quickly with less fibre breakage.

Carbonising

About 10 per cent of the Australian clip is contaminated with vegetable matter and has to go through a cleaning process — known as carbonising — to remove tenacious burrs. This is a radical treatment involving sulphuric acid. Unfortunately the acid *damages the fibres, causing about a 10 per cent wastage.* During fundamental research into the behaviour of wool protein, it was found that certain wetting agents can protect wool from acid attack. In industrial use, these agents cut wool losses by up to half. Besides, the yarn made from protected wool is stronger. It breaks less often during spinning and so less time is lost through stopping machines to rethread 'ends down'. This simple modification to the old carbonising process represents a potential gain of more than $2 000 000 a year for treating the Australian clip.

Si-Ro-Moth'd

Meals for clothes moths may have cost the world $20 000 000 a year. At any rate the progeny of a single moth can eat 45 kg (100 lb) of wool in twelve months. The Si-Ro-Moth'd process is effective, cheap, easy to apply, odourless, fast to washing and dry-cleaning, and does not alter the other natural properties of wool. The Si-Ro-Moth'd process is based on the use of Dieldrin.

Sironised

This denotes wash-and-wear (non-iron) wool garments and has given to wool the last of the 'easy-care' features. In an age when people emphasise and adopt ways and means of labour-saving in the home, discovery of ways to manufacture pure wool men's shirts, ladies' frocks, skirts and blouses, and children's clothing that can be washed in the home washing machine, hung to dry, and then worn with little or no ironing is sure to influence deeply the choice of wool as a textile fibre. Like the other processes developed by the CSIRO, a vitally important aspect of the Sironising process is its simplicity and cheapness. These virtues mean that the process can be quickly and widely adoped by the textile industry — a primary step in securing wider markets for wool products.

Si-Ro-Set

Basic research had led to a fuller understanding of the way in which protein molecules are built up into wool fibres. It was then possible to see how the properties of wool might be manipulated by chemical treatments. Some further experiments on the setting properties of single fibres suggested that chemical agents might produce permanent creasing effects in wool fabrics. They did, and it was then merely a matter of working out a simple practical method to fit the routine of garment manufacturers. In commercial practice all that is necessary is to spray the garment with a weak solution of ammonium thioglycollate just before the final steam pressing. Because the process is so simple, so cheap and so effective, the clothing industry, both here and overseas, has quickly adopted it on a large scale. It is now confidently expected that people who choose to buy all-wool clothes will lose nothing in appearance or easy-care by this choice.

Shrink-Proofing

Unshrinkable wool fabrics have long been a hope of the textile industry. There have been many previous efforts, but for one reason or another they have not been universally adopted. Usually they have been either too costly, too difficult to apply or too damaging to the wool. Fundamental research in CSIRO laboratories has now disclosed that concentrated solutions of certain chemical salts can fully protect wool from the damaging effects of the oxidising agents used in the less costly methods of shrink-proofing. Based on this finding, a new method has been worked out for commercial practice. It is simple, safe, higly efficient, and relatively cheap and should therefore find widespread favour in the textile trade. Most woven and knitted woollen fabric can be shrink-proofed in this way, the chemicals costing less than six cents per kilogram of wool treated.

Melange Printing

A large proportion of worsted fabrics are produced in composite colours such as greys and fawns. These colours are usually obtained by feeding dyed and undyed wool into the same machines before spinning. Thus, a grey yarn comes out if black-dyed wool and white, undyed wool are fed in. It is difficult to control the blending, however, and a uniformly coloured fabric is not easy to achieve. Because of this and other associated problems, another method of producing these colours was invented, known as melange printing. In this process, stripes of colour are roller-printed on to ribbon-like slivers of unspun wool and perfect blending occurs during the later stages as the wool is spun into yarn. The idea was sound in principle but the printing caused a production bottleneck because of the tedious methods involved. Melange printing fell into virtual disuse until CSIRO research on dyes and solvents permitted extensive modification and streamlining of the process. Now worsted spinners have renewed their interest in melange printing and are using the improved method to achieve excellence in the composite toning of their fabrics.

Automatic Noble Comb Control

The Noble comb separates the short and broken fibres (the noil) from the long fibres, which are then combined into a continuous sliver, known as top. It used to be thought that the ratio of top to noil depended only on the proportion of long and short fibres naturally occurring in the wool being combed. Fundamental research showed that this was not altogether true and that the ratio depended more on the comb's setting than had been thought. The practice has been to rely on experienced operators to keep the 'setover' of the comb as uniform as is humanly possible. But even highly skilled manual control permits wide variation in the ratio of the valuable top produced to the less valuable noil. To overcome this problem, a relatively simple electrical automatic control unit has been invented. Tests have now shown that the unit does its job well. Combs fitted with the device no longer need the constant supervision of skilled operators and allow faster and more uniform output. Labour costs are saved and the units are now operating in all the major Australian woollen mills.

Fellmongering

'Ribby' pelts and skin pieces have no value to the tanner, but the wool is often quite valuable. It used to be salvaged by 'pieing' — a crude and ancient method in which the skins were, more or less, set out to rot. The result was 'pie wool', recognised as a very inferior material, weak-fibred and badly discoloured. Its value scarcely offset the rising cost of labour for its production. However, a new and profitable method for recovering this sort of wool has been developed. It is based on thermal denaturation of the skin. Heating for a short time, at or above 65°C, shrinks and denatures the skin tissues, which are then susceptible to rapid digestion by bacteria. In this way the wool is recovered in excellent condition. Since all fellmongers are faced with the

Rapier loom

Stentor

Photographs: courtesy CSIRO

Circular knitting machine

Photograph: courtesy CSIRO

WORSTED PROCESS

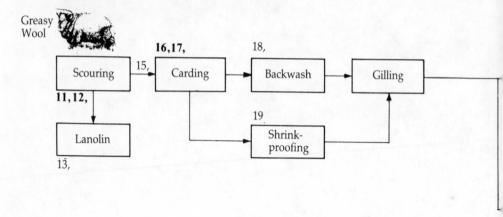

WOOLLEN PROCESS

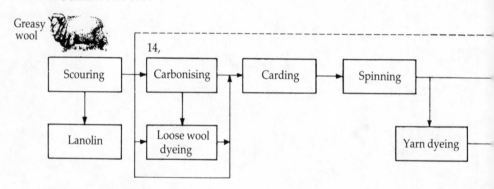

CSIRO Developments

1. 'Si-ro-mark'
2. 'Sire-sine'
3. Sheep rugs
4. Medical sheep skins
5. Fellmongering
6. Teflon coated shearing combs
7. Core sampling
8. Yield testing
9. High density bales
10. Vacuum pressing
11. Solvent degreasing

12. Aqueous jet scouring
13. Lanolin recovery
14. Carbonising
15. Rapid regain testing
16. Wool lubricants
17. Felts
18. Mothproofing
19. Shrink-proofing
20. Comb control
21. Cigarette filters
22. Whitening

23. Rapid dyeing
24. Rapid spinning
25. Pad dyeing
26. Pastel shades
27. 'Si-ro-set'
28. Setting treatment
29. Stretch fabrics
30. Marl printing
31. Detergents
32. Hospital blankets

Figure 27   Wool Processing Sequences

This flow chart shows the main steps in converting raw wool into finished fabrics by the worsted and woollen systems. The numbers on the chart refer to the list of CSIRO developments.

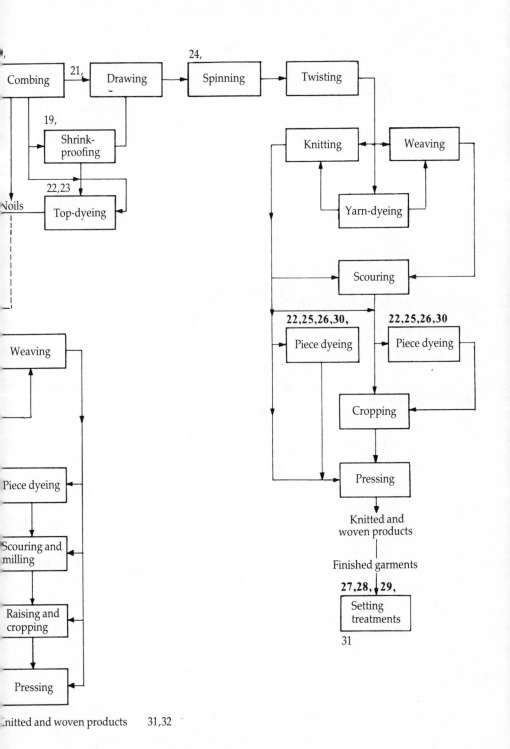

problem of treating 'ribby' pelts and skin trimmings, the CSIRO method has found universal acceptance in Australia and is being adopted overseas.

Marl Printing

Development of the process at the CSIRO Division of Textile Industry for applying marl and mottled finishes to wool fabrics has led to the construction of a full-scale machine that applies four colours simultaneously to fabric at speeds of up to five metres per minute. In this machine, the dye droplets are directed downwards by air jets from four mesh bands containing films of dye. Each band is supplied with a different colour from a dye well and roller system, which traverses the length of the band with the air jet nozzles. The machine is at present undergoing commercial trials.

Superwash

A treatment in the production of fully machine-washable knitwear in pure new wool. It is based on the CSIRO's Chlorine/Hercosett process and has proved to be highly successful.

19

SHEEP CLASSING AND MERINO BREEDING

Purchase of Rams

Special care should be exercised in the choice of the rams, since they exert such a large influence on the flock. The small flock-owner should obtain his rams from some reputable stud-breeder, where he can at least be sure of getting purebred animals, whether they are for mating for wool production, or for fat-lamb raising.

The rams used should be suitable for the district where they are required. If the breeder is satisfied that improvement is being made in his flock by the introduction of rams from a particular stud, it is advisable to continue using the same strains, as more even results are likely to be obtained than if purchases are made from different studs, even though the animals are similar in every way.

The most important qualities in the male, in addition to good lineage and high individuality, are masculinity, bodily vigour and prepotency, plus of course his wool covering, especially if he is to be used for wool-breeding. He must also be typical of his breed.

Classing of Ewes and Rams

Importance of Classing and Mating

Australia is to a certain extent dependent on the sheep-breeding industry for her national prosperity. In order to maintain and improve the standard of the Australian flocks, it is desirable that the importance and value of careful sheep classing should be accepted by all not only in merino flock and stud merino sheep, but in British breeds and their crosses also.

The standard of any flock is largely, if not entirely, dependent not only on skilful classing but also on the selection of stock of high quality and purity of breed. Ample evidence of this is provided by the evolution and improvement brought about in merino sheep from their small-framed, light-woolled primitive state to present-day standards of robust, large-framed animals, cutting heavy fleece weight of attractive wool. These results have not been obtained by chance and are entirely due to careful breeding, selection, classing and culling.

Sheep Classing

Sheep classing is in itself the study and guidance of the development of a flock, and involves the weeding out of animals showing the greatest departure from the desirable type and breeding from those most approximating the ideal. These methods must embody considerations of both constitution and covering.

Sheep culling is the persistent discarding from the flock of all animals showing undesirable characteristics, especially those showing lack of symmetry or weakness of constitution. Care should be taken to avoid overlooking formation, symmetry and constitutional vigour when striving for quality and quantities of wool.

Classing has many main objects:

1. selection and mating with a view to obtaining maximum evenness both of conformation and covering;
2. discarding from the flock the least profitable sheep;
3. effecting an improvement in the breeding quality of the flock on the assumption that the favourable type remaining will produce their quality with some degree of continuous improvement;
4. enabling the flock to be more easily managed through elimination of sheep susceptible to fly strike disease and constitutional defects causing difficulty in movement and handling.

Deciding Type and Standard

The breeder must first of all understand his country and know the type of sheep most suitable. Information of this nature can be gained by taking notice of the successful breeders in the district and observing their standards and their aims and achievements during the past.

Before deciding on the actual standard of classing and culling, it is advisable to first pen up the sheep. A portion of the mob should be caught and examined so as to become properly aware of the general standard and form a definite plan of the grades and standards which will then be made. Breeding should have a definite aim: plain and semi-plain bodied animals with the maximum of good wool forms the basic program. When making the standard, an even type of sheep should be selected to provide the foundation on which to build.

Choose the big-frame cutters of profitable type, aim at even, dense covering and at times be prepared to sacrifice a little quality for bulk weight. When establishing the objective type be particular about uniformity, correct conformation and constitution.

Sheep Classing and Culling

Each year, drafts of young ewes are brought into the breeding flock and bred from for several successive years and then sold as cast for age. The portion of maiden ewes classed into the flock will depend on the number required to replace those ewes discarded from further mating and to maintain flock numbers at a consistent level in keeping with feed and pasture available. Points to be considered will be:

1. suitable age for first mating;
2. the age of casting for age;

3. the mortality amongst the ewes;
4. the lambing percentages.

Having decided on the standard of the conformation and covering to be aimed at, the next job is to grade the flock. The time best suited for this operation is when the sheep have nine to twelve months' wool. As previously stated, it is a very sound idea first to make a thorough examination of a small portion of the mob. In dealing with young ewes before they enter the breeding flock it is advisable to cull fairly heavily and in most cases to dispose of from 25 per cent to 30 per cent — occasionally more. This heavy culling removes any borderline sheep and increases the market value of the culls.

Knowing just where to draw this borderline is a matter entirely for the classer's own judgement. The absolute best and the worst will be easily selected; however, the doubtful sheep will at times cause the novice much thought. Probably the best method is to take out three grades — tops, doubtful and culls — and to go over the doubtfuls whose distinction is less obvious when the standards are more firmly set.

A good classer will usually make three lines when classing ewes:
1. those culled for serious reasons, including the more highly hereditary constitutional faults;
2. a sale line consisting of ewes away from the general flock standards or obviously inferior;
3. the flock ewes to be retained as station breeders.

When classing or culling, always bear in mind the seasonal conditions under which the young sheep have been grown. If they have been reared under hard conditions, make some allowance for this factor. To the competent sheep breeder this point should not cause any difficulty, as the whole flock is affected to some degree, and due allowance for seasonal conditions does not complicate the task. The first requirement, as already stressed, is constitution. Look for a dense level and broad back, well-sprung ribs with plenty of lung room; an open or lightly covered back with any sign of 'devil's grip' must be avoided. The characteristics of the face and head always afford indication of conformation and covering. Look for the broad, clean and open face with wide muzzle, bright alert eyes, skin soft and velvety cream-coloured texture. Strictly cull all animals showing overshot and undershot jaw.

The weak face is small, narrow and wedge-shaped, whilst another most objectionable feature of the face is frostiness or a hard harsh growth of white hair; the latter usually indicates harsh marbly wool and at times a tendency towards roughness and hairiness of breech and neck folds. Bare and common-face sheep with no bonnet should be rejected; these are generally light cutters. Strict attention should be paid to the muffed-face sheep with bad wrinkles under the eyes and cheeks. (These can successfully be removed by the mules operation provided the sheep does not possess a woolly face.) A good symmetrical front is most desirable; a heavy irregular wrinkly front is objectionable as these types are hard to shear and the wool is inclined to be cut into irregular lengths by the shearer.

Stance is most important. The breech should be deep and wide and the legs have substance and bone; all sheep showing signs of cow hocks should be culled. Again, where possible, big, bulky, heavy-cutting sheep should be chosen with as even and dense a covering as possible. Any sheep with naturally dingy and bad-coloured wool should be discarded together with those showing sweaty face and points.

A ewe might come up possessing a good weight in covering, lacking in style and character but without a roughness and possessing a good frame. This ewe could well be retained if mated with a suitable ram. A little allowance may be made for quality of wool but under no circumstances should sheep with weak constitutions be allowed to remain in the flock.

Importance of Using Good Rams

The golden rule of the flock owner is to use good quality rams. No breeder can raise the standard of his flock above the standard of the sires he uses.

Good ewes mated with rams of lower standard will produce lambs inferior to themselves, but the progeny of lower grade ewes mated to superior rams will be an improved type from that of their mothers. No money is more wisely spent and so quickly repaid as when rams of high grade quality are purchased. However, it should not be overlooked that it is possible to use rams that are of too high a grade for the standard of ewes they are used on. The greatest mistake is to continue to purchase rams from different breeders; by doing this a wide variety of both type of wool and conformation are developed. The clip soon becomes uneven and a greater problem for both the sheep classer and wool classer to evenly grade.

The Essential Points of a Merino Ram

Suitability of Type, and Purchasing Rams
It is generally recommended that rams be bought off the same blood as the ewes to maintain uniformity of type. Any change from the blood should not be too drastic. Endeavour to purchase rams of a grade which will show an improvement in the progeny.

It is most advisable to visit the stud property and make personal selection of rams, as stud breeders have little or no idea of what each breeder requires, and further, they have a business habit of including animals whose standards are not in keeping with the draft.

Having decided from which breeder to purchase rams, the next important point is to select rams suitable for the rainfall and country on which they are to be used. The very wide range of conditions under which merinos are run makes it important to choose the most suitable types. The only choice for the open, Plains country of big distances and spacious nutritious feed is the large-frame medium- and strong-wool type. This type possesses a vigorous constitution and a covering that will stand up to the more arid, hard conditions. The ewes are excellent mothers and have a high resale value to fat lamb breeders when cast for age.

The colder climates and less abundant feed of the Tablelands will better suit the smaller-framed fine/medium and fine wool sheep.

When selecting merino flock rams, the essential features to be observed would be clean, open, soft, broad face, with good alert eyes, free from wool blindness. Perfect bite is most essential and rams that show signs of dribbling at the back of the jaw should be avoided. Be particular about substance of horns, a good, even spiral turn with ample space between the horn and face. Reject a ram showing small and weak horn growth; this is a sign of weak constitution, and chalky substance in horns usually indicates chalky and musty white wool.

Conformation

The top line should be well-sprung ribs; straight good loins are essential. Loose shoulder blades, high wither and flat shoulders or devil's grip are serious faults. Sound feet and legs with good hocks and pasterns are most desirable. Be particularly sure the rams are physically entire.

Covering

The covering must have length with good staple, well defined crimp and character of suitable quality for the particular climate. Avoid slackness but not to the extent of choosing short, dense, tight wool. The short, fatty-woolled sheep should not be tolerated. Reject very dry wools, fuzzy tips and hairy breech and necks.

Culling Aged and Faulty-Mouth Ewes

It frequently happens through seasonal conditions that breeding numbers become difficult to maintain. There are quite a number of excellent ewes among cast for age ewes which would go on breeding for two or three years. It is therefore a good plan to keep back about 10 per cent to 15 per cent and allow these to remain until the following year or two. This will help a breeder to maintain his numbers.

Reverting to good-looking ewes, a careful check should be kept to see that barren ewes be kept out at all times; they are easily picked out at lamb marking time by their extreme fat condition and careful examination will clearly reveal their identity. They should be placed in the killing paddock.

Disposal of Culls

Culls may be disposed of in different ways. They can be immediately sold on the ground, fattened and sent to the carcass butcher, or joined to a mutton-breed ram and sold to fat lamb breeders, or disposed of direct to a fat lamb grower. A station may decide to retain them as ration sheep, especially in the case of stud breeders.

There is one other use for cull ewes, often in a small number on stud properties, for the purpose of breeding lambs to use as foster mothers to rear orphan and stud lambs.

Conclusion

Results are not secured by chance but by careful classing and judicious mating, persevered with and continued in spite of many setbacks and disappointments and with a wide appreciation of the fact that only by breeding and carrying good sheep can a reasonable measure of prosperity be looked for.

Merino Breeding

Progeny testing of merino sires is of limited value. Selection of the sire on his own measured characteristics will usually lead to a greater annual increase in production when the sires being compared have been grown in the same environment. Progress will be more rapid and certain if satisfaction is based on definite measurements, such as an amount of wool produced, rather than on eye appraisal.

Increased wool weight can come from various sources, such as body size, staple length, number of fibres per square centimetre of skin, coarseness of fibre, skin wrinkles and 'point cover'. These sources are not equally profitable from the point of view of money return. Increased coarseness of fibre, for example, may increase the total wool weight, but decrease the value per kilogram. Some sources offer opportunity for quicker progress than others; variations from animal to animal in 'density' and fibre fineness, for example, have been found in some flocks to account for 60 per cent to 70 per cent of variation in clean wool weight.

Increasing the number of characters under selection will decrease the rate of progress for any one of them. Effort should therefore not be spent on a character unless it is known to be of value in increasing returns. Information on the extent to which selection for one character in the merino will influence changes in another is as yet scanty. However, research on this aspect is in progress, and some results are given later.

Let us now examine the background of knowledge from which these recommendations have risen.

Heredity and Environment

The livestock breeder's main tool for improving production in his flock is selection. Progress depends principally on his ability to select animals which will not only be superior in their own generation, but will pass on some of that superiority to their offspring. The main stumbling block is that, except in a few special cases, it is impossible to tell how much of the superiority of any individual animal is due to heredity and how much to environment. 'Heredity' comes to an animal from its parents, and in part will be transmitted to its offspring.

'Environment' includes the feeding, management, climatic conditions, diseases and so on which have influenced the animal even before birth, as well as variation due to inaccuracies of measurement or appraisal at the time of classing. None of the effects of these will be transmitted to the offspring.

Although in an individual animal the results of heredity and environment can only be separated in very few cases, research has shown that basic principles exist which can be used as a general guide. Before discussing these, let us first consider how heredity works. Heredity is controlled by units called 'genes', which are carried in pairs in the animal's cells. The inheritance of every characteristic in the animal is controlled by one or more pairs of genes. The exact number is not generally known. In a few cases, such as full black or full white colour, at least in British breeds of sheep, it is thought that a single pair of genes is responsible. However, in most characteristics which can be measured, such as amount of wool, body weight and so on, the number of pairs is probably very large.

When sperm cells are formed in the ram and ova (eggs) in the ewe, the pairs of genes split. Each sperm or egg then carries only one of each pair. When the sperms and ova unite after mating, the corresponding pair of genes in the offspring is formed by the combination of one unit from each parent. Both parents contribute equally to the offspring, but in each case it is a 'toss-up' which unit of the parent's pair goes into the offspring, just as it is for a 'head' to fall uppermost when tossing a coin.

Heredity is always controlled by genes. The methods used for predicting the probable outcome of any mating differ slightly, however, according to whether a particular characteristic in the animal is controlled by one pair of genes, or by more than one pair. Let us deal with these in turn.

Control by One Pair of Genes

The units which go to make up any pair of genes are generally of two types, and the pair may contain two like or two unlike units. If we call the types A and a, then there are two main kinds of like pairs (AA and aa) and one of unlike pairs (Aa). Let us take full colour in sheep as an example of a characteristic which is controlled by a single pair of genes. The types of genes are
1. a gene carrying a 'factor for white', as it is called; and
2. a gene carrying a 'factor for black'.
An animal may carry two genes with the white factor, two the black factor, or only one of each. Sheep with both genes alike will be white or black, according to the factor. Animals with two unlike genes of this nature are sometimes intermediate in appearance between the two extremes (a roan cow is intermediate between red and white). The sheep in this case is not grey, however, but white. The gene carrying the white factor is said to be 'dominant' to the gene for black, because it masks it.

This mechanism explains how a black lamb can occur in a flock of white sheep. If a ram is used whose gene pair for colour contains two unlike units, he will be white himself but half his lambs will receive a gene carrying factor for black. If some of the ewes also have gene pairs of unlike genes, then half of their lambs will also receive a gene carrying the black factor. Any lamb receiving a double dose of genes carrying black factors will be black. The odds in favour of a white lamb, even when both parents carry unlike pairs of genes, are 3:1.

In selection the problems confronting the breeder who has to deal with characteristics controlled in this manner are two-fold. First, what is to be done with the outstanding ram which is white in colour but has been found to carry a 'hidden' gene with the black factor? Secondly what can be done to test whether a ram carries a black factor? The answer to the first part of the problem will vary with the circumstances. Even when the ram in question meets ewes also carrying a 'hidden' black factor, only one-quarter of their offspring will be black, while half will be white, but carrying the 'hidden' black factor. The degree to which the ram can be used will depend on:
1. the proportion of ewes in the flock which carry the hidden black factor. This proportion will determine the number of black lambs likely to be born;
2. the lambing percentages usually obtained in the flock, with consequent culling level possible;
3. the extent of the ram's superiority over his fellows.

Thus the problem of what to do with the ram in any particular flock cannot be solved satisfactorily unless records are kept of animal lambing percentages and the number of black lambs born each year. From these records it would be possible to estimate the number of ewes carrying the black factor. If the number of black lambs likely to be born is small compared with the number which will be culled, then it might be safe to use an exceptionally good ram, even if he carried the black factor.

If a ram suspected of carrying hidden black factor is to be tested, the quickest way is to mate him with at least six full-black ewes. A single black lamb will be enough to prove that he does carry the black factor. He could of course throw six white lambs in such matings even if he carried the black factor, but the odds against this are 63:1. To lengthen the odds, which might be desirable in the case of a ram required for widespread artificial insemination, the number of black ewes mated should be increased.

Only a few characteristics are known to be controlled in this way by single gene pairs. There are some physical abnormalities, however, which occur from time to time and are probably controlled by single pairs of genes. These can be eradicated from a flock if the rams carrying the 'hidden' factor can be identified and eliminated. In such cases, father-daughter matings can be of assistance in identifying the ram concerned, as the animals showing the abnormality frequently do not live and so cannot be used as a test flock. If they can be kept, written records of the number of lambs born and the number of abnormalities will always be of value.

Full-colour is one of the group of special characteristics mentioned earlier, over which, as far as is known at present, heredity has full control and environment none. Such characteristics may, of course, be controlled by more than one pair of genes, and the study of them is then not as simple as the case just described.

A second and much more important group, however, consists of characteristics which are controlled by more than one pair of genes and which are also influenced by environment. This group includes most characteristics of economic importance.

Control by Many Pairs of Genes
When many genes control one character, they are still arranged in pairs. The units of the pairs are still of two types, only one type ('plus' genes) tending to increase the size

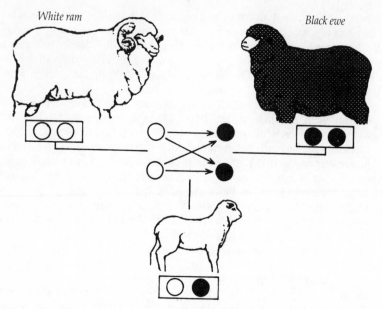

Figure 19.1   Genetic selection. All progeny will be the same (i.e., white with one recessive gene for black).

of the character concerned. The pairs still split during the production of sperms and ova, one of each pair going into the offspring on mating.

It is not possible to decide just what effect is produced by any one pair of genes. The usual assumption made is that an animal's capacity to produce some characteristics, such as a long staple, depends on the total number of 'plus' genes for staple length, which it receives from both parents, irrespective of the particular pair to which any gene belongs. This assumption seems to work reasonably well. Further research may indicate that the effects of the genes do not add up quite in this simple manner and that the particular pairs may matter. However, the recommendations based on the assumption will assist practical selection for some time.

The selection problem facing the breeder this time is not 'Does this superior animal carry a hidden undesirable factor?' 'How much of the superiority is likely to be due to heredity and how much to environment?' The question cannot be answered for individual animals. However, by examining the extent to which related animals resemble each other in any characteristic when they have been grown in the same environment, it is possible to estimate what fraction of the variation from animal to animal is due to differences in heredity. The relatives commonly used are half-brothers and half-sisters, or ewes and their own progeny. This fraction is called the 'heritability'; its value will lie somewhere from 0 to 1. It is by no means a constant figure when calculated for the same characteristics on different flocks, or even on the same flock in different seasons. However, estimates from different flocks for the same type of sheep seem to agree well enough to be useful. Values for different characteristics will differ considerably, and the methods of selection likely to give most progress will depend on the level of the heritability. These levels may be arbitrarily divided into three classes:

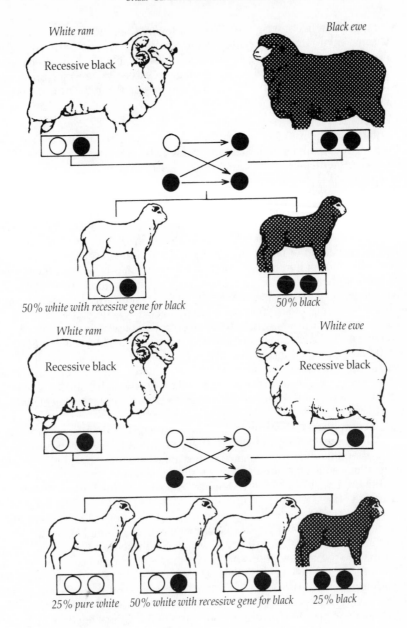

White ram
Recessive black

Black ewe

50% white with recessive gene for black

50% black

White ram
Recessive black

White ewe
Recessive black

25% pure white   50% white with recessive gene for black   25% black

Figure 19.2   Recessive genes for black

1. less than 0.15 — low heritability;
2. from 0.15–0.3 — medium heritability;
3. over 0.3 — high heritability.

For the merino, fleece weight has been estimated to have a heritability of approximately 0.3. It may even be more in some cases. Just what does this mean? A breeder chooses a ram with a fleece weight of 454 g (1 lb) more than the average of his drop, run in the same environment. He mates him to a group of ewes whose average fleece weight also exceeds that of their general flock by 454 g (1 lb). Then the fleece weight of their progeny will be expected on the average to exceed the average for progeny of the whole flock by 136 g (0.3 lb). This increase in production in the progeny is called the 'genetic progress'.

Way to More Rapid Progress

When heritability is high (0.3 or more), the way to achieve the fastest progress is to select the individual animal accurately on its own characteristics — so-called 'mass selection'. When heritability is medium or low, selection may be aided by assessing 'performance', as it is called, on more than one record on the same individual (two years' wool data instead of one, for example), or by examining the records of groups of relatives (such as progeny, or half-sisters and half-brothers). Some of these methods may increase the testing time, thereby increasing the average age of parents when their offspring are born. So when the heritability is high they may make annual progress slower, instead of faster, than it would be with mass selection.

Whatever the value of heritability, selection can sometimes be aided if the effects of specific factors in the environment are known. Twins, for example, cut less wool per head than singles even when they are adult, and some adjustment should be made for this if the two types of sheep are to be compared in selection.

Estimates for the merino show that, with the exception of weaning weight, all measured productive characters so far studied have a high heritability, with values above 0.3. The characters studied include wool weight (clean and greasy), adult body size, staple length, fibre diameter, number of fibres per square centimetre of skin and skin wrinkles. Weaning weight has a medium heritability of between 0.2 and 0.3. Characters like wool count, which are appraised but not actually measured, also have a medium heritability.

None of the productive characters which are of interest to the woolgrower, therefore, will benefit greatly from progeny testing, provided accurate comparisons can be made of the individual performance of the animals under selection. The earlier recommendations which favoured progeny testing of sires were based on the fact that the first work on heritability of fleece weight was done with the romney marsh breed in New Zealand. These and later figures for the same breed gave low estimates — 0.1 to 0.2 — and the assumption was made that such figures applied to all sheep. Recent work on the merino in Australia and the rambouillet in the United States has shown that the estimates of heritability of fleece weight for these breeds are much higher (over 0.3) and the recommendations have accordingly been changed.

Selection of sires on their own individual characteristics will only give maximum progress if measurement is made as accurate as possible, and if selection is on comparable performance. Research indicates that selection of rams by commercial sheep-classing methods of appraisal gives only one-third of the annual advance in wool cut per head which could be obtained if selection were based on actual fleece weight. Direct comparison even of fleece weights is not possible when rams of a different age are to be compared, or rams from a different property compared with home-grown rams. In the first case, some age adjustment must be made. The second case is one where a progeny test might be of value.

Selection for More Than One Character

Fastest progress in any one character will be made if selection is confined to that character alone. In some experimental merino mating groups in the CSIRO, for example in which a comparison is being made of selection for long and short staple, the progeny in the long-staple group have a staple 34 per cent longer than those in the short-staple group. When the breeder wishes to improve more than one character, progress for any one is much slower, and he has a choice of three methods of procedure:

1. tandem selection, when attention is concentrated on one character until it reaches a desired level, and then the next is dealt with, and so on;
2. selection by fixing a standard for each character, and culling all animals which fall below standard for any one character;
3. selection on a score which combines all the characters in some way.

The first method is the slowest and least efficient. The last gives the most rapid progress, but requires special computing techniques for its application. The second is probably the best where such facilities are not available. In the case of merino sheep, the situation differs slightly from that in some other animals, in that at least some of the characteristics in which the breeder is interested, such as staple length and fibre fineness, are part of his main interest — the amount of wool. The breeder may have to spend a small portion of his culling on faults such as black wool and physical abnormalities. After that his best policy is to select on total amount of clean wool, but to see that the source of increased wool is a desirable one. The definition of 'desirable' will vary from flock to flock. In general, extra wrinkles or increased coarseness are undesirable sources; increased 'density' in general is a desirable source and one which at first should lead to a rapid rise in cut per head, though with very high densities the cut per head may fall again.

When dealing with a number of characters at once, a new factor enters the picture: the extent to which the characters are associated. Is it possible to select both for staple length and fineness of fibre, or is this impossible because long staples are tied genetically with coarseness of fibre? It is necessary to emphasise that it is the genetic association which is important. A well-fed animal may grow a longer coarser fibre than a starved animal, so that in appearance length and coarseness are associated, but this does not mean that the 'plus' genes for fibre are also 'plus' genes for length of staple.

Accurate estimates of these 'genetic correlations', as they are called, require the accumulation of a great amount of data. Work on them is in progress, but the answers are not yet clear-cut. The following conclusions, however, may be drawn:

1. As 'density' (number of fibres per square centimetre of skin) increases, so does fibre fineness;
2. As amount of wrinkling increases, staple length decreases;
3. There is no definite association between staple length and fibre fineness; merino sheep may have a long staple combined with fine or coarse fibres;
4. There is no definite association between 'density' and staple length; merino sheep may have a high 'density' combined with various staple lengths.

20

FERTILITY AND MATING OF SHEEP

Fertility in sheep is usually measured by the lambing percentages, but this is not a true measure. Fertility is the capacity to produce living offspring. A potentially fertile ewe may be prevented from producing a lamb by unfavourable conditions or bad management. When the conditions are favourable, the lambing percentages are a measure of true fertility.

In all large animals, each sex has a part to play in reproduction, which begins when the male and female cells fuse. The glands which produce these germ cells are the testes in the male and the ovaries in the female. Other glands assist — the adrenals, close up to the kidneys, the thyroids in the throat, and most important of all the pituitary at the base of the brain. The front (anterior) portion of the pituitary is the control of the whole system.

The fertility of Australian flocks is said to be low as compared with those of other important sheep-raising countries. Reasons for this include:
1. high percentage of merino blood;
2. the hot dry summers;
3. the low plane of nutrition of many of our grazing pastures;
4. lack of attention, for economic and/or personal reasons, to mating and lambing flocks.

The fertility rate is one of the most important single factors affecting the gross return per flock. Where the rate of increase is high:
1. more surplus sheep are available for sale;
2. the ewe depreciation cost per lamb is lowered;
3. the range of selection is increased.

The average standard of sheep classed into the flock is raised, increasing the wool clip, the sale value of surplus stock and raising the breeding standard of the flock.

Fertility in the Ram

The male germ cells called sperms, or spermatozoa, are secreted in the testes and stored in the curled tube known as the epididymis, which lies along the length of the testes. Thence they pass up the straight tube (*vas deferens*) to the urethra or penis. A vasectomised ram is one from which a short length of the vas deferens has been cut away. Of the millions of sperms emitted at each service, only one is necessary to fertilise the egg. Although a certain percentage of rams may be permanently infertile because of disease and physiological abnormalities, it appears that one of the major causes of infertility in rams is the temporary variations which occur in the quality of the semen produced.

Typical causes of this temporary infertility have been shown by Dr Gunn to include:
1. high body temperatures produced in rams by high summer temperatures and other causes. Long periods of hot weather, over-driving, long periods of full wool, and perhaps to some extent the position and suspension of the testes and the amount of wool around the flanks and scrotum are all likely causes of high and sustained body temperatures, causing a reduction in the quality of the semen.
2. sudden changes in feed and climate;
3. poor quality of grazing, especially when deficient in vitamin A. Green pasture or good quality lucerne hay are rich sources of vitamin A, which is notably deficient in inferior hays, over-mature grass and the cereal grains.
4. prolonged over-feeding, especially on rich rations as for show, which will produce seminal degeneration;
5. fevered conditions induced by fly strike, foot rot, abscesses, etc;
6. diseased conditions or abnormalities of the testes, such as atrophy of one testis, epididymitis, hernia, etc;
7. absorption of arsenic through the skin at dipping or jetting — especially if rams are dipped in long wool, or carelessly dipped even in short wool.

With the exception of (6), completely normal sperm may not be produced by rams until two months after the cause of infertility has been removed. Other causes of low lambing percentages associated with rams include:
1. lameness due to badly formed feet and legs, overgrown hoofs, injury, etc.;
2. overfatness. Overfat rams become lazy and slothful and are disinclined to follow the ewes.
3. grass seeds, ophthalmia or pink eye;
4. insufficient rams, mating paddocks too large, and any other cause which makes it difficult for rams to contact all ewes which come on heat. Vigour and opportunity are not necessarily indications of high fertility.

A rather common cause of infertility in rams is known as 'epididymitis' (i.e. the thickening and swelling of the epididymis). This condition can easily be discovered by a manual examination through the scrotum and may disclose one or both testes to be softer and smaller than normal. Such testes are incapable of satisfactory sperm production.

Managerial Aspects

The wastage amongst rams is very much higher than it should be and includes loss of efficiency as well as a high death rate. When not required for mating, rams should be kept in securely fenced, well-sheltered paddocks, if possible away from ewe paddocks. The practice of allowing rams to run with ewes the whole year round should not be followed. The rams should be inspected regularly through the year, especially for fly strike about the head and horns, and for lameness. The rams should be yarded regularly and gone over thoroughly, special attention being paid to hoofs, testes, pizzles and head.

Fertility in the Ewe

During the breeding season the ewe will come into heat or oestrus at about seventeen day intervals. Ovulation or the shedding of the ova or egg occurs at the end of the oestrus and may last from a few hours to two or three days. Conception occurs when the ovum is fertilised in the fallopian tube by the male sperm and the fertilised ovum becomes implanted in the uterus. Normally, one ovum is shed at each heat. Rising bodily condition is conducive to the expulsion of more than one ova. Fertility in the ewe is the ability to bear progeny; ewes which bear progeny in large numbers are prolific. Ewes of British breeds are more prolific than merinos.

Due to lambing mortality and prolificacy, lamb marking percentages are not always a reliable guide to fertility. The rhythm of oestrus is not readily upset, as is fertility in the ram. It is often stated that 'flushing' will bring ewes into breeding more quickly. A rising plane of nutrition such as from poor feed to fresh green feed, tends to cause the expulsion of a greater number of ova (eggs), thus resulting in a greater percentage of twins, but does not necessarily expedite the onset or occurrence of oestrus. When carried out purposely, this is referred to as 'flushing'.

Young ewes reared on a high plane of nutrition will breed earlier than those developed slowly. There is some evidence that prolonged feeding on dry, un-nutritious feed restricts the breeding season. Over-fatness of ewes may restrict the ability to breed. On the other hand, the condition of such ewes may be due to barrenness.

Ewes reach maximum fertility at four years and remain at the same level until eight years of age. Ewes should not be mated until they are mature. The period of gestation is approximately five months (147 to 152 days). The majority of early-maturing sheep reared under good conditions can be put to the ram so that they lamb at two years. Slow maturing strains or those reared on country which does not grow a young sheep well are left to lamb at three years for the first time. Only in exceptional cases should ewes be mated as lambs. Lactation is a severe drain on growing ewes and, except under very good conditions, may reduce the mature size and the breeding life of the ewe.

Discovering the Height of the Breeding Season

A method which has proved successful to discover when the height of the breeding season occurs is to use rams which have been rendered sterile by vasectomy. Castrated rams are unsuitable because they have lost the desire to serve the ewes. The vasectomised rams are run with the ewes and coloured, with raddle, on the brisket so that they mark the ewes they have mounted. In this way the flock owner can determine when the greatest number are coming on heat. Fertile rams can then be introduced.

Managerial Aspects of Mating

The general fertility of the flock will also be raised when
1. shy or late breeders are culled regularly from the flock;
2. a reasonable proportion of the aged ewes is retained.

Theoretically, the best time to mate is during the autumn months when most ewes are on heat and excessive temperature are not likely to affect the fertility of the rams. However, other practical points must be considered and these must include:
1. the climatic conditions during lambing. Summer lambings, especially in hot dis-

tricts, are rarely successful. Lamb mortality due to heat is liable to be high. Mid-winter lambings in cold, elevated, high-rainfall districts will also be attended with heavy losses.

2. condition of pastures. In some districts, with mild winters, a winter growing season and summer drought, lambs will 'do' better if they and their mothers can have as long a period as possible on green feed. This applies especially to lambs bred for the fat lamb market, which must be finished and marketed before pastures coarsen and dry off and grass seeds fall. Winter rainfall areas with a long spring and cold winters, pastoral areas with unreliable autumns and districts with summer rainfall provide better pasture conditions for the late winter and spring lambing.

Length of Mating Period
For paddock-mating ewes at the height of the breeding season, with sound, vigorous fertile rams, six weeks should be ample — allowing each ewe two opportunities to mate. Longer matings result in an uneven drop. Stud matings may be longer, especially where valuable sires are mated to a large number of ewes by 'hand' service.

Proportion of Rams to Ewes
The number of rams per ewe depends on the type of mating and the size of the pad-docks. Normal flock practice in Australia is to join 2½ per cent of rams (occasionally 3 per cent in large paddocks or when two-tooth rams are used). In small paddocks four- and six-tooth rams will handle up to 100 ewes each; younger and older rams proportionately less. It is not usual to mate yearlings or younger rams except for pro-geny tests or special matings, and those are usually carried out under yard conditions with one to twenty ewes per ram lamb.

Rams can be made to serve many more ewes if the number of services per ewe is restricted, if the amount of travelling the ram normally does is reduced, and if the ram is fed a supplementary ration. These conditions are partly fulfilled if the ewes are yarded each evening and the ram is only allowed with them at night, being re-strained and fed during the day.

Hand Service
This economy in the use of rams can be carried a stage further in the hand service of valuable sires. The ewe flock is run with raddled vasectomised rams, which marks ewes on 'heat'. These ewes are removed twice daily, given one or two services by the desired ram and turned into another flock which is examined periodically for ewes which 'return'. It may be necessary to restrain the ewes in a suitable bail. Ewes are usually served at the inspection following that in which they were removed from the flock. Rams, which should be hand fed in a spacious exercise yard, may average up to four services per day and up to 300 ewes have been served in the one season by one ram.

21

CARE AND MANAGEMENT OF THE EWE, AND LAMBING

Careful management of the pregnant ewe will have a marked influence on the percentage of lambs dropped and reared successfully. At all times, in-lamb ewes should be handled as little and as carefully as possible. Ease them through gateways and work them quietly in the yards. Particular care should be taken to get them out of yards as soon as possible.

The nutrition of the ewe during the latter part of pregnancy is important in its effect on the milk production of the ewe and the birth weight and maturity of the lamb, details of which have been discussed in other sections. Ewes falling in body weight are especially prone to pregnancy toxaemia or twin lamb disease. In poor seasons, losses from this source may be heavy, unless steps are taken to reserve paddocks of fresh feed or to supplement the paddock feed with hay and/or grain.

Limited Australian experimental evidence suggests that the prenatal feeding of ewes need not be nearly as heavy as is advocated in other countries. This is due to the relatively high but often overestimated value of the dry feed in many Australian pastures. The type of supplementary feeding to be carried out will depend on pasture conditions, the district and the breeding policy and the fodders available, as well as the condition of the ewes. In the wheat areas of winter rainfall where crossbred lambs must be dropped early, a late break to the season will catch farmers with pastures and stubble almost depleted. A balanced ration of roughage, including silage for succulence, and concentrates, is a theoretical ideal but, except for the purpose of forcing special lambs for early markets, the economic value of such a ration over one merely sufficient to maintain body weight in ewes and prevent losses is doubtful.

In higher rainfall districts with cold winters and especially where the rainfall is of summer incidence, spring lambing is practised. The task of wintering the ewes is so difficult as to make breeding unattractive. Feed will make little growth in the cold winter and intestinal parasites, chiefly the *Trichostrongylus*, are often a further serious drain on the ewe. Two drenchings are necessary, one at two months and the second one month before lambing. Also necessary is a change of feed, to reserve paddocks of sown crops such as turnips and oats and/or supplements in the form of conserved meadow hay to build the ewes up into condition for lambing. Large quantities of bulky feeds are not suitable for ewes heavy in lamb.

About six weeks before lambing, ewes should be brought in and crutched. While the removal of excessive wool is wasteful, crutching should include a blow in front of the udder and one blow over the tail from the hocks. Rough handling should be avoided and care taken to see that ewes are not held too long at the shed. A good deal of the crutching may be done with a portable plant at strategic points on the property.

Ewes should be put into lambing paddocks some time before lambing so that they settle down before the first lambs are dropped. Where a choice can be made, some thought may have to be given to the selection of lambing paddocks. Good shelter, particularly abundant ground shelter, is more important than abundance of feed. Convenience and freedom from disturbance are important features to be sought. These paddocks should be ridden regularly to get the ewes used to inspection, to note the condition of the ewes, and to help any that may become cast.

In some districts, wild dogs, foxes, crows and other predators are a serious problem. Every effort must be made to trap and poison to reduce losses of lambs, which can be severe, especially if lambing does not coincide with that of most neighbours.

Lambing Time

The gestation period in the ewe averages about 147 days but may range from 142 to over 150. Early-maturing breeds have the shortest gestation period. Cold and un-favourable weather conditions appear to hasten labour. Low condition and fatigue appear to have a similar effect.

The care and attention that can be lavished on the lambing flock will have a marked influence on the percentage of lambs marked. Australian technique in this respect is often deplorably casual and both ewe and lamb mortality undoubtedly represent major factors in Australia's low rate of sheep increase. The reasons for this are partly traditional, a heritage from the extensive methods of cheap virgin lands, and partly economic, in that the value of the land or the value of sheep produced thereon have not been sufficiently high to warrant an extravagant expenditure on labour. However, from the frequent paddock inspection of large flocks to the intensive supervision and individual mothering of well-managed studs, lambing time involves all hands, including the boss, in long hours every day of the week. At least two inspections should be made each day. An inspection first thing in the morning is most important in order to be amongst the ewes before the crows get too daring and to relieve any ewes that may have got into difficulties overnight. Heavily grassed paddocks and those where ewes are having their first lambs will need closest attention.

The stockman will need to keep an eye open for mal-presentations, ewes that are cast, motherless lambs, flystrike, etc. Maiden ewes in low condition or small-framed ewes lambing to big rams will give the most trouble.

Lambing

A ewe about to lamb prefers to leave the flock and will attempt to scrape a bed for herself. She will appear restless, the udder is often distended and the external genitals in a flushed and flaccid condition. The ewe will lie down and strain, and the appearance of the water bag is the first result of the ewe's labouring. Often she will rise and

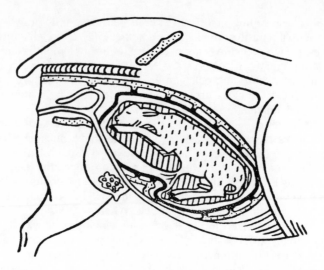

Figure 21.1　Normal presentation

turn until the bag bursts after which she will lie and strain and turn until the lamb or lambs are born. Normally a ewe will lick her lamb dry after it is born.

The correct way for a lamb to come is with its head between outstretched forelegs, though sometimes the hind legs come first. The time a ewe takes to lamb varies: single lambs will take longer than twins; large lambs longer than small lambs. It is well not to interfere unless:

1. it is obvious that the presentation is wrong;
2. the ewe has ceased to work after much trying; or
3. where a ewe refuses to settle to work after the waters have burst.

The last condition is associated with a serious condition in sheep, found to be prevalent in several areas of South Australia, and referred to as distokia.

Mal-Presentations

Causes

The causes of mal-presentation are many and varied. They are usually put down to careless handling, over-driving and crushing in gateways, and undoubtedly these are serious causes. But some flocks and breeds are more disposed to lambing troubles than others. Small ewes lambing to big rams, especially those with big heads such as the romney and dorset, are liable. Overfat ewes lacking bodily tone, or ewes lambing on bulky feed, seem prone to parturition difficulties. It is reported from New Zealand that parturition troubles in a particularly difficult flock have been reduced to small proportions by selection and breeding only from those families least liable to lambing difficulties.

It is difficult to adequately describe how to correct a mal-presentation, because one depends on touch rather than sight. As far as possible, getting one's hand right inside the womb should be avoided.

One of the commonest forms of mal-presentation occurs if one or both forelimbs is bent backwards so that the lamb appears with one leg and the head or the head only showing. In the first case, the second leg is easily brought down, taking care to turn

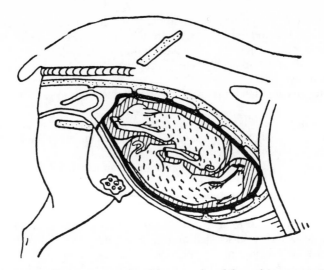

Figure 21.2   Twin pregnancy with normal and 'breech' presentation

the knee out and toe in when bringing the leg down. Often with gentle pulling as the ewe strains, the lamb may be delivered without bringing the second leg down. In the second case, the head only is presented; this is more difficult, especially as the ewe can run once the lamb's head is out. If the ewe has been labouring for long the lamb's head may become swollen. Patience is required in drawing down one leg and then the other. It is advisable not to try and force the lamb's head back into the womb. If the lamb's head is too swollen it may be necessary to remove it at the atlas joint, though a swollen head does not necessarily mean a dead lamb.

A mal-presentation that requires care to correct, and is indeed often difficult, is one in which one or both forelimbs are presented but the head is turned back. The position of the head should be determined before pulling on the legs. The legs should be pushed well back and the head brought into position by fixing a finger, a hook, or a loop over the jaw. It may be necessary to rope each leg. If delivery cannot be effected with pressure it may be necessary to remove one or both limbs. Abnormal breech presentations require more speed and some care to prevent pressure on the cord and belly. If the buttocks only or buttocks and hocks are presented (but not seen) it will be necessary to push the lamb well forward and bring each leg into position taking care not to dig the hoof into the uterus; the hock should be turned out in each case.

Complications with twins should be watched for and care taken that head and limbs brought into position belong to the one foetus. Other points to watch are as follows:
1. Always explore carefully before pulling.
2. Do not attempt to deliver a ewe before she has opened.
3. If possible, raise the ewe's hindquarters when investigating.
4. Keep fingernails short and clean.
5. Wherever possible use a mild disinfectant freely when entering the uterus.
6. If you have lambed a ewe, take the lamb to her head and hold her down until she has licked it. Then leave her lying and steal quietly away. It has been observed that arthritis is more prevalent in lambs that have been delivered and a dash of iodine on the cord may prevent much of this.

7. In intensively managed flocks and studs, much can be done to even up nature's distribution. Orphan lambs and twins from maiden ewes can be mothered to ewes that have lost their lambs or big milkers with singles by:
   (a) rubbing the lamb to be mothered in the waters of a lambing ewe and presenting it to her first;
   (b) locking ewe and lamb up together. Sometimes it is necessary to hold or tie the ewe;
   (c) skinning the dead lamb and tying the skin on to the lamb to be mothered. A New Zealand suggestion is to have a number of stakes in the lambing paddock to which ewes can be tied with a swivel lash.
8. Be careful not to put lambs older than four days on newly-lambed ewes.
9. Cold is one of the worst enemies of the newborn lamb. Sheltered paddocks and warmth for the nearly prostrate lamb will work wonders in saving many lambs.

A great deal can be written about lambing time. The best advice is to leave as little to chance as possible and to keep a watchful eye out for any abnormality, such as a ewe in trouble, a lonely, discarded and orphan lamb, or the occasional lamb thief that mothers a lamb before her own is born. The amount of supervision that can be given will, of course, depend on the carrying capacity of the country and the size of the flock. Often stricter supervision can be maintained over a small and conveniently situated flock by periodically drafting off the most forward ewes. For the stud master who has to mother lambs individually, this is almost essential. Ewes with lambs are walked off each day on to feed more suitable than the lambing paddock may provide. When circumstances permit, this separation of late ewes, lambing ewes and ewes with lambs is better for all three groups.

Ewes with young lambs drink a surprising amount of water. Keep fresh clean water available at all times.

22

LAMB MARKING

Marking calls for the utmost care in cleanliness and in the handling of lambs. It also requires careful thought, as there are many factors which should receive close attention if the small but constant losses at marking time are to be reduced. Ear-making, castration of male lambs and tail docking are the three operations involved, and they are usually performed in that order. Inoculation for entero-toxaemia, and vaccination for scabby mouth are usually carried out at lamb marking on properties where this protection is necessary.

Age at Marking

Lambs are marked according to the drop, at any time from one week to six weeks of age. A strong lamb can be marked with safety after it is a few days old, but it is usual to wait until the lamb is three weeks old. When the lambing season is protracted, some holdings have two markings — the first including the bulk of the lambs when they average a month old, and the second for the remainder, at the end of lambing. The advantages of early marking are quicker healing, less check to the growth of the lambs, and less risk of wound infection and fly strike.

The work should, if possible, be done during the mild weather; cold windy days and very hot weather are unsuitable. Blowfly strike of wounds is often a problem in hot sultry weather. The lambs should not be overheated, and the operation should be done in the morning, allowing the lambs time to 'mother up' before night. Finishing early in the day and careful shepherding are necessary precautions against lambs becoming 'mis-mothered'. Both sexes may be treated at the same time, and a useful check is obtained of the number of each sex if the tails of the male lambs and those of the females are thrown on to separate heaps.

Fat lamb breeders usually conduct several markings during lambing; it is essential that fat lambs receive the minimum check in growth and for this reason they are marked at an earlier age than others.

Choice of Site

Lambs should be marked in temporary yards erected in the paddocks — certainly never in old dusty sheep yards, where they are exposed to the risk of wound infection such as tetanus. The advantages of temporary yards are that they can be erected on a new site each year, and the lambs are on clean ground in the paddock where they are to run with their mothers, without further disturbance. The lambs do not become hot, tired and dusty by being driven to and from the yards. After the operation, provided the ewes are kept handy, the lambs soon find their mothers and settle down. If the erection of temporary yards is impracticable and it is necessary to mark the lambs in permanent yards, the following precautions against wound infection should be taken:

Clean up all sheep droppings and refuse in the yards, removing the surface soil at the same time if practicable. Water the yards well to keep down the dust before operating, and re-water during the course of the work. A new marking rail should be fitted and kept clean. Finally try to arrange for the lambs, after marking, to be dropped outside the yards on to clean grass, not into the yard.

Avoid rushing the sheep about. Deaths from haemorrhage may occur if lambs are excited and overheated when marked. Guard against dislocation of joints and bruising. Lambs should be caught around the body and not by the legs. Lamb carcasses are often rejected at abattoirs because of swollen and deformed joints, due to careless handling at marking time.

Breaking records does not pay; it is usually accompanied by rough handling and carelessness.

Castration

This is usually the first operation in the case of male lambs. There are various methods of removing or destroying the testicles of male lambs, but most commonly the end of the scrotum or purse is cut off and testicles withdrawn by a special instrument — although some operators prefer to use their teeth. This instrument may be in the form of a hook, or a toothed grip on the end of the marking knife. Whatever is used, the testicles should not be withdrawn roughly.

Another method in opening the scrotum is to slit in from side to side in order to press the testicles through. Both methods are equally effective and there are points in favour of each. It is claimed that with 'tipping', where the lower end of the purse is cut off, better drainage of the wound is effected. Those who prefer the other method of 'slitting' claim that when the wether is fattened, the cod fills with fat and so gives the animal a better appearance.

Where it is found impossible to remove both testicles, because of one not being sufficiently down to the grasp of the operator, about 5 cm (2 inches) only of the tail should be removed. This enables the owner to pick out the imperfectly castrated animals at a later date. These should be fattened and killed as soon as possible, otherwise they become a nuisance amongst the breeding stock.

On the market there are various types of bloodless emasculators, for treating the animal without cutting the scrotum. These are designed for crushing the cords leading to the testicles, and so rendering the animal sterile. Such methods, however, are much slower, require more skill and unless the instrument is in perfect order and applied properly it may not be effective. Advantages are that they cause less shock and there is less risk of tetanus and blood poisoning. There is also, of course, no risk from fly strike.

Elastration

Elastration is a bloodless method of removing the tail and castrating the lamb. It embodies the principle of completely restricting the bloodstream to the scrotum and tail by the application of a small and very strong rubber ring. The ring is expanded and applied by the use of a special instrument called an elastrator.

In castration, the rubber band is placed between the scrotum and the body and the stoppage of the circulation causes the scrotum and testicles to wither and fall off. The same applies when the rubber ring is placed around the tail at the desired position; the tail dries and eventually falls off.

Trials of this method of de-tailing and castration of lambs have been carried out but taking all things into consideration, including speed, cost, time of healing and incidence of disease, it appears that elastration has no advantage over the knife method.

Earmarking

In New South Wales the earmark — registered by the owner — is made on the right ear of ewes and on the left ear of male sheep. It is not permissible to put any other marks on the 'registered' ear. Distinctive earmarks, such as those denoting age and class of sheep, should be placed in the opposite ear. Special pliers — incorporating the owner's mark — are used.

Docking

The point at which the tail is severed is very important. Investigations have shown that overlong and very short tails predispose sheep to flystrike. This is particularly so with merino sheep. It is now an established fact that the prevention of crutch and tail strike commences at docking, and to obtain most protection the tails of female lambs should be cut at a level just below the tip of the vulva — 6 to 13 mm (¼ inch to ½ inch) — irrespective of the age of the lamb. Tails cut medium-long in this manner heal more quickly than short tails and so lessen the danger from fly-strike.

An effective method of docking is to place the knife edge on the tail at the place where the cut is to be made and then, keeping the knife pressed against the tail, pull the tail forward so that the tip is directed to the brisket and pull the knife through the tail. This method facilitates cutting the tail at the correct length, and when the tail is removed a flap of bare skin from the under-surface of the tail is left to heal back over the severed stump. The skin healing upwards over the end of the tail means that there is no wool-growing skin left opposite the vulva in a position where it is likely to become soiled with urine and so attract flies.

It will be seen that two factors are therefore of major importance in the docking of lambs' tails — the length of tail, and the way in which the tail is cut. Only careful workers should be allowed to carry out this operation.

Care of Instruments

Just as it is essential to work in clean surroundings, so it is important to use clean instruments to minimise the risk of infecting the wounds. Reliance should not be placed on the false hopes that antiseptics will kill all germs on a dirty knife. Knives and other instruments should be of a type which can be sterilised by boiling for at least 10 minutes before marking. The most popular type of knife is one with the blade

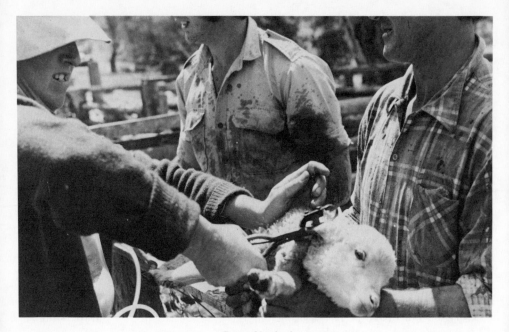

Earmarking lambs

Photograph: courtesy NSW Department of Agriculture

and handle all in one piece, without any hinged joints or cracks in the handle to harbour germs.

Throughout marking, the instruments should be cleaned often and kept in a tin of antiseptic solution when not in use. Any chemist will be able to suggest suitable preparations for the purpose. Although many owners who have taken little precaution to prevent infection escape undue losses, the risk is always present and a high death rate from wound infection occurs from time to time if these precautions are not observed.

After Treatment

As a rule, it is recommended that no dressing be applied to the wounds after marking. If the work has been carried out under clean conditions and the lambs dropped on fresh pastures after marking, the wounds heal quicker and better without any dressing.

It is a mistake to expect that wounds contaminated with dirty instruments will be sterilised by the use of some antiseptic. More often than not, wound dressings, as commonly used, irritate the end of the tail to the crutch area, including the scrotum in the case of wethers. The ideal dressing is one which repels flies.

After years of observations 'Borocit', as developed by the CSIRO, still appears to be the outstanding dressing for this purpose. It contains citronella, one of the best fly repellents, and also boracic acid which prevents the spread of flystrike. This dressing is harmless to the skin and does not affect healing to any extent. It should be applied with a swab at the rate of up to 150 lambs per 4.5 litres. It is preferable, however, to avoid using dressings and to mark at a time when flies are not active.

Infections

Many bacterial infections are likely to cause trouble at marking time. Some germs may simply infect the wounds, causing them to suppurate, so delaying healing and making the wounds more susceptible to fly strike. Other germs, such as tetanus and malignant oedema, may infect the wounds and produce powerful poisons, which kill the animal. Some infections may enter through the wounds and affect the joints, causing difficulty in movement and lameness. Another type of infection produces 'cheesy glands' or caseous lymphadenitis; prevention of losses from these infections depends largely on good management.

It is stressed that marking in old contaminated sheds and yards, dropping the marked lambs into dusty or soiled yards and paddocks and using unsterilised instruments should be avoided.

Mules Operation

Young merino sheep from the lamb to the hogget stage are more predisposed to fly attack than they are later in life, and it is desirable that the maximum protection be afforded from the earliest possible age.

The mules operation gives lasting protection, although it is more difficult to do on lambs. The operation can be done successfully on the ewe portion of the drop at marking time, if they are sufficiently well grown. It is not desirable to perform the operation on the wether portion at this time, as it imposes excessive shock on the animals. In many cases the tail operation is carried out at marking time and the remainder of the mules operation completed subsequent to healing of the marking wounds. It is essential that the mules operation be carried out when no bush flies or blowflies are about.

23

WEANING AND CARE
OF YOUNG ANIMALS

Weaning

The practices and problems associated with weaning and the care of weaners vary from district to district. In every one of these districts, weaner management is important. Too often growing and developing weaners are forced to accept hardships not even imposed on older sheep, yet from the weaners are drawn the future flock. The better they can be grown out consistent with their value as breeders, the more useful and long-lived we may expect them to be.

The average age for weaning is about four and a half to five months. Shearing and weaning are often arranged to coincide, though there are, of course, exceptions to this, as when ewes are shorn in the spring and lambs weaned in autumn. Fat sucker lambs, of course, are consigned to market or slaughter straight off their mothers.

In certain hard seasons it may pay to ration any good food to young lambs and relieve the ewes of lactation early. Some graziers prefer to allow lambs to wean themselves. Undoubtedly this eases the check to the weaner lambs. On the other hand, it is harder to control the condition of the breeding ewe and in most circumstances ewes are better for relief from big lambs drawing on them before they go to the ram. There is some danger, too, of ewe lambs being served.

The weaning paddocks need to be chosen with some care. Preferably they should contain adequate shelter and good water and have a good supply of feed that is soft and sweet rather than coarse and stemmy. Bad grass seed areas should be avoided as far as possible. In any case, every measure possible should be adopted to control trouble from seed. Hard grazing with cattle or wethers, sometimes before the weaners are turned out, is helpful. In lieu of good clean pastures and where conditions are suitable for it, there is nothing better in the agricultural areas than a paddock containing Wimmera rye grass and a legume. The Wimmera rye competes successfully with barley grass, holds its green longer and is fine and palatable.

If it can be managed it is a wise plan to run the ewes and lambs together for a short time in the weaner paddock before the ewes are removed. Lambs may then start their independent careers in familiar surroundings. Whether this can be arranged or not, weaners should be inspected daily until they settle down and those lingering along the fences brought in to water. A few old sheep put with the weaners will help them to settle. Crossbred or British bred lambs require paddocks with good fences.

On the inside country where it is often necessary to wean either on to dry summer feed, or in the autumn on the cold winter pastures, supplementary feeding will often help considerably. If the lambs become accustomed to self-feeders before weaning and while still on their mothers, they will settle down better and receive less check.

Amongst the greatest dangers to rearing weaners successfully are grass seeds, ophthalmia, malnutrition and internal parasites.

Malnutrition

Malnutrition will stunt the growth, reduce the wool-producing capacity, delay breeding and reduce the profitable life of the sheep. Undernourished weaners are much more susceptible to the effects of stomach and intestinal worms. Forced feeding and heavy condition are not necessary and unless abundant good grazing is available may not be economic. In fact, a 'store' period may be necessary and do no harm provided reasonable growth and strong vitality are maintained. In some districts the dry unbalanced pastures of summer are limiting factors. In others, especially areas of summer rainfall and cold winters, to bring weaners through the winter without loss is a problem so serious as to prevent successful breeding altogether. In these districts improved pastures and winter fodder crops have proved their worth.

Overstocking, through economic necessity or ignorance, leads to malnutrition from eating out the best feed. Although this applies with equal force to all classes of sheep and possibly to the whole property, it is interesting to note that some authorities suggest a stocking rate in which four weaners are equivalent to three dry adult sheep. Since the weaner must grow as well as maintain condition, such an assumption and others like it should be accepted conservatively. Remember, too, that quality of feed is more important for growth in young animals than for maintenance and production in adult sheep.

Grass Seeds

Seeds of barley grass and the spear, or corkscrew grasses, and geranium working into the skin and eyes of weaners can cause serious loss of condition through irritation and blindness. The need for care in selecting a suitable paddock has been stressed. If paddocks free from grass seed cannot be secured, fencing off, harrowing, burning and mowing tracks all suggest themselves as measures affecting the paddock. Shearing, wigging, hand feeding, frequent inspection and handling to remove seeds from eyes, a swim through a dip or stream of water or non-poisonous disinfectant will help the weaners.

Pink Eye or Ophthalmia

This is an infectious disease of the eye which produces a characteristic pink film over the eye and profuse watering. Most cases recover in due course; in others an abscess is formed which may damage the eye. The disease is most prevalent in spring and summer and is usually present in seeded mobs, but sheep free from seeds are just as susceptible. Outbreaks are usually worst in weaners; some degree of resistance is set up, and though outbreaks occur in older sheep, they are not as a rule so severe.

Affected sheep lose condition and may die from starvation and thirst. Badly affected sheep should be collected in a vehicle and run in a small paddock or shed with feed and water handy. Remove all grass seeds, and if numbers and time are convenient, irrigate the eye daily with 2 per cent zinc sulphate solution in water. However, most untreated cases recover just as quickly.

Parasitism

The subject of internal parasites will be covered in more detail later. Weaners, however, are particularly susceptible to worms. Overstocking affects adversely the nutrition of the sheep, rendering it more susceptible to infection, and also increases the concentration of worm eggs in the pasture. Rotational grazing, providing fresh nutritious feed, also results in many worm eggs drying off during the spelling period. Drenching after rain and humid conditions in the summer for the stomach worm and regularly through winter for the small intestinal worm and fluke will be necessary where these parasites are present.

The Ewe After Weaning

The merino ewe in an average season will have little milk left by the time the lamb is weaned and, put on short commons for a while, little harm may be done. With deep-milking, crossbred-rearing fat lambs for market on lush pastures, the danger is greater. Fat lambs are often sold at three to four months, in many cases in drafts according to weight and condition. It is therefore difficult to draw off immediately the ewes in heavy milk whose lambs have been so suddenly weaned. If they can be separated, so much the better, to save feed for lambs yet to be finished. But always watch out for ewes with distended udders. They can be yarded on short rations and then some milk, but not all, can be drawn off by hand.

24

SHEARING, CRUTCHING
AND WIGGING

Shearing

Shearing is one of the busiest times of the year on a sheep property and considerable preparation is necessary before this annual event. Most of the sheep in Australia are now shorn by machines. The machine is a metal comb over which is driven a cutter, the whole being held in a handpiece. A system of flexible gut encased by various joints makes it possible to reach the various parts of the sheep. The power is usually supplied by an internal combustion engine or electric motor which drives an overhead shafting, the machines being driven off this shafing. A great number of electric machines are used in Australia, though they are not so common as the overhead shafting system.

Preparation of the Woolshed

Some days prior to the start of shearing, the machines should be given a trial run to see that the engine, belting, shafting, overhead gear and machines themselves are in thorough working order. The handpiece should be checked over to see that everything is satisfactory. The trial run of the machinery has an additional advantage in that it dislodges a large amount of dust from the walls and roof of the building, and is the first step in cleaning up the shed, which is necessary for two reasons:
1. as a precautionary measure against disease in the sheep; and
2. to ensure that the wool does not pick up unnecessary dirt and foreign matter.

Although a woolshed should not be used other than for the purposes for which it was built, namely shearing, crutching or general handling of sheep, nevertheless, and especially on smaller properties, it is frequently used between shearings for various purposes, such as the storing of grain and even chaff, and the drying of sheepskins — a most undesirable practice.

Sometimes accumulations of crutchings and dead wool occur in the shearing shed, all contributing to rendering it unsanitary and necessitating a thorough cleaning up of the building before shearing. The highest degree of cleanliness possible should be maintained in the shearing shed, and in all places in which sheep are handled. After allowing the machinery to run for some time, during which the opportunity can be taken to sweep down cobwebs and to tidy up generally, the whole shed should be thoroughly swept out. This process should be followed by washing down with hot water and washing soda — 227 to 454 g (½ to 1 lb) of washing soda to 4.5 litres (1 gallon) of water — and a hard broom used to assist in the removal of grease, especially on the shearing board, and in the sweating pens and catching pens. A carbolic or equally efficient disinfectant solution should now be sprayed over walls and floors of that portion of the shed used for sheep. The chutes require special attention, and should be scrubbed with hot water and soda and then antiseptic solution.

Cleanliness in the counting-out pens in which sheep are placed directly after shearing, and with many fresh cuts in the skin, is of the greatest importance. Accumulations of wool and sheep faeces should be removed and, if possible, 8 to 10 cm (3 to 4 inches) of the surface soil should be removed and replaced with clean soil or gravel. Concrete pens are the most hygienic and save a lot of trouble. The best types of pens are those which do not run under the building as they are more easily kept clean, and are exposed to the disinfecting action of the sun.

If sheep-droppings are allowed to accumulate in the counting-out pens, they dry and are then ground to powder by the sheep's feet and so blow on to the open wounds. For this reason, attention should also be paid to the branding race, and any other small yards where recently shorn sheep are held, and all manure cleaned up. Further, sheep manure should not be allowed to accumulate under the sweating pens, as so commonly occurs. Old sheep-manure is a constant source of disease infection. During shearing there will be little time for work of this nature, so it should be attended to beforehand.

Repairs should be made where necessary to yards, gates, pens, doors and broken rails so that everything is in good order for the start. Overhaul the machines as soon as shearing or crutching is over so that any repairs necessary can be effected and supplies of duplicate parts ordered. Wool packs and bale clips should be ordered in plenty of time so that no delay will occur due to shortage of them. The wool press must be overhauled and the wool scales checked for accuracy, and to see that no weights are missing. Check over the stencils and hang them on the wall near the scales. Hang a bag near the press for short ends of twine. The wool-rolling tables should be thoroughly cleaned and repaired, if necessary, and canvas tacked round the sides to keep the locks under the table. Clean out the wool bins and make adequate provision for receptacles for the pieces.

Very good light in the wool room is essential if the woolclasser and his team are to do the best work. It may be possible to improve the lighting by putting in one or more windows in the roof. When building a woolshed, the question of lighting should receive careful attention. This is of particular importance where shearing takes place in the late winter or early spring when the days are comparatively short and the light fades towards the end of the afternoon.

How the Shearing Is Carried Out
It is usual on stud properties to get the stud sheep shorn before the start of the general shearing. Picked men are put on to do this work. Shearers are engaged months ahead so that the owner may be reasonably sure of having a team when shearing is due to

A modern shearing shed featuring raised shearing board, circular wool tables, fluorescent lighting and electric hydraulic wool press. This one is the Allawah *woolshed, Illabo, NSW*
Photograph: Courtesy Wool Harvesting Project, Australian Wool Corporation.

commence. The owner may do the woolclassing himself, or he may engage a classer; and the shed hands are generally station staff unless the shearing is done by contract.

A good deal of shearing is done by contract and shearing contractors relieve the sheep owner of a good deal of work. The contractor engages a team of shearers and shed hands, and often the woolclasser and machine expert also. These contractors arrange a 'run' of sheds, so that their team of men will have as much work as possible during the season. A good 'run' may start in June in the Far West and, working towards the coast, finish on the Tablelands in November, or even later.

An expert is necessary to attend to the machinery, and to the grinding or sharpening of the combs and cutters. Shearers and shed hands are paid according to award rates which, of course, vary from time to time, and the owner must provide suitable accommodation in accordance with regulations. A good shearer will shear over a hundred sheep in the day, but the number shorn depends a good deal on the class of the sheep, some being much easier to shear than others. Special rates are paid for stud sheep.

The order in which the sheep are shorn depends on circumstances appertaining to the property. It is preferable to start with the young sheep — weaners — two-tooth and so on. This not only assists in the control of caseous lymphadenitis ('cheesy gland disease') but helps the woolclasser to blend his lines of wool by using limited space in the woolshed. The ewes and lambs are generally shorn last, and to avoid the lambs suffering a check it is advisable to shear in such a way that they are not more than a day away from their mothers. It would be preferable, in order to minimise risk from the forementioned disease, if the lambs were the first sheep shorn at the annual shearing.

The different flocks are yarded and any drafting necessary is done so that the sheep go into the shed in even lines. It is a great advantage if the yards can be watered to prevent dust entering the fleece and blowing on to the wounds of recently shorn sheep during drafting and other necessary yard work. When laying out a shearing shed and yards it is desirable to place the latter so that prevailing winds take the dust away from the shed. If this is not possible the design should be such that shorn sheep leave the shed without passing the dusty yards in which unshorn sheep are being worked.

Plenty of cover for the unshorn sheep in case of wet weather is desirable on any property, either in the shed or even under it. Shearers will not shear wet sheep because of danger of illness, and further there is a risk of fire in the bales from spontaneous combustion if wet wool is pressed. When filling the shed at night, or even when putting sheep under cover on account of threatened rain, care should be taken, especially with young sheep, that they do not crowd into a corner, as they are likely to smother. Dogs should not be allowed loose near the shed at night as they may disturb the sheep and cause them to crowd together.

Sheep should be kept in the pens and yards for as short a time as possible after shearing. Every endeavour should also be made to get them away from the shed in time for them to have a feed before dark. This will often prevent deaths if a cold night follows.

During the shearing of the sheep, many small skin cuts, and sometimes severe wounds, are inflicted on the sheep. Some of these are unavoidable, but many are due to the shearer racing for big tallies, or to incompetent or careless shearing, or to lack of supervision. It is not practicable to treat with some antiseptic solution every wound made on a sheep during shearing, but an attempt should be made to dress any large wounds. The application of an antiseptic solution to as many wounds as possible, if properly carried out, will tend to minimise the risk of wound infection. The commonly used 'tar stick' is of doubtful value, apart from which the use of tar on the wool is very deleterious and should be avoided. Whilst carbolic or similar antiseptic solutions can be kept handy in containers on the shearing board and are usually applied with a swab consisting of a piece of rag on a stick, the method of application leaves room for much improvement.

A far superior method of application of the antiseptic solution is to spray it on with either a hand apparatus like a duco sprayer, or with a compression apparatus run from the power plant. A mixture of one part phenol to six parts of water is suitable for use in this manner and ensures that the wound is sprayed with a fine spray of antiseptic solution which is generally preferable to the use of a contaminated swab stick. The hand apparatus can also be used in the branding race or yards. A still more scientific treatment of shear cuts likely to find favour in future is the dusting on of sulphanilamide powder with a handblower.

Comb, cutters, and shears should be frequently cleaned and disinfected. For this purpose a 2 per cent caustic soda solution of 28.3 g (1 oz) caustic soda to 1.42 litres (2½ pints) of water is best. Combs and cutters should be threaded on a wire and shaken in a vessel of plain water and dried. They may afterwards be oiled in the usual way. The caustic soda solution is more effective if warm. Whenever it is known that shears or combs and cutters have been in contact with pus from a discharging abscess, or when an abscess has been opened, they should not be used again until thoroughly disinfected.

There is room for considerable improvement in the manner in which sheep are shorn generally in Australia. Better shearing in its wider sense means not only less wounding of the sheep, but also the adoption of measures to reduce the infection of wounds by greater cleanliness of the shearing shed and yards, particularly the shearing-board and counting-out pens, and by keeping down dust in the yard as much as possible in the manner already indicated.

Crutching and Wigging

Crutching consists of shearing wool away from the breech, over the tail and down the back of the hind legs, thus making those parts the same as in recently shorn animals. It is carried out mainly to prevent blowfly strike in the breech region of the sheep, but also for reasons of general cleanliness, especially in the case of ewes just prior to lambing. The majority of owners crutch sheep only once per year and as close to lambing time as possible. Some owners, however, make a practice of crutching twice a year, either in the spring or summer and again in the autumn, the dates depending on the time of shearing and lambing. One crutching at or near mid-season has value in addition to blowfly protection, by preventing the occurrence of excessively dirty wool at shearing.

The whole idea behind crutching, as far as the blowfly is concerned, is to render the breech clean and allow it to dry out thoroughly. When this is brought about, an inflammatory condition of the skin and bacterial decomposition are not so prone to occur and the breech area is not then attractive to the blowfly. In the case of plain breeched sheep of good formation, that is, wide behind, or those sheep on which the modified mules operation has been carried out, or the Manchester treatment applied, crutching is not necessary, as the accumulation of urine and sweat are not so marked in these animals. Nevertheless, one crutching at least of such sheep should be undertaken for the reasons already given. When a thorough system of jetting is carried out in conjunction with a single crutching, there is not the same necessity for a double crutching.

To be carried out properly and effectively, the wool should be shorn close over an area extending from above the tail, down each side of the breech, (including well outside the pin bones), to include the crutch (behind the hind legs) and down the back of the legs to the hocks. Simply to cut away dung-soiled and matted wool below the breech (which is really only 'dagging') is not sufficient. It is much more important to shear the wool from over the folds each side of the breech and their extensions to the crutch and down the legs. Crutching is carried out chiefly on ewes, although at times wethers and rams are also crutched. It is frequently necessary to shear the wool away around the prepuce of wethers and rams ('ringing') to prevent this area being fouled with consequent bacterial decomposition, and fly strike. The heads of rams may be shorn in March (when the flock ewes and rams are usually crutched) in order to minimise later summer and autumn attack on rams' heads. Other sheep are sometimes 'wigged' to overcome wool blindness, and if grass seed is troublesome.

Shearing Sheds, Yards

Wool Sheds and Yards

The placing of yards and sheds and the situation and relation of fences, yards, floors and races, buildings and equipment are important in relation to the handling of sheep and wool. Good arrangements can improve visibility in the shed and make supervision easier, thus assisting shearers, shed hands and classer. It also leads to greater ease in the use of equipment and facilities available. Mechanical devices such as gate

POINTS FOR GOOD SHEARING

A good shearer is skilled and, like others in skilled professions, his competence will build up over many years. Improvement of his skill level will make his work easier and of better quality and will improve his status in the industry and the community.

Highly skilled shearers approach their profession intelligently and constantly analyse their performance to improve the quality and speed of their work.

The following points should be considered by anybody who wishes to improve their shearing skills.

1. POSITIONING AND CONTROL

Positioning and control have been combined because each depends on the other. A flat surface is required for high quality shearing. This is created by correct positioning and control. Good control of sheep also permits correct blow placement. Correct positioning of both sheep and shearer is necessary to ensure that the blows are made where required.

2. BLOW PLACEMENT

Correct blows are essential for good shearing and correct positioning and control permits this. It is vital that blows be run on as flat a surface with as full a comb as possible. Great care must be taken not to run out into the wool through taking blows too far. Follow the recommended pattern carefully and always count the blows in each section to ensure that you do not become careless and run too many. Concentrate at all times on keeping the bottom tooth firmly on the skin. Run blows accurately even if a little slowly. A quick return, keeping the handpiece close to the skin, saves a lot of time. Keep a flexible but firm grip on the handpiece.

3. FITNESS AND STAMINA

Besides demanding considerable skill, shearing requires a high level of physical fitness and stamina. These can only be built up over years of experience and only then by being aware of the physical and mental demands of the profession of shearing. It is essential that a shearer start each season at a high level of fitness and some preseason preparation is needed. Since shearing imposes heavy strains on the body, it is worthwhile to do a series of warming-up exercises before starting in the morning and after lunch. It is also important to plan the day's work so that regular rest periods can be planned throughout the day. In hot weather a regular intake of cool water is essential, rather than large intakes on the hour or at the end of each run.

4. TEMPERAMENT

A good shearer controls his sheep and his temper and does not exhaust himself fighting troublesome sheep. It is essential that shearers learn how to control their sheep with a minimum of effort. Time spent learning to control sheep will help prevent frayed tempers and assist with quality workmanship.

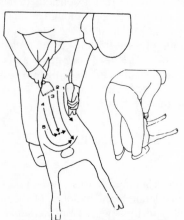

1. Avoid running into the fleece wool when shearing the belly. Stretch the skin with the left hand and pull the wool through the comb. Approach the pizzle with a horizontal blow.

2. Straighten the leg by pressing your left hand into flank. Place left hand over teats of ewe when running blow around crutch.

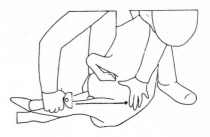

3. Stretch skin with left hand. Run blow to clear flank.

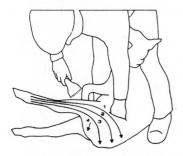

4. Press left hand into flank to keep leg straight. Run blows to within a comb's width of backbone.

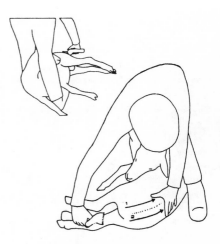

5. Keep right shoulder and heel of handpiece both well down.

6. Clear topknot and above left eye.

7. A clearing blow is followed by a blow up the side of the main neck fold, then clear the left face. The second main blow follows the first and continues around the back of the head to the base of the right ear. The top of the left leg is now cleared, the left hand stretching the skin.

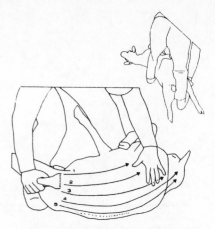

8. Two more main blows complete the leg. Do not run them too far. Then clear the shank.

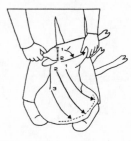

9. Keep right shoulder well down and heel of handpiece below comb entry point. Start blows with full comb. Finish this section on backbone.

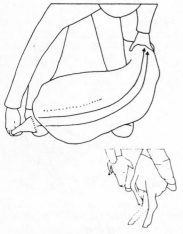

10. Bow sheep around left leg and push head firmly down. Run two full blows.

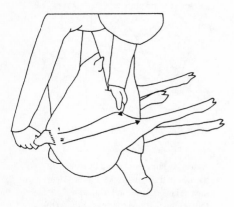

11. Shear carefully down neck one wrinkle at a time. Do not run blows past line of shoulder.

12. Heel of handpiece below point of entry. First blow across top of leg and down to clear brisket. Second clears top of leg.

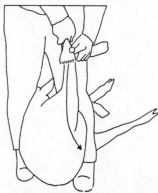

13. Pick up leg and clean shank. Blow down leg runs towards flank and sets up line for last side.

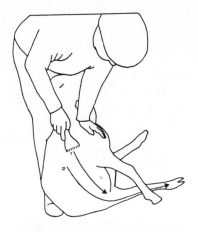

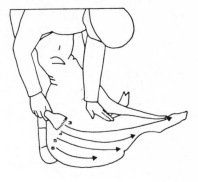

15. Step back slowly and stretch skin of leg with left hand, pressing into flank to keep leg straight. Start blows with full comb.

14. Pick up the head from between your legs. First blow runs to flank, second right out the leg.

Figure 24.1   High quality shearing method for merino sheep
Source: Australian Wool Corporation

catches and automatic fasteners can be important operational aids and should be considered in relation to general management.

The average economic life of sheep is up to eight years with six to eight shearings. Sheep may be physically affected, temporarily or permanently, by faults in yard and shed design and operation. They may be disabled or injured by straining or crushing or their skin may be damaged by wool tearing out. The greatest risk of loss is death from exposure to unfavourable weather in the first few days after shearing.

Shearing Sheds and Yards
Where new sheds are to be built or existing sheds altered, the factors to be considered in the overall arrangement of shed and yards are as follows.
1. size of property;
2. number of sheep to be handled;
3. location of shed and yards in relation to:
    (a) the property;
    (b) highways and towns;
    (c) each other;
    (d) water supply;
4. orientation in relation to:
    (a) climatic conditions;
    (b) protection from sun;
    (c) lighting;
    (d) topography and slope;
5. availability of labour to operate yards and shed;
6. type of shed;
7. types of machinery and static equipment;

8. protection of sheep from weather before and after shearing;
9. storage facilities and methods;
10. handling equipment and aids;
11. off-season use of shearing shed and yards.

The location of shed and yards should be decided after considering the property as a whole, and the maximum possibilities of economic operation. Location close to other buildings is not always an advantage. The latitude and elevation of a property determine to a large extent whether general conditions are hot or cold. Prevailing winds must be considered also. If cold country, sunlight to provide warmth and light may be an advantage; if hot, too much sunlight may be a disadvantage and buildings or yards should be oriented accordingly. If sheds are near the homestead they should not be built on the windward side. Wise orientation may make it necessary to cover or partly cover yards against sun and weather. In Australia, southern is the best natural lighting as it is even and avoids the disadvantages of glare from the direct rays of the sun.

The natural features of a site can often be used to advantage in yard and shed construction. Advantage can be taken of variations in ground levels, as by building the yards so that sheep work uphill and enter the shed at a point where the shed floor level is nearest the ground. In this way steep sheep ramps are avoided and construction is simplified and cheapened. Sometimes the natural levels are such that space can be provided below the shearing floor for storage of sheep.

As the floor of the elevated end of the shed is used for wool storage, the height above ground level should be fixed to simplify the loading of bales to road vehicles. It should be possible to arrange this floor so that the bales may be dropped on to the waiting lorry and then the upper rows rolled on. By working sheep in yards to a high point of entry, all wool movements after shearing can continue on a level plane or downhill. The force of gravity can be used to aid handling operations.

Sometimes simple aids can be used to control sheep. One used on a Western Australian property is based on the observation that when sheep go backwards they shuffle and do not lift their feet as readily as they do in going forward. Sheep are hindered from moving back in the races by placing a piece of timber about 8 × 5 cm (3 × 2 inches) in section across the race floor and with each end against the vertical side-members of the race.

Shearing Sheds

Shearing sheds should be erected on a convenient site with good drainage and the best possible design. They cost a good deal to erect and mistakes in construction or design may prove expensive. The main thing to watch when building a shed is that there is inside it plenty of natural light early and late in the day. Have the board, tables, bins so arranged that the wool goes ahead all the time. If it is to be an end-loading shed, which is the best type, have the entrance end (i.e., where the sheep come in) facing the west; as the shed is usually filled last thing at night, the sun is then behind the sheep, and by the time more are required next day the sun will be well up.

If possible the shed should be on a rise; not only will it be better for drainage purposes, but sheep yard better uphill as they can see those ahead of them.

Inside the shed the following points should be watched:
1. Sheep should not have to cross the shearing board.
2. A shearer should not have to step over fleeces to reach the catching pen.
3. It should be possible for fleeces to be picked up without the picker-up getting in the way of the shearers.
4. Sufficient space should be allowed to house enough sheep for at least a day's run.

Cleanliness should receive important consideration in the shed, to prevent infection, blood poisoning, tetanus, etc. caused by a dusty, dirty shed and yards. Painting the inside of the shed white is a sound idea. The woolclasser should have a clear view of the fleeces while they are in the bins and when they pass into the press. In designing and building the shed it should be remembered that it is not advisable to use posts that are part of the building to carry the machinery, unless the shed is of unusually solid construction and the posts extremely heavy and sunk well into the ground.

Accommodation
The accommodation for shearers and shed hands is a matter that is governed by law in Australia and Acts of Parliament making it compulsory to provide a minimum of what is termed 'proper and sufficient accommodation for shearers' are in force in the various states of the Commonwealth. Buildings, therefore, must be designed and constructed to comply with all the provisions of the Act under which the owner is liable.

Yards
Yard layout varies from property to property. A common arrangement includes holding paddocks, receiving yards, forcing pens or yards, drafting races, classing,

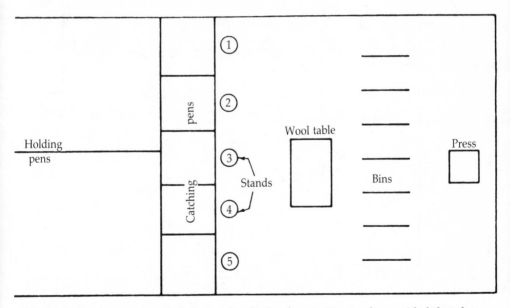

Figure 24.2   A recommended design for sheds of up to six stands provided that the bins and press are portable. By having the press directly behind the classer, fleeces in the major line could be placed directly in open bale and bins could be arranged so that minor lines (or the major line when pressing was actually taking place) could also be close to the classer. The stands face the classing table and the average time taken in walking from stand to table is 6–7 seconds. Access to counting-out pens is by reverse chute under the shed.
Diagram: courtesy Australian Wool Corporation; from the AWC Technical Report, Objective Measurement of Wool in Australia, October 1973.

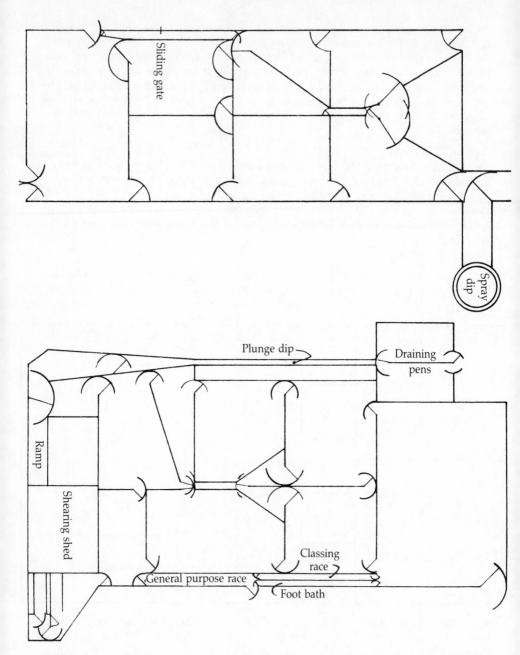

Figure 24.3   Glen Innes Agricultural Research Station has been using the yards from which this plan was drawn for a number of years. The flock is just under 1000 sheep.

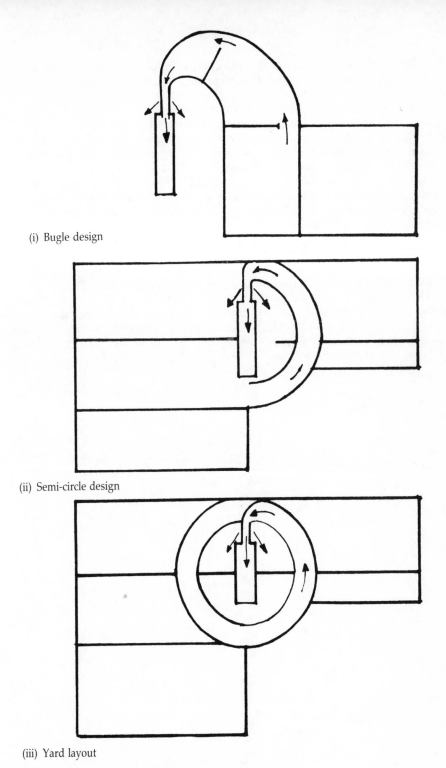

(i) Bugle design

(ii) Semi-circle design

(iii) Yard layout

Figure 24.4   Some improved layouts for drafting yards.
Diagram: courtesy Australian Wool Corporation

drenching and branding races. Well-constructed yards of good design are invaluable in the saving of extra work when handling sheep for routine operations, including drenching, inoculating and classing.

Some yards include facilities for dipping sheep, such as swim-through bath or spray type. Other yards include facilities for lambmarking and the mules operation, although lambmarking should be carried out in temporary yards. The mules operation is a surgical job frequently done in the vicinity of the shearing yards. As each sheep is handled separately, special cradles are used to hold the sheep.

The yard surfaces should be carefully watched where dust or mud may form. The yards can be watered to prevent dust rising. In districts where rain is frequent, it is a good plan to have the yards bricked or constructed of concrete to offset the mud nuisance. Grading of yards is important. Because of limited groundfall it is sometimes difficult to grade yards effectively, but some measure of drainage should be aimed at by establishing local falls, or by building up the ground level before fences are erected, or building dished gutters running into rubble drains.

Yards and sheep can be protected by encouraging the growth of trees, and many properties have trees planted in yard corners. Sometimes the trees are protected by drums or by corner barriers placed across the angle of the fence. Boarding off the corner of the yard to protect the trees serves another purpose: it keeps sheep from jamming into corners, and yards with the corners cut off work better than yards finished with right-angle corners.

25

PASTURE IMPROVEMENT AND MANAGEMENT, FODDER CONSERVATION AND SUPPLEMENTARY FEEDING

Pasture Improvement

Pasture improvement may be defined as the improvement of natural pasture by sowing seed of selected strains of legumes or grasses combined with fertiliser application.

To establish a good pasture and maintain a balanced pasture composition the following practices are also necessary:

1. stock management;
2. sub-divisional fencing;
3. siting of watering points to prevent localised overgrazing;
4. periodic removal of stock to allow recovery and setting of seed.

Scientific pasture improvement in Australia dates from the beginning of the century. Significant progress did not begin until the 1920s, however, when sowing of subterranean clover and topdressing with superphosphate first became commercial practice. High prices for rural products, good seasons, and reduction of the rabbit population were contributing factors which resulted in an upsurge in area under sown pastures in the late 1940s. Under a variety of incentive schemes, pasture improvement continued to be popular during the 1950s and 1960s. The area under sown grasses and clovers in the early 1970s exceeded 25 million hectares, which was more than three times the area twenty years earlier.

The distribution of sown pastures in Australia is limited by various climatic factors. Rainfall, temperature and seasonal variation in the length of daylight are of particular importance. The accompanying map indicates that, in rough terms, the 30 cm (12 inch) isohyet is the inland limit of sown pastures in Western Australia, the 38 cm (15 inch) isohyet in the south-east of the continent, the 50 cm (20 inch) isohyet in Queensland,

and the 76 cm (30 inch) isohyet in the Northern Territory and northern Western Australia. Distribution of rainfall, soil fertility and extremes of climate are also limiting factors.

Soil Deficiencies

Most soils in Australia are deficient in phosphorus and nitrogen. Superphosphating of natural pastures does offer some improvement in this regard. In some areas deficiencies have also been found in major plant nutrients such as sulphur, calcium, potassium and magnesium. Perhaps the most dramatic evidence of how far deficiencies of plant nutrients may limit production lies in the work with minor trace elements such as molybdenum, zinc, copper and boron. A few kilograms per hectare, even a few grams in some cases, of the appropriate compound applied to a soil lacking the element concerned will satisfy the requirements of pastures for years and will often make the difference between a complete failure and a highly productive pasture. The development of the Ninety Mile Desert, now known as Coonalpyn Downs, in South Australia is one of the outstanding examples of the use of copper and zinc.

The Degree of Acidity or Alkalinity of Soil

The degree of tolerance of different pH levels in soils varies quite widely among different species of pasture plants. The bacteria associated with legumes of the temperate regions, such as clovers and medics, are particularly susceptible to acidity.

Nitrogen Fixation

Grasses and legumes fulfil a complementary function in providing a nutritive pasture for the grazing animal. Legumes provide a high level of protein feed and contribute to soil fertility and nitrogen content. Grasses make use of nitrogen and fertile conditions promoted by legumes to produce a bulky and plentiful supply of high quality feed.

Legumes are able to obtain nitrogen from the air in association with bacteria of the genus *rhizobium* which penetrate the roots and form small cysts or nodules in the root

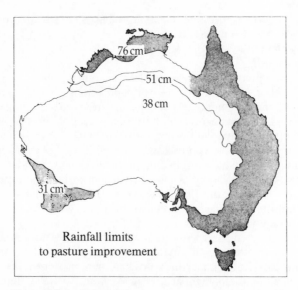

Figure 25.1   Rainfall limits to pasture improvement

tissues. These bacteria live within the nodules and obtain their food supplies from the plant and from the nitrogen in the atmosphere. Nitrogenous compounds excreted by them meet the nutritional needs of the legume host and are in large part conveyed to the soil through the grazing animal. Rhizobium often occur naturally in the soil, but appropriate strains for particular species are not always present and therefore it is a wise precaution when sowing legumes to apply a suitable culture to the seed before sowing.

Seed Selection
1. Select a suitable variety for climate and environment.
2. Use only tested seed: various Commonwealth and state Acts of Parliament exist to ensure that all seed entering trade conforms to minimum standards.

Each state Department of Agriculture has a seed-testing laboratory for administration of the various Acts and these provide a testing service for a small fee to farmers and seed merchants who want a report on the quality of their seed. Farmers using seed produced on their own properties are not obliged to comply with the regulations under the *Seeds Acts*, but clearly it would be in their own interest to ensure that the seed was of good quality.

Sowing Methods
1. Seed may be sown directly on unploughed ground by sod-seeder (cuts furrows through sods with discs or heavy tynes and deposits seed in them, like a combine drill).
2. Seed may be broadcast from a farm vehicle or aircraft.

Economics
Pasture improvement is a good investment in the long run, although increased profits are not immediate. It involves not only an initial capital outlay, but also a continuing higher level of annual cash expenditure to maintain the pasture and the greater stock numbers required.

Pasture Improvement and the Future
1. It is possible to raise stock-carrying capacity threefold and more in many areas.
2. In future it may be undesirable to continue to build up live stock numbers indefinitely, but rather from time to time cash in on the rising fertility of the soil by growing crops.

What to Sow
- *Phalaris tuberosa, perennial and annual rye* grasses and *lucerne* can be grown in almost any type of soil and combine high carrying capacity, drought resistance and supplementary fodder value to a high degree.
- *Cocksfoot* is another valuable winter green species. It can be grown on poorer soils than rye grass and is very useful in Tablelands districts.
- *Wimmera rye grass* is suitable for the Slopes and portions of the Plains. Thrives on fertile free working soils providing valuable winter feed. Can be grown in association with lucerne or subterranean clover.
- *White clover* is a valuable pasture of perennial habit and should be used combined with *Phalaris tuberosa*, perennial rye and cocksfoot in Tablelands districts.
- *Red clover* is an excellent supplement to white clover because it will grow during summer and is more resistant to dry spells than white clover.
- *Subterranean clover* is suitable where rainfall is insufficient for white clover and for building the fertility of large areas of country in Tablelands and inland districts.

The Effect on Wool

CSIRO trials at Shannon Vale Nutrition Station and elsewhere have shown that improved pastures increase fleece weight, giving wool better length and improved yield. Handle and colour are not affected, and it has been further demonstrated that fine and superfine wool can still be produced without becoming coarser in fibre diameter.

Pasture Management

The principal object of improving pastures is to increase the carrying capacity of the holding. This may be achieved, once the improvement work has been undertaken, by gradually increasing the number of stock carried, while on the other hand the elimination of losses due to drought and to shortage of winter feed in cold districts, and losses due to disease caused by malnutrition, will further assist in increasing the numbers of stock.

Sown pastures should be managed so that succulent feed and clean paddocks will be available for lambing ewes, for 'topping off' lambs, as a tonic for sick animals, and to provide grazing during the periods when the natural pastures are depleted. With good management, as the areas under improved pastures increase, a gradual increase in the number of stock that can be safely carried on the holding will be obtained without harm to the pastures.

An efficient system of pasture management should be aimed at maintaining a good supply of succulent pasturage to be grazed at the stage of maximum feeding value, at prolonging the life and productivity of the pastures — especially the more palatable and nutritious plant species — at providing a change of feed for stock and ensuring a supply of feed in the seasonal periods of shortage which occur in every district. Rotational grazing followed by treatments to maintain soil fertility, such as top-dressing, harrowing, mowing if necessary and practicable, and re-seeding, may all enter into the system.

Handling Sown Pastures

Sown pastures may be stocked as soon as the plants have made sufficient root growth to withstand grazing. It is preferable to commence by turning in a fairly large number of stock and leaving them in the paddock for a short time only; they must be removed as soon as the growth is shortened back. Regulated grazing in the early stages encourages the plants to stool out and make stronger growth. Stock should not be turned on to young grass and clovers if the weather is wet and the ground boggy and soft, for the trampling under these conditions will kill out many of the plants; when the land settles down and becomes firm, the liability of plant damage from this cause is reduced. As the plants develop, the stocking periods can be lengthened, until eventually normal grazing can be carried out; however, this should be rotational. During the early grazing periods it is advisable to observe closely the grazing behaviour of the stock, and in the event of their paying too much attention to comparatively slow-growing grasses, they should be removed for a time.

Preserving Native Grass

Natural pastures composed of such plants as the Wallaby (*Danthonia* spp.) (*Andropogon* spp.) and the Panic (*Panicum* spp.) grasses should also be carefully managed in order to preserve the most useful species and to prevent the growth of weeds and useless grasses.

Where useful native grasses are still in existence on a property, they should be encouraged to seed and spread by properly regulated grazing and by avoiding overstocking. Paddocks of native grasses showing signs of seeding should be rested or lightly stocked until the seed has ripened and fallen after which normal stocking may be resumed. A system can easily be evolved whereby a number of paddocks can be allowed to seed each year, so that the grasses will thicken up considerably. The majority of the useful native grasses are commonly known as 'root' grasses by graziers, and their presence adds considerably to the drought-withstanding capabilities of a property.

In regulating the movement of stock, the usual practice of graziers is to study the condition of the animals; if found to be in poor condition, they are moved to another area. However, there is a strong objection to this system from the point of view of efficient pasture management: the condition of the pastures should receive more consideration. If it is found that the better quality grazing plants have been grazed close and are being defoliated severely, stock should be moved despite the fact that they may be in good condition. Non-observance of this principle is to a great extent responsible for the deterioration of many valuable areas of native grasses.

The Subdivision of Paddocks
No hard and fast rules can be laid down with regard to this matter, as the conditions vary so much in different districts and even on the same property. The aim should be to provide sufficient paddocks to control the grazing completely, so that an even growth can be maintained in each. With controlled stocking the pastures can be fed off when at their maximum feeding value, that is, when they are providing short, succulent growth high in protein content and there is no waste such as that which is associated with more mature but less palatable and less nutritious growth.

Paddocks that are too large result in stock having to travel long distances for food and water, which is particularly undesirable for fattening, as much of the food consumed is then used to supply energy for unnecessary walking. This point is of importance to those contemplating fat lamb raising. Large paddocks are also responsible for a certain amount of erosion, as the animals in their search for food tend to traverse definite tracks, which become bare of grasses. In hilly country these bare tracks form channels along which water flows, finally resulting in erosion.

The expenditure entailed in laying down large areas of sown pastures on well-prepared land is considerable, and although returns amply justify the outlay, some pastoralists do not have the capital available for this work on a large scale. By choosing the most suitable soils and situations, however, it is remarkable what excellent results can be obtained from small areas of sown pastures when used in conjunction with larger areas of natural pastures, and graziers with limited capital should proceed on these lines.

The Influence of Different Kinds of Stock
Sheep are very selective in their feeding, preferring the low-growing fine plants, and neglecting the tall-growing, coarse species. Over-grazing by sheep will therefore result in the 'eating out' of the finer grasses and will encourage the development, seeding and spreading of the coarser types.

Horses are 'harder' on the pastures than sheep, being even more selective in their grazing habits and concentrating on patches of the finer grasses until they eat these out. Pastures used entirely for horses will quickly revert to coarse grasses and other unpalatable pasture plants. Cattle are less selective in their grazing than sheep and horses, and pay considerable attention to the coarser growth. Advantage can be taken of this fact, and cattle used in conjunction with sheep and horses to maintain an even balance in growth between the fine and the coarse plants in the pasture. 'Stores' are naturally of more use in this connection than fat sheep.

Overstocking is probably the greatest contributing factor in pasture deterioration in New South Wales, for in many cases the overstocking has been due to rabbits as well as stock, and this, combined with the effects of drought, has resulted in many pastures becoming denuded of the original covering of natural grasses. Our uncertain seasonal conditions make it rather difficult always to be on the safe side with the number of stock carried, and occasionally trouble cannot be avoided.

Sown pastures may often be stocked at the rate of forty-nine to seventy-four sheep per hectare (twenty to thirty sheep per acre) without being overstocked, and so long as they are removed and the numbers reduced before the plants are nibbled too close, no harm will be done. If, however, the farmer persists with this rate of stocking it will not be long before the most palatable species disappear, followed in time by the less palatable species. As the grasses and clovers become eaten out, unpalatable plants and weeds will obtain a hold, until finally they will assume control, to the detriment of the area in stock-carrying capacity and drought resistance. The dominance of the wire or three-awned spear grasses (*Aristida* spp.) in some localities is due entirely to overstocking and drought.

Overstocking, particularly by close-feed animals such as sheep, prevents the development of renewal shoots; and heavy grazing followed by dry weather will invariably result in many plants dying.

Continued understocking of sown pastures is likely to lead to the stronger growers becoming dormant, to the exclusion of the finer types, and to loss of feed on account of the grasses seeding and becoming more or less harsh, unpalatable and innutritious. Finally, weeds and annual grasses grow and seed unheeded, and assume control.

Fodder Conservation

In those districts where crops and pastures can be grown and properties are of moderate size to allow economical feeding, a reserve of hays and grains should be maintained for feeding stock during temporary or prolonged feed shortages. To a large extent these reserves of fodder will be cereal hays or grains and meadow hay. Where conditions allow, as much as possible should be conserved in the form of leguminous roughage. Rich in protein and vitamins, legume hay or silage is invaluable in balancing rations and keeping stock healthy.

In many districts of light rainfall, fodder conservation is impracticable. Unreliable seasons, the lack of farming labour and the huge reserves needed to maintain the large numbers of sheep make such a scheme impracticable. Some fodder conservation may be carried out by buying up stocks of grains and hays in good seasons, but the main solution to the problem will depend on careful stocking. Faced with the onset of drought conditions, the owner has several courses open to him:
1. reduce stock numbers by disposing of sheep;
2. find agistment for some or all of his sheep;
3. hand feeding.

If normal stocking is light and sheep disposed of according to the feed available, some financial loss is sustained by selling on a falling market, and the grower may not like unloading in case the drought is only short. On the other hand, the sheep retained will do better, cut more profitable fleeces, and will produce and rear lambs if rams are joined, while the pastures will receive a spell and recover more quickly when rains come. More clips from properties where this practice has been carried out during recent droughts have been outstanding on show floors where they have been displayed.

Food Elements Required by Sheep

The materials required by sheep are proteins, carbohydrates, fats, fibre, mineral matter, vitamins and water.

- *Proteins* are nitrogenous compounds and their main functions are:
  1. building and repair of tissue;
  2. provision of material for secretions (e.g. hormones, internal secretions, glandular products and digestive juices);
  3. tissue stimulation — body development, wool growth.
- *Carbohydrates* are starchy and sugary food elements. They are all nitrogen-free compounds. They produce heat and energy in sheep and are an important source of fat in the carcass.
- *Fats* are different to animal fats. The latter are used to supply bodily heat and energy and excess in the food is stored as body fat.
- *Fibre* supplies bulk to feed — excess causes digestive disorders, as may too little.
- Mineral matter. Minerals enter into the formation of bones, teeth, tissues and blood and play an important part in digestion and other processes. Required in relatively large quantities are calcium (lime), phosphorus (phosphoric acid), sodium, potassium and chlorine. Required in relatively small quantities are iron, iodine, sulphur, manganese, magnesium, silicon, fluorine, boron, copper and cobalt.
- *Vitamins* are complex substances that occur in very small quantities in various foodstuffs of vegetable and animal origin. In absence of knowledge of exact composition in early days of research, letters were used to denote each vitamin (A, B, C, D, E, etc.).
  Vitamin D deficiency is found in weaners on succulent green pastures during winter months.
  Vitamin A deficiency occurs in lambs reared without access to green feed. The symptoms of night blindness are revealed if the lambs are driven after dusk.
- *Water* An ample supply of good water is essential. A good deal of water is supplied by green grass. Average sheep on dry feed require 4.5 litres (1 gallon) per day.

Supplementary Feeding

Supplementary feeding of sheep is an important phase of sheep management and should not be confused with drought feeding. There are many occasions when supplementary feeding of sheep is desirable. Sheep on improved pastures where feed is green and succulent require a certain percentage of roughage. At times this can be supplied by giving them access to paddocks of unimproved natural pastures, but this is not always practicable. Under these circumstances cereal hay or grass should be made available to the sheep, or chaff in self-feeders should be placed in the paddock to provide a supplementary feed to the rich grazing. A strip cut across a paddock and allowed to dry will overcome this to a certain extent.

Supplementary feeding checks worm infestation and a ration should be supplied to ewes and to lambs grazing on natural pasture shortly before weaning. It should be continued to the weaners for several months afterwards — this helps the weaners over the danger period of worm infestation. The other advantage of supplementary feeding is that it keeps young sheep growing. With larger holdings, supplementary feeding is more difficult, and more reliance must be placed on rotation of paddocks, lighter stocking and top dressing of pastures. It also increases the amount of wool grown on the sheep.

Self-feeders

The provision of self-feeders in the paddocks allows sheep to balance their own ration. These feeders should contain chaff, oats, maize, wheat, etc., and be available to the sheep at all times. They are available from stores and agents or can readily be made on the property.

The main points to observe are:

1. that they are big enough, in order to reduce the labour of filling to a minimum, yet not so large that they cannot be conveniently moved to other paddocks;
2. that the internal construction is such that the chaff and grain run freely;
3. that they are weather proof.

The Cost of Feed

Many feeds may be used for supplementary feeding of sheep, but cost is most important. The various grains, linseed nuts and other nuts containing a combination of crushed cereal grains and protein-rich foods such as blood or meat meal, peanut meal, coconut or linseed meal have all been tried. Cereal chaff is perhaps the most suitable and common feed for supplementary feeding; the addition of grain — particularly oats — is an advantage, on account of the higher protein content.

If supplementary feeding is to be carried out, not only the cost of feed must be considered, but also the manpower and cost of handling the feed. The cost of supplementary feeding need not be very great if the feed is grown on the property and the troughs, etc. are homemade. Having these troughs on the property also lessens the cost of putting out the feed. Experience has shown that any costs incurred are more than compensated for by the greater return of wool, lambs, lessened mortality and greater commercial value of the sheep.

Conclusion

An important point in favour of supplementary feeding is the fact that where it is practised, the advent of drought conditions will make the eventual cost of feeding through that period very much less. This is owing to the fact that the sheep have been trained to eat at feeders, and there is no risk of leaving feeding too late — when it is frequently impossible to pick up lost ground.

26

MINERAL DEFICIENCIES

Many minerals are required by the animal to maintain its normal physiological activities and normal health. Deficiencies may occur because the soil itself is lacking in some particular mineral, or if it is absent it may not be available, and hence the herbage in turn becomes deficient. Stock grazing on such herbage may in turn develop symptoms of mineral deficiency. While calcium and phosphate are the two minerals required in greatest quantities (and at times deficiency of these may occur), by far the more common deficiencies are those that are referred to as 'trace' or 'minor' element deficiencies. These refer to elements that the animal requires in only very small or 'trace' quantities, such as copper, cobalt and iodine.

Cobalt has been found to be essential for sheep and areas of deficient country have been defined in most states, particularly South Australia. Areas deficient in copper also exist in most states and are more extensive. These two deficiencies do exist together, most commonly in coastal areas. In localities where there is virtually no cobalt, sheep quickly show signs of malnutrition, losing their alertness and becoming lethargic. As a consequence their wool suffers and shows a dead, ill-nourished appearance. On copper-deficient country, sheep become anaemic and lose their appetite, but the most striking symptom in mature sheep so affected is a marked depreciation in the quality of their wool. Their wool changes from the evenly crimped character of the normal fleece to a straight, shiny hairlike growth, referred to as 'steely' wool.

Another condition exhibited in pigmented wools is a banding of pigment, dark bands alternating with light bands in a somewhat irregular pattern. Yet another symptom invariably associated with advanced copper deficiency is the ataxia which affects young lambs born of copper-deficient ewes. These lambs are unable to coordinate the movements of their limbs, particularly their hind limbs, which collapse when they attempt to walk, so death accounts for a considerable number. But the most serious source of monetary loss is the general lack of thrift in the animal with failure to produce wool of good quality and weight.

The best method of combating these trace element deficiencies is to improve the soil condition which is simply done in the case of cobalt or copper, but is better administered in lick form (as iodine salt) in the case of iodine deficiency. About 3 kg (7 lb) copper sulphate and 140 g (5 oz) cobalt sulphate per .405 ha (acre) usually applied as copperised or cobalticised superphosphate has corrected soil deficiencies.

Licks designed to overcome minor deficiencies are generally unsuitable because some animals will not take them and some take them in excess, which can be dangerous in the case of copper. Where top dressing is not practised, copper and cobalt deficiencies in stock can be overcome by drenching the stock, but it must be emphasised that, wherever possible, the soil conditions should be corrected, for where the pasturage contains adequate minerals, it almost invariably follows that the stock obtain adequate for their needs.

27

DROUGHT FEEDING

I t is known that drought occurs, although somewhat irregularly, in pastoral areas, and as a result heavy losses have been experienced in the sheep industry. For the purpose of sheep husbandry, in this section, drought is regarded as a temporary or prolonged period in which climatic conditions reduce the nutrients available to all animals below those necessary for maintenance. The best insurances against drought are:
1. adequate water supply;
2. reserve stocking;
3. fodder conservation.

Water Supply
Water supply must be adequate to supply needs of stock. Sheep will only travel limited distances to water, so that if the number of watering places is limited, feed shortages may occur and drought conditions prevail. Owing to restricted areas being eaten out, sheep will 'do' better if water is clean and fresh. Greater care can be exercised if sheep are watered at troughs. Low muddy tanks and dams are a source of danger in drought.

Reserve Stocking
The effects of drought may be delayed if stocking is regulated below full carrying capacity, so that pasture and fodder plants can be spelled for long periods to allow recovery. By this means plants last longer into a dry spell and recover more quickly. Further, adequate soil cover is provided which protects the soil from eroding winds, maintains the surface layers and assures quicker germination of seed and ultimate recovery. If light stocking is carried out within reason, the smaller number of sheep

carried will produce heavier fleeces and more lambs, and will maintain such condition as to carry them well into dry periods. In those districts where conserved fodder cannot be grown and stored in bins and stacks, it is fundamental to preserve and keep in healthy, recoverable condition the edible shrubs and pasture species.

Fodder Conservation

In those districts where crops and pastures can be grown and properties are of moderate size to allow economical feeding, a reserve of hays and grains should be maintained for feeding stock during temporary or prolonged feed shortages. To a large extent, these reserves of fodder should be cereal hays, grains or meadow hay. Where conditions allow, as much as possible should be conserved in the form of leguminous roughage. Rich in protein and vitamins, legume hay or silage is invaluable in balancing rations, keeping stock healthy.

In many districts of light rainfall fodder conservation is impracticable. Unreliable seasons, the lack of farming labour and the huge reserves needed to maintain the large numbers of sheep make such a scheme impracticable. Some fodder conservation may be carried out by buying up stocks of grains and hays in good seasons, but the main solution to the problem will depend on careful stocking.

Indications for Drought Feeding

The woolgrower whose country is drought-stricken may consider one of several methods of meeting the situation. These include:
1. moving of sheep to agistment;
2. feeding edible scrub;
3. hand feeding;
4. selling part or all the stock;
5. allowing the sheep to take their chance in the paddock.

Seeking Agistment

Suitable agistment areas are not always available, and even if they do exist it is often impossible to move sheep to relief pastures; again, rents for agistment are usually high during drought periods. If relief country can be found, stock routes are short of feed and trucking difficulties are accentuated. Only strong, grown sheep should be sent away.

Feeding Edible Scrub

This or agistment is usually the cheapest method. However, the more valuable edible trees and shrubs such as kurrajong and mulga are restricted in their distribution and there are large areas of Australia where scarcity of top feed makes scrub feeding impossible. When these circumstances arise, the owner has to decide which of the other three methods he will adopt and to what extent he will use the one selected. Accordingly, in making a decision about hand-feeding, consideration has to be given to:
1. Which sheep will be fed?
2. What is the market value of the sheep?
3. What is likely to be the market value of the sheep at the end of the drought?
4. For how long is it probable that feeding will have to be undertaken?
5. What is the availability of foodstuffs, and are supply, transport and labour assured?
6. What will be the cost of feeding?

The answer to (1) is obvious — the young breeding ewes must be given first preference. From them the flock can be rebuilt when the drought breaks; (2) and (3) depend on conditions at the time. If the prospect is for a well-sustained wool market, it is likely that the sheep will maintain their value. If, however, the wool market falls suddenly, sheep would tend to lose value quickly.

Hand Feeding

Sheep should be hand fed in as small mobs as possible. It may pay to separate poor doers and timid sheep and feed them separately. The question of weaning lambs must be considered. If sufficient feed is available, lambs will do better if left with their mothers. If feed supplies are limited, it may pay to separate the lambs from their mothers and even slaughter the wether lambs. Thus the best feed and conditions can be reserved for the growing ewe lambs. They will receive less knocking about if they don't have to compete with older sheep for food, while the dry ewes can often survive on the barest minimum.

Choosing Drought Rations

The energy value of the various feeds shown in Table 27.1 is based on the metabolisable energy (ME) contained in each feed and is expressed in megajoules (MJ) of ME per kilogram of feed corrected to 90 per cent dry matter. For example, grain with 90 per cent dry matter contains an average of 12 MJ of ME per kg.

Table 27.1   Average Nutritive Values of Various Feeds.

Feed type	*Energy value (corrected to 90% dry matter)		Crude protein (% of dry matter)
	Average	Range	
Grains			
Corn (maize), grain sorghum, wheat, barley and oats	12.0	10.5–13	8–16
Grain by-products			
Sheep nuts	10.0	8–12	
Wheat pollard	10.0		15
Wheat bran	8.5		15
Oat bran	7.5		8
Rice bran	10.0		6
Roughages			
Lucerne hay	9.0	7–9.5	15
Clover hay	8.5	8–9	13
Pasture hay (mostly clover)	8.5	7.5–9	11
Pasture hay (mostly grass)	7.0	6–8	8
Bush hay	6.0	5–7	
Good cereal hay	8.5	7.5–9	8
Poor cereal hay	6.0	5–7	5
Oat, barley or wheat straw	5.0	4–6	4
Lupin hay	8.5	7–9	15
Protein rich meals			
Meatmeal	10.0		40–55
Fish meal	10.5		55
Meat and bone meal	9.0		40
Peanut meal	11.0		42
Cottonseed meal	9.5		41
Linseed meal	10.5		30–35

Table 27.1   Average Nutritive Values of Various Feeds (cont.)

Feed type		*Energy value (corrected to 90% dry matter)		Crude protein (% of dry matter)
		Average	Range	
Coconut meal		11.0		21
Sunflower meal		10.5		40–45
Safflower meal		10.0		45
†Succulents	†			
Pumpkin, swede, turnips	10	12.5		13
Potatoes	20	12.5		10
†Green fodder and silage	†			
Lucerne and clover green feed	25	9.5	8–10	16
Lucerne and clover silage	25–30	8.0	7–9	15
Young oat, wheat, barley green feed	25	8.5	7.5–9.5	10
Maize, sorghum, millet green feed	25	8.5	7.5–9.5	4–6
Maize, sorghum, millet silage	25–30	8.0	7–9	4–6
†Waste products	†			
Citrus pulp	10	12.5		10
Molasses	75	12.5		6
Poultry manure	90	Nil		Equivalent to 30–50
Poultry litter	90	Nil		Equivalent to 20–30

* Energy values are expressed in megajoules of metabolisable energy per kg of feed.
† Values for energy for waste products, succulents, green folder and silage are for the dried material only. To calculate how much to feed, calculate as dry material then divide by DM per cent to get wet weight, for example, citrus pulp:

$$\text{need } 350\,\text{g dry} = \frac{350 \times 100}{10} = 3500\,\text{g wet}$$

This system is the most useful for drought feeding calculations because it takes into account the digestibility of feed in relation to its total energy content and gives an accurate assessment of the energy actually available to the sheep. Feed with a high energy value (grain) is more digestible than low-energy feed (poor hay); therefore a greater proportion of the total energy is available to sheep from grain than from poor hay. For example, a 35 kg sheep would require 430 grams of grain per day or 1000 grams of poor hay — more than twice as much feed with half the energy value.

The energy requirements of sheep vary with different physiological states (pregnancy, lactation) and with weather conditions. Liveweight and age also influence energy requirements; larger and older sheep need less feed in relation to liveweight than smaller and younger sheep. For example, a mob of dry sheep with average liveweight of 40 kg *do not* require twice as much feed as a mob of weaners at 20 kg liveweight.

All the variables in feed quality and sheep requirements have been included when calculating the various tables. The feeding levels include an allowance for the energy expended in normal walking and will be sufficient to maintain sheep at the selected liveweight. If more feed is provided, sheep will stabilise at a higher weight; with less feed, sheep will lose weight and, if this is too severe, they will die.

The total amount of any feed a sheep can eat, and digest, largely depends on the

Type of sheep	How often to feed	Type of feed	Crude protein requirement %	Essential mineral supplements	Vitamin 'A': Note — This is supplied by green hay, scrub and yellow maize
Dry adult ewes and wethers.	Once weekly — not more than twice weekly	Grain alone is satisfactory. Keep hay supplies for cold weather.	6	Ground limestone with grain	Not needed for first 12 months of drought
Pregnant ewes	Every 3rd day	Grain alone is satisfactory; some good hay for cold weather as required	8–10	Ground limestone with grain	Supplement before mating or lambing if intake was nil for 8 months.
Ewes and lambs	Every 2nd day	Grain plus hay is better than either grain or hay alone	10–12	Ground limestone with grain	Supplement at lambing if no intake or other supplement for 8 months
Weaned lambs fed for survival	Daily, trough or self-feeders	Equal weight of grain, and hay; protein meal with poor quality hay.	10–12	Ground limestone with grain	Supplement at weaning if intake was nil from birth.
Older weaners for survival	Every 2nd day in trough or self-feeders	Grain alone satisfactory. Wheat can be used with safety, some hay in cold weather.	10	Ground limestone with grain	Supplement if no intake for 6 months
Weaned lambs fed for maximum growth	Every day in trough or self-feeders	Equal weight of grain and hay (preferably oat grain and lucerne hay). Some protein meal with poor quality hay.	15	Ground limestone with grain. Salt when drinking water is low in sodium.	Supplement at weaning if intake was nil from birth
Rams	Twice weekly	Grain and good hay	8	Ground limestone with grain. Salt when water is low is sodium.	Supplement 2 months before mating if no intake for previous 4 months.

digestibility of that feed. Feed which is readily digested is consumed in greater quantities than feed of low digestibility.

Energy value and digestibility of feed are closely related, and the high-energy feeds (for example, lucerne hay and grain) are generally more digestible than low-energy feeds (for example, poor pasture hay and cereal straw).

Consumption of low-energy feed is so severely restricted by poor digestibility that sheep are unable to eat sufficient of feed such as cereal straw to provide their basic energy requirement for maintenance. However, the amount of lucerne hay that dry sheep can consume will provide more than sufficient energy for survival.

To prevent losses from exposure in time of drought, adhere to the following guidelines:

- Keep the most susceptible sheep in the most sheltered area.
- Use hay to provide additional energy. Sudden increases in grain feeding could cause grain poisoning and deaths, but an additional ration of hay can be fed with safety.

Because some of the energy in hay is more readily available than that from whole grain, sheep will obtain heat from hay sooner. The less digestible portions of hay will supply energy and heat for a longer period than grain.

How Often to Feed

Experiments and observations on properties show that feeding dry, grown sheep once or twice weekly can give results at least as good (and often better) than daily feeding. However, pregnant and lambing ewes and young weaners require more frequent feeding.

Frequency of feeding must depend on several factors — type of sheep, type of feed, availability and capacity of troughs and self-feeders, numbers to be fed, whether feeding in distant paddocks or in yards, possible loss of feed to kangaroos and birds, and weather conditions.

28

CROSSBREEDING AND
FAT LAMB RAISING

There are three types of lamb production, broadly in two groups: intensive and intermediate.

Intensive Lamb Raising

1. Lambs from first crossbred ewes and a downs breed ram. These ewes are produced from the type of breeding described in 'lambs from merino ewes' (see (3) below). Intensive lamb raising is a specialised type of production to produce lambs for home consumption. This section deals with this aspect of the industry.

Intermediate Lamb Raising

2. Lambs from grade corriedale or first-crossbred ewes using corriedale, border leicester or romney marsh rams.

3. Lambs from merino ewes using border leicester rams. This type of breeding is usually carried out in marginal country. In favourable seasons a large number of these lambs are marketed as fats; otherwise they may be shorn and taken to irrigation areas for fattening. The ewe portion of this cross is always in demand by lamb breeders for use as dams, and are described as border leicester first-cross.

These two intermediate groups produce large numbers of marketable lambs when pastures are favourable. This lamb-raising in conjunction with woolgrowing or crossbreeding is carried out in the areas marked on the map (Figure 28.2) as intermediate lamb areas.

The wide variation of climate and rainfall in New South Wales makes it necessary to use a number of breeds. Lamb raisers in New Zealand can confine their breeding mainly to romney marsh ewes and southdown rams for marketing a standard type of lamb. Most lamb raising areas in New South Wales are in the winter and non-seasonal rainfall belt where pastoral conditions are at their best during late winter and spring. From the middle of October, summer temperatures, pasture growth and quality declines on the Slopes and in the Riverina.

In these areas, grass seed such as barley grass (*Hordeum leporinum*), corkscrew grasses (*Stipa* spp.), and wire grasses (*Aristida* spp.) is formed in November and may become a problem in ewes and lambs. Unless provision is made for grass seed-free paddocks, the lambs must be marketed by the end of October. Otherwise, they are likely to receive a check, and it will be necessary to carry them on. These are the type of lambs that are marketed between February and May. In grass seed areas lambing is often timed for late autumn and winter to take advantage of the favourable pastoral conditions. Marketing begins in July, increasing to a peak in October–November and decreasing sharply in December. These lambs are designated in the trade as 'winter' or 'spring' lambs.

Production of summer lambs is restricted to the Tablelands and the irrigation areas, where green feed in summer can finish lambs, and grass seed is no problem. These lambs are usually marketed in January, February and March. This form of production will extend with the development of the various irrigation schemes along the western and southern river systems, and with increased pasture improvement on the Tablelands areas. Lamb marketing is now fairly evenly distributed throughout the year, but peaks of production in October–November are the result of a favourable winter and spring. In June, July and August, quality lamb is sometimes in short supply and prices may rise.

Type of Lamb Required

Although New South Wales does not export any large quantities of lamb, the standard of production is based on export requirements. These standards should be maintained so that lamb surplus to local requirements will be suitable for export. Export lambs are graded into three grades, each of which includes four weight ranges — 9 to 13 kg (20 to 28 lb), 14 to 16 kg (29 to 36 lb), 17 to 19 kg (37 to 42 lb) and 20 to 23 kg (43 to 50 lb) frozen weights.

Conformation

The ideal lamb carcass should have a maximum amount of meat with a minimum of waste in the form of bone and fat. A blocky carcass with the meat on all four quarters is required, but emphasis should be placed on the fleshing of hindquarters and loins. When hanging on the hooks the carcass should show a definite 'U' shape between the hind legs, not a 'V'. The market demand is for lightweight cuts, and lambs weighing 14 to 18 kg (30 to 40 lb) dressed weight are the most suitable.

Finish

The carcass should have a high percentage of lean meat covered with a fine layer of fat. Lack of this fat covering causes drying and 'blueing' of the carcass during storage, but excess fat is wasteful and lowers carcass value. Lambs should not become overfat because of late marketing. Fat is developed successively as kidney and caul fat then fat between the skin and flesh followed by 'marbling' fat, or fat dispersed between the muscular fibres.

Quality

Quality in the carcass is indicated by the grain and distribution of the fat. A lamb properly finished will show a good even covering of fat, and marbling fat evenly spread throughout the meat. One of the most important factors in the finish, this can be obtained only by correct breeding and feeding and avoiding any check to the growing lamb.

Quality in the live lamb is shown by general appearance and finish. Lambs which have received a check and have recovered lack the bloom and desirable shape of lambs properly fed. This is evident to experienced buyers.

Growth and Development

The importance of nutrition in the production of quality lambs cannot be over-stressed. Knowledge of growth and development and correct feeding at all stages will produce the desired result in the finished lamb.

After birth, the lamb's growth rate and body development follow a well-defined order: development of bone (or the skeleton), growth of muscular tissue (or lean meat), development of fat.

When a lamb is born, the bones of the head and legs are well developed. They continue to grow, and reach their maximum size before other parts of the body. The essential organs — the heart, digestive organs, lungs, brain and internal tissues — also develop early. Skeletal development is rapid, but muscular tissue develops slowly, and fat is formed later as maturity approaches.

The most valuable parts of the carcass are the hindquarters, which are usually last to develop. Should the lamb receive any check, the hindquarters suffer first, nourishmen going to the vital organs and skeleton. In the improved British mutton breeds, however, especially through selection, early development of the hind quarters has been secured. It is this, amongst other early maturing characteristics, which make the British downs breeds of sheep so valuable for lamb raising.

Factors Affecting Development

In sheep there are early and late maturing breeds, the British downs breeds representing the early types, and the merino the late. Intermediate to these are the English long-wool breeds and crossbreds. A knowledge of these breeds is necessary to select suitable stock for lamb breeding, where early maturity is so important.

Birth Weight

Birth weight will depend on breeding-nutrition of the ewe during gestation, age of the ewe, sex of the lamb, and whether a single or multiple birth. The last six weeks of pregnancy are important, when good feed is needed even more than during early pregnancy. The individual weights of twin lambs are about 16 per cent less than single lambs. The combined weight of twin lambs is over 60 per cent higher than singles.

Growth rate indicates that, depending on breed, each .454 kg (1 lb) of birth weight above 4 kg (8 lb) will mean between 1 and 1.5 kg (2.5 and 3.5 lb) at twenty weeks of age. For example, if a 4 kg (8 lb) lamb weighed 39 kg (85 lb) at twenty weeks, then a 5 kg (10 lb) lamb at birth would weigh 40 kg (90 lb) to 41 kg (92 lb) at twenty weeks. A good birth weight for singles, depending on breed and management practices, is 5 kg (10 lb). Lambs heavier than this may cause trouble at lambing.

Rate of Gain in Weight

Rate of gain is important in determining the marketing of the lamb. Normal rate of gain is rapid during the first six to eight weeks and will vary from 2 to 2.5 kg (4 to 5½ lb) per week, after which there is a decrease to 1.5 kg (3 lb) and under per week. For the first six to eight weeks the lamb depends on milk for its nourishment, then progressively more of its food comes from grazing. At this stage it is very sensitive to the quantity and quality of the feed received. Poor nutrition will quickly cause slowing-down of the rate of weight gain, and may even stop it. Recovery from such setbacks is slow and the lamb does not attain its former desirable shape or bloom until a much later age.

The ewe's maximum milk yield is reached about the third week after lambing, then it steadily decreases. At ten to twelve weeks after lambing, the yield has dropped to about 25 per cent of the maximum yield, and only supplements the needs of a fast-growing lamb.

Mature ewes and ewes rearing twins produce more milk than young ewes. Twin lambs each receive about two-thirds the milk single lambs get and, with lighter birth weights, this explains their slower growth. They need preferential treatment if early marketing is desired. Thus good management of the pregnant ewe is essential. The greatest demands on pregnant ewes are in the last six weeks of gestation. During this period they should gain in body condition, as distinct from body weight. The foetus and membranes together weigh about 9.5 kg (21 lb) and most development takes place in the last six to eight weeks of gestation. Quality of feed is most important at this stage. Good pastures or fodder crops are the best feed for the prelambing build-up and for subsequent high milk yields.

The effect of nutrition on milk is shown by the following work carried out in New Zealand on 200 romney marsh ewes of fixed ages (Table 28.1).

Table 28.1   Nutrition and milk yield

Group	Average yield [g (oz.) per day] for ewes with:	
Nutritional plane	Twins	Single
High plane before lambing and during a 12 weeks' lactation	1698 g (60)	1274 g (45)
Low plane before lambing and during a 12 weeks' lactation	877 g (31)	623 g (22)

Bad nutrition will increase the death rate of new-born lambs, slow up the growth rate of lambs, decrease wool cuts, and increase the proportion of tender wool. It will also increase ewe deaths from nutritional diseases, such as hypocalcaemia, hypomagnesaemia and pregnancy toxaemia.

Selection of Ewes and Rams

The parent stock should have the essential qualities required to breed quality lambs. Good management and feeding cannot overcome bad or unsuitable breeding stock. The essential qualities required for ewes are: favourable mutton qualities, early maturity, high fertility, high milk production and good wool production. The suitability of breeds and crossbreeds available in New South Wales is as follows.

Merino

Pure merinos are unsuitable for lamb production compared with other breeds and crosses; they are slow in maturing, low milk producers, give low lambing percentages, and their lambs are poor in shape and depth of carcass. The merino is nervous, easily excited and frightened, and is a more selective feeder than the crossbred. Merino ewes, however, are readily available, early mating, and they give high wool returns.

British Breeds

British breeds are divided into two distinct groups, the long-wools, and the short-wools, or downs breeds. These two groups, because of their low wool return, and because they are not readily available, are used as pure breeds for production of rams. They are valuable for mating to the merino or to crossbreds for production of lambs.

Long-wool Breeds
The long-wool group consists of the lincoln, English leicester, border leicester and romney marsh — the wool sheep of England. In New South Wales the rams are used to mate with merino ewes to produce crossbreds. The crossbreds are valuable in the industry for the breeding of prime lambs.

Short-wool or Downs Breeds
Short-wool or downs breed sheep are noted for their inherent mutton qualities. There are many different breeds in this group and some of them have been introduced over a period of time. The main breeds at present in New South Wales are the dorset horn and southdown.

Crossbreds
The merino ewe, when mated to one of the British long-wool breeds, produces an excellent type of crossbred ewe for breeding lambs. This crossbred ewe combines the mutton qualities, early maturing and high lambing percentages of the long-wool breeds with the high wool production of the merinos. Crossbred ewes available are border leicester × merino, romney marsh × merino, dorset horn × merino.

Border Leicester × Merino
Border leicester × merino sheep are bred in the intermediate lamb-raising areas and are noted for being vigorous sheep, able to move about in the hotter climate. They fatten readily; their clean heads and legs are not troubled by grass seeds to the same extent as other breeds; they give high lambing percentages; they are very good mothers and will continue to nourish their lambs under adverse conditions; and they give a good marketable fleece of medium-fine quality crossbred wool. These ewes mated to downs breed rams produce high numbers of early maturing marketable lambs.

Romney Marsh × Merino
The romney marsh × merino, along with the border leicester cross, is the most popular crossbred in New South Wales. It is usually bred in the higher rainfall areas but has a tendency to 'dry off' under adverse conditions. It is more compact and blocky than the border leicester cross, and its main features are a very good mutton conformity and ability to combat wet conditions. It can fatten readily, particularly as a hogget, and gives a better wool return than the border leicester.

Lincoln and English Leicester
The lincoln and English leicester were used in the early stages of the industry in New South Wales, and the English leicester × merino is still used in the irrigation areas. Compared with the romney or border cross they are much slower maturing and the wool return is considerably less. Neither of these two breeds can compete with the border leicester or romney marsh under New South Wales conditions.

Dorset Horn × Merino
The dorset horn × merino cross is becoming popular in the hotter parts of the state, particularly in the irrigation areas. This cross is prolific, early mating and early maturing — especially when mated to the dorset horn ram. The wool return from this cross is less than from the other crosses and the rate of lambing deaths has been higher than in other crosses. The survival percentage is good because of many multiple births.

Crossbred ewes are usually purchased when they are 1 to 1½ years old, where most breeding is in the zone marked 'Intermediate' on the map (Figure 28.2). The general practice is to buy a proportion of young ewes each year as replacements. These ewes are often in short supply and competition is keen, so many breeders purchase them as weaners (nine months of age) and so ensure their requirements. However, stocking these young ewes while they are maturing for breeding poses problems. Lamb breeders normally do not breed their own crossbred ewes, as a flock of merino ewes and British long-wool rams — and the country necessary to run them — would be needed. The average lamb-raising property of 200 to 600 ha (500 to 1500 acres) is suitable for one flock only. The specialised lamb raiser on a highly developed property prefers to carry the maximum of crossbred ewes.

Australian Breeds

Corriedale

The corriedale is the result of crossing English long-wools with the merino and breeding to a type from selected progeny. The ewes are suitable dams for lamb raising but will not give the growth rate in lambs from first-cross ewes. The wool return is higher than from crossbred ewes.

Polwarth

Polwarth are a mixed merino comeback. Because of mostly merino blood, this breed will mate early, but it is slower to mature and will not produce the quality of lambs of a crossbred ewe. Their wool return, however, is considerably better than the crossbred.

Choice of Rams

Short-wool British rams are most suitable for production of prime lambs, as they are early-maturing and have the desirable depth and quality of flesh required for meat production. Rams over three years old should not be relied on.

It is suggested that 25 per cent of the rams should be young. Of the many breeds, experiment and experience have shown that the dorset horn and southdown are the most satisfactory. The dorset horn ram will give the earliest marketable lambs, and the southdown the desirable export quality lamb. Other breeds used are the ryeland, shropshire and suffolk.

Marketing

Marketing of lambs in New South Wales is carried out by direct paddock sales and marketed when they are ready. The tendency to keep lambs for a price rise or for the extra weight is not in the interest of the breeder as the big, heavy lamb is usually over-fat. Price per kilogram drops sharply for overweight and over-fat lambs. With variation of weights and age amongst the lambs, lots should be marketed as sufficient lambs become available.

If necessary, lambs should be dagged and cleaned around the crutch and kept off green feed some hours before trucking; otherwise they are likely to foul one another and get very dirty which detracts from their appearance in the sale pen.

Marketing of lambs in New South Wales is carried out by direct paddock sales to wholesalers, or auction sales at country or metropolitan saleyards. Weight and grade at abattoir marketing, which is the most satisfactory, is not used to any extent in New South Wales.

In New Zealand, practically all lambs go direct from farms to abattoirs and are sold on a weight and grade basis. New Zealand, with some forty abattoirs, consistent seasons, stability of numbers and even type of lamb, has developed this method of marketing. The advantages of this method of direct marketing ensure full market value to the breeder of quality lambs. The opening of country abattoirs, increased

An example of top quality prime lambs

Photograph: courtesy The Land

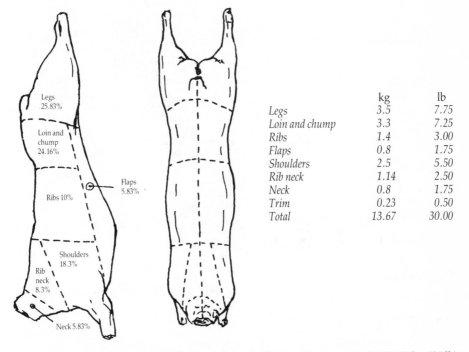

	kg	lb
Legs	*3.5*	*7.75*
Loin and chump	*3.3*	*7.25*
Ribs	*1.4*	*3.00*
Flaps	*0.8*	*1.75*
Shoulders	*2.5*	*5.50*
Rib neck	*1.14*	*2.50*
Neck	*0.8*	*1.75*
Trim	*0.23*	*0.50*
Total	*13.67*	*30.00*

Figure 28.1   Conformation and dissection of a lamb. Carcass weight 13.67 kg (30 lb).

irrigation and pasture improvement are making conditions better for selling on a weight and grade basis.

Breeders aim to market their lambs by the time they are three and a half to four months old and dressing from 14 to 18 kg (30 to 40 lb). Marketing between these weights gives the breeder some scope to meet the fluctuations of the market which occur because of supply and demand. When lambs are plentiful, the lighter and more shapely quality lamb will always give better financial returns. Lambs over 18 kg (40 lb) dressed weight are becoming too big and the increased weight is mainly fat.

In lambs weighing between 14 and 16 kg (30 to 36 lb) dressed weight, there is very little wastage and fat is at a minimum. This type of lamb is always in demand as it provides the popular small joint with a maximum of flesh to fat and bone. Lambs should not be dirty, as this detracts from their appearance in the sale pen. All sheep and lambs should enter abattoirs free from dags and soilage and in a clean and attractive condition.

Care during the drafting and loading of lambs is necessary to prevent bruising and other injuries. The use of sticks and biting dogs, and grasping the wool, cause most of this kind of damage. Buyers allow for wastage of this nature in their estimates and

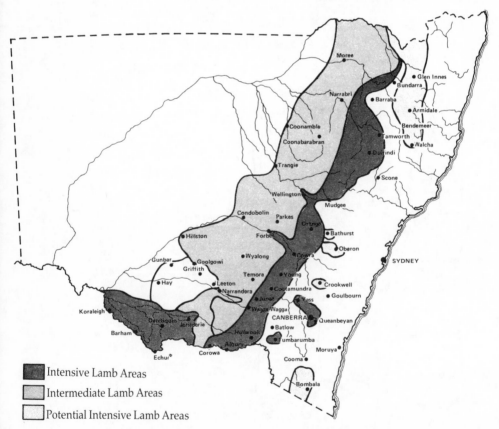

Figure 28.2   Lamb-raising areas in New South Wales

will not give maximum prices for lambs badly handled. Lambs should not be loaded straight into railways trucks after being driven or after a long road trip. They should be given time to settle down in the yards. If hot, water the lambs before loading into railway trucks.

Overloading trucks must be avoided; otherwise, lambs get down and are unable to get up and are trampled and knocked about by the other lambs. A yard 5.26 × 24.45 m (17 ft 3 inches × 8 ft 2 inches) is useful for estimating the number of lambs to fill a sheep van deck.

Shearing Lambs
Shearing is seldom necessary or desirable when lambs are marketed direct from the ewes because shearing spoils the appearance of lambs and involves bruising. Buyers are not eager to buy shorn lambs for this reason.

29

ANATOMY AND PHYSIOLOGY
OF SHEEP

The sheep is a ruminating or cud-chewing animal and, in common with other ruminants, it has four stomachs. Distinctive features are its ability to produce wool, the cleft upper lip, which enables it to graze closely, and the special glands (interdigital glands) between the claws of the foot. While, as a group, sheep have a very wide environmental range, they have certain characteristics in common and, as in other animal groups, there are a number of different types which have been developed under domestication for special purposes. Thus in Australia we have wool types and mutton types and, intermediate between these, dual-purpose types.

Anatomically we can distinguish several distinct systems, each with their own specific physiological functions. Although in these notes greater attention is paid to the digestive system, it does not follow that any one system is any more important to the animal than another.

Externally we usually distinguish four parts of the body, the head, neck, body or barrel, and legs.

The head
The head is very strongly formed, as its bones are flat, short and thick. The cranium encloses the brain and the face encloses the cavities of the mouth and nose. There are several sinuses or 'air spaces' in the bones of the head and while this special construction of the head gives added protection to the brain, these sinuses may be a source of trouble from the nasal bot fly. In addition to the skull proper, there is the lower jaw which works from a joint near the base of the ears. It is important to note that in the sheep the upper jaw contains molar teeth only, and the lower both molars and incisors, the latter biting against a hard pad in the upper jaw. The eruption of the per-

manent teeth serves as a means of determining the age of a sheep between rather broad limits and while age marking is a precise means, a mention of the normal eruption would be to the point at this stage. The complete set of teeth numbers thirty-two, with eight incisors in the lower jaw and twelve molars in each jaw.

When born the lamb usually has no teeth. Within a week after birth, the milk teeth or temporary teeth appear in the front lower jaw and by the time the lamb is two months old these, eight in all, have erupted. At twelve to nineteen months, the two central incisors are replaced by two permanent incisors and the animal is spoken of as being a two-tooth. At eighteen to twenty-four months the two milk teeth, one on each side of the central incisors, are replaced by permanent incisors (middle incisors), the animal then being a four-tooth. At twenty-three to thirty-six months the next two permanent teeth (laterals) erupt and we have the six-tooth, and finally at twenty-eight to forty-eight months the last two permanent teeth (the corner incisors) erupt, the sheep being eight-tooth or fresh full-mouthed. First, second and third temporary cheek or molar teeth erupt at birth or soon after and the approximate times for the appearance of the permanent molars are:

- fourth molar — three months
- fifth molar — nine months
- sixth molar — eighteen months

The first and second temporary molars are replaced by permanent teeth at about 24 months.

The state of the teeth is the accepted rough guide as to age, but the teeth cannot be considered as an accurate means of estimation, for their condition depends to a great extent on the type of country on which the sheep have been grazing, and upon breed. Climate also is believed to have an effect, but this factor would be mainly reflected in the nature of the feed available. Well-nourished sheep usually have much

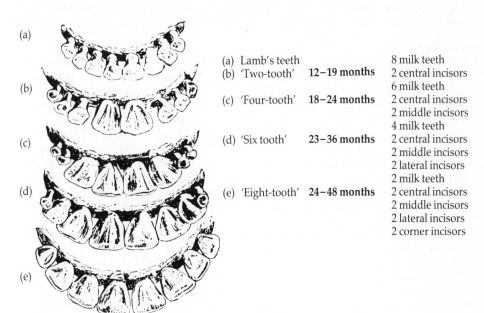

(a)	Lamb's teeth		8 milk teeth
(b)	'Two-tooth'	**12–19 months**	2 central incisors
			6 milk teeth
(c)	'Four-tooth'	**18–24 months**	2 central incisors
			2 middle incisors
			4 milk teeth
(d)	'Six tooth'	**23–36 months**	2 central incisors
			2 middle incisors
			2 lateral incisors
			2 milk teeth
(e)	'Eight-tooth'	**24–48 months**	2 central incisors
			2 middle incisors
			2 lateral incisors
			2 corner incisors

Figure 29.1   Appearance times for teeth

better teeth than those under hard conditions and animals that are forced to graze closely on sandy or quartz country will show early signs of wear. Sheep fed a prolonged calcium-deficient diet develop dental abnormalities which can be avoided by feeding a supplement of finely ground limestone.

After the full-mouth stage at four years, it becomes more difficult to tell the age of the animal, but as a guide, it can be remembered that at five years the two central incisors usually show a space between them. At six years, these two teeth have become broken and the two adjoining incisors will separate. The sheep is then described as being 'broken-mouthed'. However, it is not altogether rare for a sheep to have a sound mouth at twelve or even fourteen years, so that it can be readily appreciated why the conditions of the mouth are but a rough guide to age.

The neck

The neck is short and powerful, particularly in the male. The gullet is large and this, together with the sheep's method of feeding, makes choking a rare occurrence. The thyroid gland in the neck is sometimes found to be enlarged (particularly in lambs) and this condition is indicative of iodine deficiency. Besides the windpipe and gullet, the neck also carries the carotid arteries and jugular veins which carry blood to and from the head.

The body or barrel

The barrel or body houses the vital organs, the digestive tract and the generative system. The backbone or vertebral column traverses the length of the barrel and in the thoracic region thirteen pairs of ribs — in some cases fourteen — curve downwards from the backbone uniting with the breast bone by means of cartilages which become increasingly hard with advancing age. The thirteenth rib is not attached to the breast bone but is 'floating', as is the fourteenth, if present. The ribs, backbone and breast bone form a bony cage housing the lungs and heart, separated from the belly or abdomen by a thin sheet of muscle tissue — the diaphragm.

The legs

The legs are relatively short and strong, the length varying with breed, the less improved breeds having longer legs than the more improved mutton types. It is interesting to note that the length of the cannon bones in the legs is closely correlated with the depth of fleshing in the carcass, for example, the short cannon of the southdown breed is associated with deep fleshing of the leg and other parts of the body.

The hoof is cleft into two claws and is lubricated from the interdigital gland. The ability of the sheep to forage is dependent upon its ability to get about, making its legs and hooves of special importance. Bent and twisted legs (e.g. cow hocks) and feet and poor horn can always be regarded as features on which an animal would be culled. Good hard horn is associated in the breeder's mind with resistance to foot-rot, as is dark colour, although there does not appear to be any factual basis for this latter contention.

The several systems referred to earlier are:
1. skeletal system;
2. muscular system;
3. nervous system;
4. circulatory system;
5. respiratory system;
6. digestive system;
7. lymphatic system;

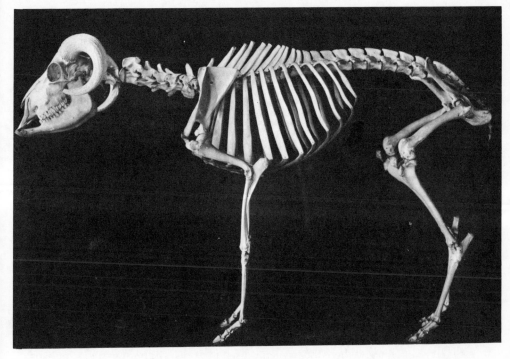

Skeleton of the sheep

8. urinary system;
9. generative system.

Skeletal system
The skeletal system is the bony framework of the body serving for protection and support of the softer and more delicate parts of the animal body. It consists of a number of bones which are composed chiefly of calcium carbonate and calcium phosphate.

Muscular system
The muscular system is made up of the muscles of the body and is really the 'flesh' or meat. They are attached directly to, or through tendons to, bones and it is by their contraction and elongation that movement is made possible. The majority of the muscles are voluntary (i.e. they are under control of the animal's will) but there are some muscles such as those of the heart and intestines, over which the animal has no control, these being referred to as 'involuntary' muscles.

Nervous system
The nervous system is centred in the brain and from this and the spinal cord, which passes right down the backbone, numerous nerves are given off which pass to the various parts of the body. Nerves are agents for conveying sensations and transmitting to the brain all the impressions made on the sensory organs by which consciousness is attained. They stimulate the muscles and thus indicate movement.

Circulatory system

The circulatory system controls the supply of blood to all parts of the body, the heart being the centre of the entire system. The heart is a four-chambered muscular organ that functions as a powerful pump. Each side has two chambers; the upper chambers, or auricles, act as receiving chambers; the lower, or ventricles, acting as delivery chambers. Purified blood returning from the lungs passes to the left side of the heart and is pumped to all parts of the body through the arteries. Near the heart the arteries are large, but soon branch, decreasing in size progressively until each tissue is rightly supplied with small, capillary vessels. These capillaries in turn unite to form larger vessels, the veins, which return blood to the right auricle. The right ventricle pumps the venous blood to the lungs for purification.

Respiratory system

The respiratory system begins with the nasal cavities, the entrance to which is formed by the nostrils. The cavities are connected through the pharynx and larynx to the wind pipe, or trachea, which extends down to the lungs. The lungs occupy most of the chest cavity, are soft and spongy and are divided into various lobes. The chief muscle of respiration is the diaphragm, which causes the lungs to expand and contract allowing air to be drawn in or to be expelled, the lungs thereby functioning somewhat like bellows. In the lungs air charged with oxygen is brought into close contact with blood and a gaseous exchange takes place. The red colouring matter of blood, the haemoglobin, has a great affinity for oxygen which is taken up and the gas carbon dioxide is displaced to be passed out of the body on exhalation. The blood assumes the rich colour characteristic of arterial blood and passes to various parts of the body via the heart where the oxygen is used to oxidise, or burn tissue, supplying heat and energy to the body. The by-product, carbon dioxide, passes into the blood eventually to be expelled in the lungs.

Digestive system

The feature of the digestive system in ruminants is that it is designed to cope with fairly rough and bulky types of food and it provides for a double mastication. These features are developed to a greater degree in cattle than in sheep, hence the fact that cattle can deal with roughage more successfully than can sheep.

Starting at the mouth, food passes down the gullet to the stomachs, which are four in number and account for approximately two-thirds of the total capacity of the gut. The four sections are known as the rumen, reticulum, omasum and abomasum.

1. The *rumen* or *paunch*, which serves as a storehouse for food, is the largest and has a capacity of approximately 14.1 litres (three gallons). It occupies almost the whole of the left side of the abdominal cavity and extends to the right side. All solid food passes into the rumen and from here it is regurgitated to the mouth for further mastication when the animal is at rest (chewing the cud). While in the paunch, various types of microflora are active in breaking down the food material.
2. The *reticulum* is to the right side of the rumen and is the most forward of the four stomachs. The lining membranes are arranged in honeycomb formation and give a characteristic appearance. Here food is still further reduced to a more finely divided state. By muscular movement, the animal has the power of passing food from the gullet either into the rumen or into the reticulum.
3. The *omasum*, or Bible, lies to the right side of the abdomen and is characterised by the lining membranes being raised in folds (likened to the leaves of a book) between which the food is subjected to a grinding action.
4. The *abomasum* is the true stomach inasmuch as it is here that we find the various

digestive juices being secreted. The gastric or digestive juice contains several ferments which have specific effects in breaking down the food materials. Thus it can be seen that the function of the rumen is largely one of a storehouse. While the action of the reticulum and omasum is mechanical in nature, that of the abomasum is chemical.

From the abomasum food passes into the intestines which can be divided into two sections, the small intestine, which is about 24 m (80 ft) long and runs from the abomasum to the second section, the large intestine, which terminates at the anus, and is about 5 metres (18 ft) long. The first part of the small intestine is folded on itself and is distinguished as the duodenum. It is there the bile from the liver and pancreatic juice from the pancreas enter the intestine. These secretions have important functions in the digestive process.

Small elongated bodies, villi, which are richly supplied with blood vessels, project into the intestines from the intestinal walls and their function is to absorb food nutrients passing through the bowel. The small intestine is more richly supplied with villi and consequently most absorption takes place in this organ. As the food passes through the intestines it becomes progressively less liquid until the waste material at the latter end of the large intestine is, in the healthy animal, voided as relatively solid matter.

The liver plays an important part in the digestive system as it not only secretes the digestive juice bile, but it also functions as a storehouse of nutrient material (mainly glycogen or animal starch) which is liberated in the bloodstream as required. It is also the main organ for the removal of waste products from the blood which are later filtered out and eliminated through the kidneys.

The purpose of the spleen, which lies close to the stomach, is not fully understood, but it has an important function in supplying the colourless corpuscles (leucocytes) to the blood. The supply of red corpuscles is a function of bone marrow.

Lymphatic system
The lymphatic system consists of a series of thin-walled vessels which drain from all parts of the body to two central vessels. The lymphatics play an important part in digestion by taking up nutrients and carrying them into the tissues. Numerous lymphatic glands are distributed throughout the lymphatic vessels and these act as filters arresting the products of bacterial activity when these invade the lymph system. (You have, no doubt, experienced swelling of the lymphatic glands in the armpit or in the groin following injury or infection in the arms or legs respectively.)

Urinary system
The urinary system consists of the two kidneys, the ureters or vessels, passing from the kidneys to the bladder, and the urethra which passes urine from the bladder out of the body. As indicated previously the kidneys function as filters separating nitrogenous waste materials from the blood mainly in the form of urea.

Generative system
The generative system comprises the organs of reproduction. In the male these consist of the two testicles which are enclosed in the purse or scrotum, and the organ of copulation or the penis. From the testicles two tubes (the vasa deferentia) pass to the vesiculae seminales which in turn passes to the urethra, which traverses the length of the penis. We find associated with the vesiculae certain accessory glands the main function of which is the secretion of semen, a diluent for the spermatic fluid which is produced by the testicles.

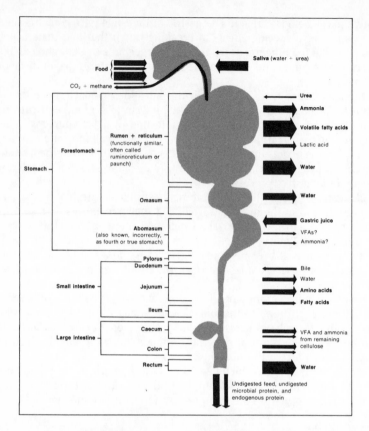

The various parts of a sheep's digestive tract are shown in the picture above, and some of the functions of each are illustrated diagrammatically (top). The digestive organs of all ruminants look very similar and function in much the same way.

In the female the generative organs consist of two ovaries which have connection with the uterus or womb through two tubes, the Fallopian tubes. The ends of these tubes nearest to the ovaries are extended into cup-shaped funnels and eggs expelled from the ovaries pass into the tubes via these funnels. The uterus opens through the cervix to the vagina which opens, in turn, to the exterior through the lips of the vulva. The urethra opens into the floor of the vagina and urination takes place through the vulva. The lower portion of the vulva is pointed and projects outwards serving to clear urine from the breech region. If this tip is cut off or damaged during shearing or crutching, urine does not clear the breech, setting up conditions that will encourage fly strike.

At mating, spermatozoa from the male are deposited in the upper portion of the vagina near the cervix and they pass through the cervical canal to the uterus. Fertilisation (i.e. the union of a sperm with an egg) normally takes place in the upper regions of the uterus or in the upper third of the Fallopian tubes.

Ova or eggs are normally shed from the ovaries towards the end of the heat period, so that hand service, when practised, is more successful if mating is delayed for about 24 hours after the onset of heat. Just prior to the heat period a breakdown of the uterine wall takes place and if fertilisation occurs, the egg attaches itself to the wall in a somewhat parasitic fashion and development of the embryo commences. The period of gestation in sheep ranges from 147 to 152 days (five months), but there is quite a variation within breeds. The mammary glands, or udder, are two in number, each being supplied with a teat, the function of the glands being to secrete milk for the nourishment of the lamb or lambs. Varying numbers of rudimentary teats are frequently present.

30

EXTERNAL PARASITES OF SHEEP: BLOWFLY STRIKE, MULESING, JETTING, DRESSING

Historical

Blowfly strike in sheep has been known to occur for several centuries. Tusser (1557) in his treatise, *One Hundred Good Points of Husbandry*, recommended the drying out of struck areas by the application of dust. It was recognised, even in those days, that moisture played an important part in rendering an animal susceptible to fly strike.

For many years very little thought was given to combating the blowfly pest, because the seriousness of fly strike had not manifested itself to any great extent. Even towards the end of last century, when the pest first made its appearance in Australia, struck sheep were regarded as no more than an accident. However, from 1900 onwards the seriousness of the position gradually became more apparent. Flockmasters began to realise that they were faced with heavy losses as fly wave after fly wave occurred over increasingly widespread areas. By 1909 it was evident that there was cause for alarm. Since then, however, the scientist and the sheep breeder have combined successfully to control losses due to fly strike.

Economical

The control of blowfly strike imposes a heavy burden on sheep breeders. Many thousands of sheep die each year as a result of it. The wool yield of many thousands more is greatly reduced and many fleeces are ruined. The blowfly is thus directly responsible for the loss of millions of kilograms of wool to the sheep industry.

The work involved in the control of blowfly strike calls for constant vigilance, periodical mustering, crutching, jetting, and the treatment of affected sheep. The annual cost of these measures and the direct loss caused by blowflies amounts to a colossal sum of money.

In addition to damage to fleeces, however, there are several other indirect losses from blowfly strike. For example, the general health of sheep is affected by fly-worry, and blood poisoning often results. There is a lowering of fertility in affected sheep, and lambing percentages suffer in consequence. Crutch-struck ewes are often too sick or irritable to suckle their lambs. Mustering tallies are short, due to struck sheep being missed and left in the paddock.

Blowfly Strike in Sheep

The Green Blowfly — Its Life Cycle

The green blowfly (*Lucillia cuprina*) is responsible for up to 90 per cent of all strikes. It breeds almost entirely on the living sheep. There are four stages in its life cycle: the adult blowfly; the egg; the maggot; and the pupa. When the female green blowfly strikes a living sheep, egg masses are laid on moist susceptible areas of the fleece where, after twenty-four hours, young maggots emerge from the eggs. If the area remains moist for a further twenty-four hours, the young maggots establish themselves by scratching and tearing into the skin of the sheep with their powerful mouth hooks. In two or three days the maggots are fully grown. They are smooth or hairless in appearance and are frequently the only ones present in a strike wound. They then drop off the sheep and crawl into cracks in the ground or under debris, where they change into a resting stage known as the 'pupa'. After some days the young blowflies emerge from the pupa cases and are able to force their way up through earth or out from under debris by means of bladder-like sacs on their heads. Once in the open, the life cycle is rapidly completed.

Other Flies

Several other flies also attack sheep, but are of less importance than the green blowfly.

The two *common brown blowflies* mainly live on carrion, but in the cooler climates also strike the living sheep. Only about 10 per cent of strikes, however, are caused by these flies. The maggots of the brown blowfly are hairless, like those of the green blowfly.

The *blue-green blowfly*, which causes secondary strike, is found in the hotter climates. It only attacks sheep after other flies have established a primary wound. Its maggots are hairy in contrast to those of the green and brown blowflies. Like the brown blowfly, it chiefly breeds in carrion. Secondary strike by the blue-green blowfly is usually of extreme severity and, if not treated, the affected sheep may die.

The *black bush fly* causes irritation to wounds and keeps them open to attack by other flies.

All blowflies breed rapidly in warm weather and their power to reproduce is immense. Over 2000 young blowflies have hatched from the eggs laid by one fly. The fact that the green blowfly breeds almost entirely on the living sheep accounts for failure of trapping to effectively minimise blowfly strike in sheep.

Causes Predisposing to Fly Strike

Sheep are not flystruck by chance. First of all blowflies must be active and areas on sheep's fleece must be attractive to them. Conformation of an animal may predispose to the retention of moisture, which is the main essential for a strike to become effective. The part must remain moist for 48 hours for young maggots to hatch and establish themselves. Moisture on soiled portions of the fleece is particularly attractive to flies and a sheep is thus rendered susceptible to strike. One of the main factors in keeping wool and skin moist is length of wool, particularly in the breech area. The length of the tail, too, plays a most important part in predisposing to strike.

A 'fly wave' may occur at any time of the year, if climatic conditions are suitable, but spring and autumn are the two seasons most favourable for blowfly breeding. The effect of rainfall and temperature on fly strike can well be seen in showery weather with consequent high atmospheric humidity and temperature during fine intervals. At such times, particularly in flush seasons, avoid excessive yardings, as this may cause staining and fouling of the fleece when a sheep lies down in badly drained or dirty yards. Fleeces wet by long grass or otherwise are easily fouled by manure from scouring sheep packed together in yards.

Types of Fly Strike

1. *Head strike*: This occurs chiefly in rams when the wool and skin at the base of the horns are struck. Animals with horns set close to the poll seem to be struck more frequently than others. Some degree of protection against head strike is obtained by jetting. In the treatment or the application of preventative measures in rams for any strike, however, veterinary advice should always be sought, for certain jetting or other mixtures, if absorbed, may have an adverse effect on a ram's fertility.
2. *Pizzle strike*: This is not uncommon. Such strikes frequently spread to and involve the sides of an animal. Pizzle strike can be controlled by ringing and jetting.
3. *Body strike*: Occurs mostly on the back, withers or sides of an animal. In young sheep it usually starts on the withers or other points along the back, shoulder or neck. Body strike occurs after prolonged wet weather and is frequently associated with fleece rot or mycotic dermatitis. It is more likely to occur in sheep carrying six months' wool than in recently shorn sheep. Sheep of either sex are affected to an equal degree. In this type of strike, bad conformation may predispose a sheep to strike, for high shoulder blades, broad withers with depression between the shoulder blades and 'pinched' behind the withers ('devil's grip') all favour the retention of moisture.
4. *Breech strike*: Includes all strikes in the region of the crutch or tail: 90 per cent of strikes in sheep occur here. Ewes are most liable to breech strike because of the wetting of this area with urine. The degree of wrinkledness of the skin of the breech in the merino varies considerably. The more wrinkled the breech, the more likely it is that moisture will be retained there. Saucer tail, wrinkles, etc. lend themselves admirably to the retention of moisture, especially when the wool grows long.

The Control of Blowfly Strike in Sheep

Early detection of a struck area saves a lot of damage, both to the health of the sheep and to the fleece, and helps to keep in check the green blowfly population.

Treatment of Affected Sheep

Struck sheep are treated by first removing the wool over and around the struck area. Care should be taken to follow up all pockets of maggots, as these soon develop into a more extensive strike if missed. Wool should be clipped from an area of 4 to 5 cm around the strike and the dressing applied over the whole of the clipped area. It should be well rubbed in to make sure that it penetrates down to the skin.

Do not use dressings such as tar or other substances that will not subsequently scour out of the wool. Also avoid dressings that are irritants as they delay healing. The longer a strike wound remains unhealed the more likely it is to be re-struck. The use of a mild dressing ensures that the skin remains soft and pliable, and that it heals in a few days. The use of an irritant dressing leaves the skin inflamed and thickened for a long time and with severe dressings it may die.

Probably one of the most commonly used dressings for fly strike over the years has been a 5 per cent solution of copper sulphate (bluestone) in water. Its effectiveness is probably largely due to its lavicidal and antiseptic action; it renders the wound dry and clean and for a time unattractive to the blowfly. The organo-phosphorus insecticides such as diazinon are now widely used in preparation of blowfly dressings.

In order to assist in reducing the green blowfly population, discarded maggots from struck sheep should be destroyed wherever practicable. Untreated crutchings from struck sheep, if dumped in a heap, are ideal for the continuance of the life cycle of the green blowfly.

Jetting
Struck sheep can be treated by jetting, but care is required as there is a risk of badly struck animals being poisoned by the absorption of certain jetting mixtures.

Prevention of Blowfly Strike
Jetting is more often used as a prevention than for the actual treatment of affected sheep. It has certain advantages, but success depends on careful attention to detail. Wool in all cases must be properly saturated. Jetting programs need careful arranging for each individual property so as to coordinate shearing, crutching, ringing and jetting at the necessary intervals. To be most effective it should be commenced at the beginning of a 'fly wave' in the spring and repeated as often as necessary until the heat of summer ends. It is required again a month or two after crutching and is repeated at intervals until the winter or, in exceptional circumstances, until shearing. Diazinon gives good results and is most effective if well applied by jetting to thoroughly saturate the fleece.

Marking
Prevention of blowfly strike starts at marking time and care should be taken with every lamb to see that the tail is docked to the correct length. A tail should be severed just below the tip of the vulva in ewe lambs and the skin on the under-surface left long so as to reflect back over the stump when healing takes place. Smooth skin over the end of a tail not only assists in prevention of strike, but it helps to keep a tail 'dag' free.

The modified tail operation for stretching the bare skin of the tail is best done when marking, as it not only saves time when mulesing, but is less severe on the sheep than when done concurrently with mulesing. The amount of skin removed from the wool-bearing surface of the tail depends on the individual lamb and the method used. Advice on this operation should be sought from a veterinarian before the operation is carried out on a large scale.

Shearing and Crutching
Dry wool is not struck, and on this fact most present methods of prevention of breech strike are based. They aim to dry out the area and free it from stain. The most common methods of doing this are by shearing or crutching. The latter must be well done to be really effective.

Shearing or crutching and ringing are only temporary measures at the best, for as the wool grows their effectiveness fades. They should, however, be so co-ordinated as to ensure that the wool in the region of the breech and pizzle is always short when spring and summer rains, followed by sunny days, bring conditions favourable for blowfly 'waves'.

Selection and Culling

Shearing or crutching protects both plain and wrinkled sheep, but because wrinkly sheep are very much more predisposed to strike, the importance of the protection given is greater for wrinkled types.

Studies of the inheritance of wrinkledness have shown that while either the very plain or the very wrinkled types can be maintained, the very nature of the inheritance makes it extremely difficult to hold intermediate types.

Mules Operation and Modification

Artificial de-wrinkling of the breech of sheep is now practised, and sheep are made plain breeched by surgical or chemical means. These methods aim primarily to stretch the bare area around the vulva and anus and thus de-wrinkle the part of the sheep that is most susceptible to fly strike.

The mules operation, as developed by the CSIRO, has given excellent results in the prevention of breech strike. The operation requires skill, but it can be performed quickly once an operator has become proficient. The results in many cases have been astounding. For example on one property only 120 strikes were recorded in 10 000 young mulesed ewes from shearing to shearing. They were crutched once. On the same property 8000 old untreated ewes showed 10 800 strikes over the same period, despite a double crutching.

The operation is not just a matter of snipping off sufficient skin wrinkles to produce a plain breech, but, irrespective of the degree of wrinkledness, follows a definite pattern. The first cut is made in the direction of the A to B (Figure 30.1 (i)), that is, it starts at a point about 4 to 5 cm above the level and to the side of the base of the tail and follows the direction of an imaginary line towards the point of the hock (D, Figure 30.1 (i)) of the opposite leg. To commence this cut, the shears are held in the right hand with the back of the hand to the sheep and the thumb on the blade to prevent lapping. Using the thumb and forefinger of the left hand, the skin is picked up at a point about 3 cm below and to the side of the tail (i.e. at point A, Figure 30.1 (i)). No tension is placed on the skin at this time. The shears are now pressed firmly on to the body of the sheep with the blades open to the full extent. As soon as the cut is started, the maximum tension is put on to the skin with the left hand. As the cut comes up over the pin bone, the right hand is raised to keep the shears flat on the skin. At this stage the shears will be pointing to (D, Figure 30.1 (i)) the hock of the opposite leg and the cut is brought along to the level of the 'bare area', where a new 'bite' is taken to ensure a good wide cut at this point. After passing the level of the bare area at B (Figure 30.1 (i) and (ii)), the cut is turned up the inside of the crutch along the line B to C (Figure 30.1 (ii)) and finished off.

The operation is completed by removing a corresponding piece of skin from the other leg. There should be 25 mm of skin between the cuts. The ends of the finished cuts at A and C (Figure 30.1 (i) and (ii)) should be sharply pointed and the widest part of each should be opposite the tip of the vulva at B (Figure 30.1 (i) and (ii)).

A good deal of success in mulesing is dependent on the sheep being held in the right position, which is greatly facilitated by the use of crushes or cradles. The mules operation gives a remarkable degree of protection against breech strike, but it should be carried out correctly to avoid mistakes. Faults most frequently found are:
1. cutting into the bare area, which produces an uneven stretching of the skin;
2. removal of insufficient skin;
3. commencing the cuts too low down and leaving too much skin between them;
4. commencing and finishing the cuts on different levels.

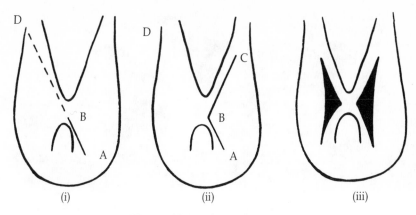

Figure 30.1   The mules operation

Very wrinkly sheep require more skin removed than plain ones. If there is any length of wool on the breech area it is essential to crutch before mulesing. A small percentage of merino flocks are mulesed as lambs at marking time, but the majority are done at the weaner or hogget age. Young sheep are lighter to lift and handle, and heal rapidly. Once healed, an animal is protected for life against crutch strike; therefore, the younger they are done the better. Complete healing takes about three weeks.

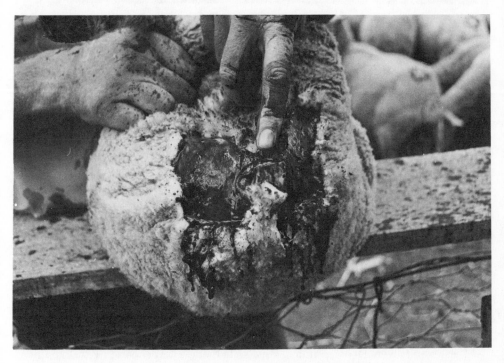

Completed radical mules operation. Although very severe, the wound heals within two weeks and gives sheep lifetime protection from strike.

Mulesing lambs at marking time

Photographs: courtesy NSW Department of Agriculture

There is no need to apply any dressing at the time of the operation, and after-care of treated sheep is simple. They are merely turned out to pasture and left undisturbed for one week.

Properly organised, mulesing should not take longer than three or four days each year, even on big properties. Much more time than this would normally be spent on fly control where sheep are not mulesed. The most suitable time to undertake mulesing is immediately after the first frosts, because blowflies are not then active. Sheep can also be done in mid-summer, providing that the small bush fly is not too prevalent. Mulesed sheep are actually less susceptible to strike than naturally plain breeched ones as shown by the number of strikes recorded below.

Strikes per 100 sheep	
Mulesed wrinkly sheep	1
Untreated plain breeched sheep	14
Untreated wrinkly sheep	86

These results were recorded from a large-scale trial over a period of twelve months.

Other methods of bringing about artificial de-wrinkling of the breech of the sheep are practised. They aim to destroy areas of skin by application of caustic agents. The treated area forms a slough which scabs over and eventually drops off, leaving scars which, in shrinking, stretch the bare area. Healing takes somewhat longer than when surgical means are used, but the final result in most cases is quite good.

Summary
Correct tail docking and mulesing give lifelong protection to sheep against breech strike, and the animals so treated are easier to crutch. These two methods are recommended to wool-growers as a practical answer to the blowfly menace.

The Sheep Ked
The Sheep Ked (*Melophagus ovinus*) is widely distributed throughout Australia in the higher rainfall areas, with the exception of Queensland, where it occurs only in the southern districts. It is thought that extreme dryness and heat are unfavourable to it, as it is seldom found to any extent in the drier inland areas.

It is far more commonly known as a 'tick', but unlike the true tick the ked has six legs and is very distinctly divided into head, thorax and abdomen, whereas the tick has eight legs, the thorax and abdomen are fused together and the head is not conspicuously distinct. The ked could be described as a wingless, blood-sucking fly, about 6 cm long, with a leathery body, reddish or greyish brown in colour, and covered with short spiny hairs. Its mouth parts are adapted for piercing and sucking blood. When feeding, it practically buries its head and proboscis in the skin of the sheep.

Life Cycle
The whole of the ked's life is spent on the one host unless dislodged. There is no evidence in support of the statement that the ked breeds in scrub and timber. This parasite does not lay eggs but deposits larvae which are attached to the wool of the sheep by a glue-like substance. These are then covered with a soft white membrane which becomes brown and hard in about twelve hours. After about nineteen to twenty-four days the young keds emerge almost fully grown and immediately work through the wool and attach themselves to the skin of the sheep. Young female keds are sexually mature in five days, and males in ten days from their emergence from the pupa. The female does not extrude the first larvae for up to twenty-three days, after emerging from the pupa, the minimum being thirteen days. She deposits eight to ten larvae at the rate of about one every six to eight days. A ked's life on a sheep may extend to four or five months, but it is commonly only three months. Under favourable conditions, keds which survive more than four or five days off their host would be in the minority, although they have been found alive for up to forty-two days.

As is the case with most forms of parasitism, the effects of the ked are most severe on animals that are low in condition. Due to the intense irritation set up by these parasites, considerable damage is done to the fleece by the animal rubbing, biting and scratching and also by excreta from the ked staining the wool ('tick stain').

Treatment
In all cases the aim should be eradication of keds from the flock. This can be achieved by effectively dipping all sheep in an appropriate insecticide. Dips effective against keds are based on rotenone (the active principle of derris root) or the organic phosphorus compounds, Diazinon®, Mankor® or Asumtol®. These preparations not only destroy all adult keds, but possess a residual effect which destroys keds that will hatch from what were pupae at the time of dipping.

For maximum efficiency sheep should be dipped as soon as shearing wounds have healed, preferably about two or three weeks after shearing, and certainly no more than six weeks. Early treatment has the advantage that at the time the sheep carry few adult keds and practically no pupae.

The chlorinated hydrocarbons (for example B.H.C., Aldrin and Dieldrin) were highly effective against keds, but their sale for use on stock is no longer permitted because of the problem of carcass residues.

The Body Louse (*Damalinia Ovis*)

The body (biting) louse (*Damalinia ovis*) is the most common of the several species of lice which are parasitic upon sheep in Australia. It is a small wingless insect, the male being about 1mm long and the female slightly longer. The general colour is white, but the head and thorax are darker. The abdomen is divided into segments, each of which has a median band of dark colour. It has the mouth parts adapted for biting and sucking.

The louse, like the ked, spends its whole life on a host and only leaves to transfer to another animal, which it does very rapidly when sheep are in contact with one another. The eggs are referred as 'nits', have a somewhat barrel-shaped appearance, and are laid at the rate of about two every three days. They are attached to wool by a glue-like substance extruded by the female. In nine to ten days the eggs hatch out, the young lice — almost invisible to the naked eye — attain maturity in about twenty-one days and commence to lay eggs in about twenty-four days. Due to the smallness of these parasites, the sheep would appear quite clean to the casual observer unless there is evidence of rubbing caused by irritation. These lice may be found on any part of the body, lying close to the skin and their small size and hard, flattened bodies make movement through the fleece quick and easy.

Like the ked, infestation is heaviest on sheep in poor condition and intense irritation is caused, resulting in the sheep biting and pulling the wool and rubbing against posts and logs. It is not unusual to see badly infected sheep with large areas of wool rubbed off and actual sores caused by rubbing and scratching. Lice are more prolific and detrimental than keds and create a serious economic menace to the wool industry, a few lousy sheep being able to infect a whole flock. Unlike the ked they live on scurf and other skin and wool products and are not blood suckers.

Treatment

Lice are easily killed with insecticides at low concentrations, but their eggs are very resistant. Consequently an insecticide applied by any method should persist in the fleece long enough to kill the young lice after they hatch out. There are many insecticides which kill by contact, those mostly used being Diazinon®, Mankor® and Delnar® (all organic phosphates) and rotenone.

The Sucking Louse

The sucking louse, of which there are two species — the foot louse (*Linognathus pedalis*) and the face louse (*Linognathus ovillus*) — is not of great economic importance. The foot louse is larger than the face louse, has a short, blunt head and a wide oval-shaped body and is blue-grey in colour. It is to be found around the coronet and back of the leg. At times when the sheep are heavily infested, they are to be found around the scrotum and belly of rams, causing considerable uneasiness due to the testes becoming swollen. It is interesting to note that sometimes infestation is so slight that it will only affect perhaps 2 to 3 per cent of the flock.

Life Cycle

Although this has not been fully worked out, it is thought that the sucking louse begins laying eggs at twenty-two to twenty-six days old, and the eggs hatch in about

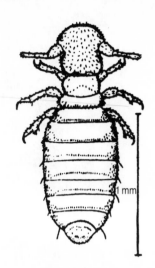

Engorged female sheep ked   *The body louse*

seventeen days. Egg laying occurs at the rate of about one a day. Sucking lice, like the biting louse, are limited to a specific host, the sheep.

Linognathus ovillus (the face louse), the second of the species, is usually found in colonies on the face or nearby parts of the body, but it is not very common in Australia. Its presence can be easily detected by the immense masses of eggs or 'nits' matting the wool together. Being a blood sucker, this parasite can cause loss of condition combined with rubbing and biting of the parts by the sheep, causing damage to the fleece.

In the treatment for this parasite, the egg masses should be cut off and burnt, and as this parasite is so uncommon, complete shearing and destruction of the wool might be desirable if large numbers of sheep are not affected.

Treatment
To kill foot and face lice a contact poison such as one of the organic phosphate preparations, is necessary. For foot lice, a foot bath containing an efficient contact poison such as an organic phosphate should be used.

The Itch Mite
The itch mite (*Psorergates ovis*) was first recognised in New South Wales in 1940 by H. B. Carter BVSc. The mite is invisible to the naked eye and lives in the superficial layers of the skin. It causes a complaint known as dermatitis which of late years seems to be spreading. It is found anywhere on the body and causes a slightly scaly appearance on the skin. The wool on the affected area looks unusual, becoming very white, dry and tender. As in the case of the ked and louse, the irritation causes the animal to bite, scratch and rub itself with consequent damage to the wool.

The life cycle of this parasite is as yet undetermined. The mite apparently spends its whole life on the sheep and passes through various phases of protracted development from egg to mature mite.

Treatment
Ordinary dips are not effective in the control of the itch mite, lime sulphur solutions being the only known control. These solutions will not wet the wool readily unless a

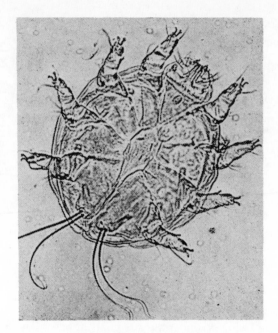

A female itch mite, 250 times life size

Photograph: courtesy CSIRO

Sheep heavily infested with body lice

Photograph: courtesy Department of Agriculture

wetting agent is added — Agral 3 or 'Wetsit' at the rate of 170 g to every 455 litres improves the wetting properties of the solution.

Other External Parasites

Other external parasites of sheep in Australia include:

1. a mite known at *Chorioptes communis* var. *ovis*, which causes foot mange in sheep as well as horses;
2. a mite of the *Dermodes* spp. which causes follicular mange of sheep; and
3. another mite *Trombicular sarcina*, which causes trombidioses of sheep. This mite has only been recorded on the black earthy country of the central highlands of Queensland. They are particularly active after summer rains.

31

DIPPING SHEEP

Although there are various types of dips, plunge and shower dips are still the most efficient and, if used correctly, can be relied on to give a total kill of lice and keds at a single dipping.

Shower or spray dips are easier to operate than plunge dips, and in fact can be worked by one person. They have the advantage of functioning on a lower volume of dipping fluid than plunge dips, as the fluid used is collected in sumps, strained and reused. There is also less risk of the wool becoming dip-stained.

Most insecticides can be used in this type of dip, with the exception of lime sulphur. The latter is poisonous if inhaled and there would be a risk of sheep inhaling the spray during dipping.

Dipping Fluids

A wide variety of compounds is currently registered for external parasite control in sheep. There are two main groups:
- organophosphates,
- synthetic pyrethroids.

Arsenic-based chemicals had been used in plunge and shower dips. However, they were withdrawn from the market at the end of 1986, and their use is now illegal. They are no longer acceptable because of their potential to cause residue problems in the scouring liquors from wool-processing plants.

In Chapter 30, dealing with external parasites of sheep, insecticides are named which are suitable for the eradication of particular parasites.

Points to be Watched in Dipping

The usual order of dipping is to put the killers, wethers and store sheep through the dip first, then the ewes, next the lambs and finally the rams.

First, it is wise to follow to the letter the manufacturer's directions for mixing the dip. Careless mixing is found most often in the case of powder dips and on no account must the dry powder be poured into the bath. To obtain the correct strength of solution it is necessary to know exactly the capacity of the dip and for this purpose a measuring stick marked off in 455 litres (100 gallons) is invaluable. When using powder dips it is essential to keep the wash constantly agitated during the dipping operation — the movement of sheep alone not giving sufficient agitation.

During dipping, no scum should be allowed to accumulate on the surface of the wash. A clean wash will not deteriorate appreciably on standing for a few weeks; dips of the non-poisonous type deteriorate if allowed to stand too long. A very foul wash deteriorates and is also liable to stain the wool, as will an insufficiently mixed wash.

If a dip is used too long it becomes so charged with suint from the fleeces of the sheep that it becomes, in effect, a very dirty scouring bath and it then tends to remove some of the yolk from the wool which is left in an unattractive dirty condition. At present there is no accurate test of when to discard the wash. The quantity of wash required depends upon the number of sheep to be dipped and the density and length of their wool. Lambs may carry out only a litre of dip each, whereas the average crossbred sheep might average two to three litres with two months' wool growth. Quantities in litres can be calculated from:

$$A + 3 \times X$$

where A = quantity of water in litres necessary to bring dip up to level; and
X = number of sheep to be dipped.

Insufficient immersion is another cause of unsuccessful dipping. It is necessary to thoroughly saturate the skin of the sheep so that the time of immersion is dependent upon the length, density and condition of the fleece. Low temperatures and hard water both tend to retard the penetration of the dip; high temperatures are liable to induce scalding. Increasing the strength of the wash will not compensate for a short immersion, but only lead to a greater risk of scalding. Tight packing of the sheep, etc., in the narrow swim-through bath will compress the fleeces and help to retard penetration. The action of swimming and movement thoroughly opens all the staples and gives maximum penetration. As they swim the sheep should be checked occasionally — particularly in a straight-swim bath — and pushed gently backwards. While in the bath, each sheep should have its head immersed twice, and it is usual to give each animal at least one minute's immersion.

Although not usually expected, any sheep that require it should be dagged prior to dipping, and it is a frequent practice to trim the feet at the same time. It is a good plan to leave the sheep in the yards for a period, to empty before going through the bath. The dipping of thirsty sheep heated from driving, or animals with a full belly, should be avoided. Hot sheep have the pores of the skin opened and scalding is likely to result. Wounds in the skin from dog bites, nail tears, etc. are likely to be infected.

Rams in the pink, and all sheep in high condition should not be dipped if it can be avoided. Care has to be taken to assist rams through the bath and a bridle is a very helpful aid. They should be ducked only once. Rams should never be put with the ewes less than seven days after the ewes have been dipped. It is not desirable to dip unweaned lambs because of the risk of mismothering. If ewes and lambs have to be

The shower dip

dipped they must be separated and the lambs kept away from their mothers for two or three hours until the ewes are quite dry.

Sheep should not be packed too tightly in the draining pens or they will chafe and rub together, increasing the risk of scalding and it is desirable that they dry in the shade. On no account must sheep start at once on a long drive when still wet with dip.

Causes of Losses

1. General rough handling of sheep in getting them into the bath.
2. Extensive wounds such as shear wounds, dog bites, nail tears and abrasions.
3. Sheep being unknowingly in poor health — weak lungs, worm infections and so on.
4. Few sheep are lost from actual poisoning from swallowing the dip.
5. Mechanical pneumonia from wash entering the lungs and ordinary pneumonia.

Don'ts in Dipping

1. Don't dip in extreme heat or cold or in wet weather.
2. Don't dip during the hottest hours of a hot day.
3. Don't dip when the sheep are hot, tired or thirsty.
4. Don't drive sheep immediately before or for several hours after dipping. Yarding overnight without tight packing is permissible provided the animals are not in a cold situation.
5. Don't guess the amount of bath water.
6. Don't shed sheep overnight or on the day of dipping. Allow them to dry in a paddock sheltered from cold winds.
7. Don't dip rams in a flushed condition and always handle them individually in the bath.
8. Don't subject sheep to violent entry into the bath.
9. Don't dip so late in the day that sheep cannot dry before sundown.
10. Don't fail to burn used dip packets and to exclude stock from access to dip containers.

Spray Races

The traditional spray race is different from the new aerosol applicators as the sheep are sprayed with insecticide as they walk through a race. The spray race should be used only off the board. The dip wash is not reticulated and only one chemical is registered for use through some spray races. While spray races are portable and have labour-saving advantages, saturation of the wool can be insufficient to kill all lice, especially if the sheep are poorly shorn or run through the race.

Backline Treatment Off-the-Board

This new treatment has many advantages over traditional methods, particularly for small flocks. The chemical is applied either with a hand gun as a stripe or spray from shoulders to rump or, using a special spray race, as a broad band misted on by aerosol.

Most methods require little capital expenditure and no extra labour is required. Treatment should be applied within twenty-four hours of shearing. The chemical contains a coloured dye, which allows treated sheep to be readily identified. However, there are no fleece staining problems. The treatment is quick to administer and no water supply is necessary. No extra musters are required. Backline treatments take up to six weeks to kill all lice on all sheep. This allows a few weeks' grace for stragglers to be mustered, shorn and treated.

Backline treatments are unsuitable for treating lambing ewes or ewes with lambs at foot without special precautions. The treatment is mainly suited for body lice and will not eliminate face or foot lice.

32

INTERNAL PARASITES

Practically every sheep harbours some worms. Disease conditions occur when the number of parasites increases as a result of suitable weather conditions, over-stocking and overcrowding, malnutrition or other cause of lowered resistance. The different parasites have different effects on the sheep, live in different parts of the animal, have different seasonal occurrences, and may require different drenches to kill them. It is therefore important to know which species occur in any particular flock or district.

It is practically impossible to eradicate worm parasites of sheep, and control meas-ures are aimed at keeping their numbers low enough to avoid economic loss. How-ever, it must be kept in mind that quite light infestations with certain parasites such as black scour worm will check growth and wool production long before the sheep shows any obvious symptoms.

The common internal parasites of sheep are: large stomach worm, black scour worms, nodule worm, large-mouthed bowel worm, large lungworm, tapeworms, liver fluke and hydatid.

Large Stomach Worm (*Haemonchus contortus*)

The females of this parasite are about 25 mm long and are red with a white spiral twisting — hence they are often referred to as 'barber's pole' worm. The males are smaller and are reddish-brown in colour. These worms are blood suckers and are to be found in the abomasum, especially in young sheep and breeders, although sheep of all ages are affected.

This is a summer parasite, the eggs of which are picked up in the spring, through summer into early autumn. Outbreaks are related to the incidence of rainfall, a fall of

10–12 mm or more, accompanied by dull humid weather, being followed by increased infestation of sheep from parasite hatching from eggs on the ground. The effects of infestation quickly develop so that sheep should be treated about three weeks after good rains, to kill the developing worms before they have commenced to lay eggs. If rains and dull weather persist, treatment should be repeated at intervals of three weeks until the weather becomes hot and dry, or cold and dry. It is also advisable to give a final treatment three weeks after a period of wet weather has ended to prevent a heavy infestation later on.

Black Scour Worm (Several Species of *Trichostrongylus*)
These are hair-like worms, light pink in colour and about 5 mm long. They occur in the first third of the small intestines and one species occurs in the fourth stomach of sheep. They are parasites of the cooler months and are picked up in autumn, winter and early spring.

Nodule Worm (*Oesophagostomum columbianum*)
This worm is about 1 cm in length and creamy white in colour. It is a summer parasite, depending on adequate rainfall during late spring, summer and early autumn to enable the eggs to hatch. Larvae are picked up from pastures during the warmer months but they develop slowly in the sheep and symptoms may not appear until the late autumn, winter, or even early in the spring following infestation. To reduce contamination of the spring pastures and help to protect spring lambs, treat all sheep, particularly breeding ewes late in August. If infestations are severe, treat all sheep in winter and young sheep in autumn.

Large-mouthed Bowel Worm (*Chabertia ovina*)
This parasite is most common in areas of winter rainfall, particularly southern New South Wales, Victoria, Tasmania, the south-west of Western Australia and the south-east of South Australia. The worm damages the lining of the crown gut, causing inflammation and thickening of the bowel wall. It is picked up in the cooler months of the year and its effects are usually seen in the late winter.

Large Lungworm (*Dictyocaulus filaria*)
This is a creamy-white thread-like worm, up to 8 cm long, found in the air passages of the lungs of sheep. Ill-effects are usually seen in late winter, particularly when feed is scarce. Young sheep are the most severely affected. Infection occurs during late autumn, winter and early spring, and is usually thrown off rapidly when the spring growth of feed appears. The characteristic symptom is a cough, but even a few worms are quite sufficient to be harmful and will lead to coughing, whereas severely affected sheep may be too weak to cough.

Tapeworms (*Moniezia* and *Helictometra*)
These are segmented, flat, white worms up to 8 and 9 m long. They are conspicuous and the segments are easily seen in the droppings. Tapeworms seldom affect sheep adversely; however, if lambs and weaners are on a poor diet and are heavily infested they may lose condition.

Live Fluke
Liver fluke in the adult stage is a flattened leaf-like parasite found in the bile duct of the liver. Each fluke produces many eggs which pass down the bile ducts into the intestines and thence to the exterior. The eggs develop and hatch into larvae only under very wet conditions and they must find a special fresh-water snail (*Lymnaea tomentosa*) in which to develop. The parasite multiplies in the snail and after about two

months emerges, attaches itself to a grass blade or water weed and encloses itself in a protective case. After it has been swallowed, in drinking or eating in swampy areas, it penetrates the wall of the small intestine and bores its way into the liver, where it spends some ten weeks, growing meanwhile and causing considerable damage. It then enters the bile ducts where it grows to maturity.

Acute fluke disease results from heavy infestations by wandering young flukes. Chronic fluke disease results from the damage caused by the adult flukes in the bile ducts. Black disease can occur following light infestation, therefore it is most unwise to permit sheep to feed in swampy places as they may pick up young flukes which have been there for many months.

For the control of fluke, drench with carbon-tetrachloride and for the control of black disease, vaccinate.

Hydatid (*Echinococcus granulosus*)

Hydatid is a tapeworm, which in its larval stage forms cysts or bladders filled with clear watery fluid. The adult worm is about 4 mm long, and is found in the small intestine of the dog and other closely allied animals, but does no harm to the animal. However, this parasite when fully developed is passed out with the dung of the dog and the intermediate host (sheep, ox, pig, man) becomes infected by swallowing the eggs. In the small intestine the egg shell breaks up and the worm is set free to burrow through the wall of the bowel into a blood vessel. It is then swept along in the blood until it reaches a suitable organ, such as the lungs or liver. Here it settles, grows, changes in form and, after several months, becomes a hydatid cyst, from 1 to 15 cm in diameter, which if eaten by a dog develops into tapeworms, thus completing the life cycle.

Hydatids is a disease that could be eradicated by following simple and easily

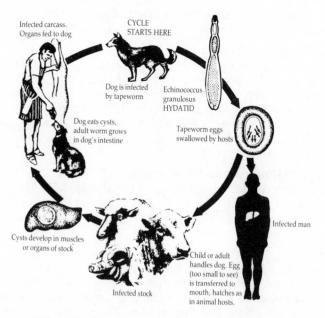

Figure 32.1   Life cycle of the tapeworm

Table 32.1 Spectrum of activity drenches registered and currently available for control of internal parasites of sheep

Chemical group	Trade name	Barber's pole worm	Other gut roundworms	Lungworm	Tapeworm	Liverfluke
Group 1 — Broad spectrum benzimidazoles and probenzimidazoles	Thibenzole® (G)	Yes	Yes	No	No	No
	Ranizole®	Yes	Yes	No	No	Yes
	Telmin RLT® (G)	Yes	Yes	Yes	Yes	No
	Panacur®	Yes	Yes	Yes	Yes	No
	Valbazen®	Yes	Yes	Yes	Yes	Some
	Synanthic® (G)	Yes	Yes	Yes	Yes	No
	Systamex®	Yes	Yes	Yes	Yes	No
	Rintal® (G)	Yes	Yes	Yes	Yes	No
Group 2 — Broad spectrum imidazothiazoles (levamisole and morantel)	Nilverm®	Yes	Yes	Yes	No	No
	Ripercol®	Yes	Yes	Yes	No	No
	Nilzan®	Yes	Yes	Yes	No	Mature fluke
	Citarin®	Yes	Yes	Yes	No	No
	Exhelm E® (G)	Yes	Yes	No	No	No
Group 3 — Narrow spectrum organophosphates	Rametin HLV®	Yes	No*	No	No	No
Group 4 — Narrow spectrum salicylaniliides and nitrophenols	Seponver®	Yes	No	No	No	Yes
	Ranide®	Yes	No	No	No	Yes
	Fascol®	Yes	No	No	No	Yes

(G) = Registered for use in goats.
* = Recommended dose effective against black scour worm in lambs.
® = Registered trade name.

effected precautions. The weak link in its life cycle is that it must spend part of it in the dog. Consequently dogs should not be permitted to eat raw offal and carrion and periodic dosing with arecoline hydrobromide is very satisfactory in destroying any parasites already carried. 'Droncit' is another new and completely effective drug now available on the market. However, through carelessness and ignorance, the disease is very common, freezing works records indicating that up to 80 per cent of the cattle and 70 per cent of the sheep slaughtered have been affected.

In view of the fact that the disease is transmissible to man, it is all the more important that the parasite should be killed out. Children should be discouraged from fondling dogs, and they should wash before eating; also any uncooked vegetables, such as lettuce, should be thoroughly washed before use.

Methods of Control

Rotational grazing and spelling should be carried out by moving drenched sheep to a paddock which has been spelled for three or four weeks, as they immediately pick up further infective larvae if they are returned to the same paddock. Reduction of stocking by making more feed available also helps. If sheep numbers cannot be reduced over the whole property, at least give the breeders and young sheep more room, as weaners particularly are most susceptible to worms. Grazing well-fed sheep develop and maintain resistance to worms. Grazing crops, improved pastures, spelled pastures, rotational grazing and conserved feed are all means for maintaining adequate nutrition.

Treatment

When sheep are showing evidence of worm infestation the ideal form of attack is a three-pronged one. It consists of:

1. the use of a drench that will remove the species of worms believed to be infesting the sheep;
2. the transfer of the sheep after drenching to a clean paddock to reduce the risk of reinfestation;
3. the provision of good quality pasture or supplementary feed to assist the sheep to overcome the effects of the worms.

In this section particulars are given of the principal drugs used for the removal of worms. Each owner must select the one that appears most suitable to his requirements. Considerations which govern the choice of a worm drench include where the property is located, the time of the year, the species of worms likely to be present, the cost of the drench and whether treatment is for curative or preventative purposes.

When conditions are such that sheep are building up infections, the treatment of choice may be a drug that removes a high proportion of larval worms, as well as adults. Under other circumstances a drench that removes only adults may suffice. Some drenches will remove a wide range of worms. They are usually referred to as broad spectrum drenches. Others are restricted in their activity to one or a few species of worms.

The barber's pole worm is the most readily removed of the worm species. Where this parasite is involved, the use of the more expensive broad spectrum drugs is not warranted. Similarly, broad spectrum drenches are not needed if only the nodule worm, or the nodule worm in combination with the barber's pole is involved. Particulars concerned commercially available treatments are given in Table 32.1.

33

SOME COMMON SHEEP DISEASES

Anthrax
Symptoms: Affected sheep are found dead. Blood comes from nose, mouth and back passage. *Important*: No post-mortem.
Immediate action: Vaccinate all contact sheep with anthrax vaccine (Sterne-type). Burn carcasses.
Prevention: Vaccinate all young sheep annually.

Black Disease
Symptoms: Affects grown sheep. Found dead on camps. Carcasses decompose rapidly. Post-mortem shows yellowish-grey areas on the liver.
Immediate action: Vaccinate all contact sheep with black disease vaccine.
Prevention: Vaccinate weaners annually by December and again twelve months later. Keep fluke snails under control.

Blackleg
Symptoms: Affects lambs and grown sheep after birth, lamb marking, shearing, dipping, mulesing and inoculation. Affects ewes after difficult lambing and hoggets grazing on turnips. Causes sudden death with signs of blood poisoning. Causes swelling of the head of young rams.
Immediate action: Move flock into a fresh paddock. Treat sick sheep with penicillin (½ to 1 million units for four days). Give 1 million units of penicillin to all ewes assisted at lambing.
Prevention: Vaccination prevents blackleg. Two annual inoculations required for young sheep. To prevent losses in newborn lambs, vaccinate ewes two weeks prior to lambing.

Blood Poisoning

Gas gangrene (malignant oedema). Caused by gas gangrene group of organisms of *Clostridium* species.

Symptoms: Causes deaths after shearing, marking, dipping, dog bite, crow pick and lambing. There is usually a soft doughy swelling around a cut.

Immediate action: Treat affected sheep with penicillin. Dose ½ to 1 million units. Repeat at twenty-four-hour intervals for four days.

Prevention: Prevent infection through open cuts by keeping standard of hygiene high at shearing, marking and dipping.

Cheesy Gland

Symptoms: Forms chronic abscesses in lymph glands, specially of shoulder, flank, g oin and sometimes in internal glands. Abscesses rarely burst. Contain yellowish-green pus, consistency of cheese.

Immediate action: Remove affected gland from carcass which is then fit to eat.

Prevention: A germ living in dust of sheep yards causes 'cheesy gland'. Keep down dust at shearing to prevent contamination of shearing cuts.

Copper Poisoning

Symptoms: If due to overdosage with bluestone, deaths occur within twenty-four hours. If due to chronic poisoning (from grazing on pastures near copper mines or on subterranean clover pasture), jaundice and coffee-coloured urine most noticeable symptoms.

Immediate action: No antidote for acute copper poisoning. For chronic poisoning on subterranean clover, feed lick of one bag of salt, one bag (63kg — 140 lb) gypsum, 454 g (1 lb) sodium molybdate or feed 227 g (½ lb) lucerne hay per head sprayed with 2.7 kg (6 lb) Glauber's salts and 113 g (¼ lb) sodium molybdate per 91 kg (200 lb) hay.

Prevention: Do not use excessive quantities of bluestone for topdressing pastures or for control of fluke snails. Prevent clover dominance by topdressing late in autumn. Apply 113 g (4 oz) molybdenum per .405 ha with superphosphate.

Epididymitis

Symptoms: Softening and slight enlargement of the tail of the epididymis are early symptoms.

Chronic symptoms: Tail of epididymis hardened and enlarged. Most cases due to *Brucella ovis* infection. A non-specific type also occurs.

Immediate action: Aureomycin shown to be effective but costly. Only practicable for very valuable stud rams. Vaccinate all unaffected rams with Strain 19 plus *Br. ovis* vaccine.

Prevention: Keep unmated rams separate from mated rams. Seek veterinary advice on advisability of vaccination. Vaccinate ram hoggets with Strain 19 plus *Br. ovis* vaccine. Vaccinate purchased rams not previously vaccinated at least eight weeks prior to mating.

Foot Abscess

Symptoms: Two types, both cause extreme lameness. Top type — no visible abnormality early. Foot is hot. Only one claw affected which is painful on squeezing. Pus breaks out at coronet. Heel type is worst type. Foot swells on one side. Pus breaks out between heels. Foot may swell above coronet.

Immediate action: Toe type — open with a knife at junction of inside wall and sole. Releases pus. Heel type — no effective treatment. Try injection of 500 000 units penicillin in early cases. Remove affected flock to dry ground if possible.

Prevention: No specific method of prevention. Keep sheep out of yards in wet weather. Keep feet trimmed. Foot abscess is not contagious.

Foot Rot

Symptoms: Starts with inflammation between claws. Wall and sole become underrun with foul-smelling material. Both claws are affected and this is highly contagious.

Immediate action: If only in one mob keep it isolated. Must decide whether to attempt eradication or just keep under control. *Treatment of individual sheep*: Remove all affected horn and stand in 5 per cent formalin for 15 minutes or paint with 'Chloromycetin' or 'Cetafoot'.

Prevention: Isolate all sheep coming from foot rot areas until feet have been examined.

Eradication: Only start when there is ample feed to allow spelling of paddock.

1. Concentrate all sheep in one part of property and spell remainder for two weeks.
2. Separate clean sheep, doubtfuls and affected sheep.
3. Once treated, do not allow clean sheep to travel over same ground.
4. Use 5 per cent formalin in foot baths and treat all sheep.
5. Thoroughly pare feet.
6. Repeat treatment in seven days.

Foot Scald

Symptoms: Mainly affects lambs and young sheep. Starts with scalding of skin between claws. Underruns sole. No foot-rot smell.

Immediate action: Stand in 5 per cent formalin for ten to fifteen minutes.

Hypocalcaemia

Symptoms: Prevalent among ewes in lamb especially if grazing on green oats. Move with stilted gait at first then develop staggers. Go down and drag along on hind-quarters. Die in twenty-four hours. Occurs after untrucking fat sheep, also if fat ewes in lamb are held in yards overnight.

Immediate action: Give immediate injection of calcium borogluconate, 80 cc 25 per cent solution, 50 cc 40 per cent solution. Respond by getting to feet in half an hour.

Prevention: Keep ewes in lamb grazing on green oats under close observation. Remove them immediately a sick one is noticed. Do not starve ewes overnight in yards if close to lambing.

Mycotic Dermatitis

Symptoms: Causes hard, dry lumpy crusts over face, ears, legs and back of lambs and weaners. Firmly adherent to skin. Called 'lumpy wool'. May not be noticed in wool until sheep handled. Affects newborn lambs all over body causing numerous deaths. Causes wart-like scabs on the face and lips of adult sheep.

Immediate action: Treatment not highly effective. Dipping in bluestone 454 g to 227 litres (1 lb to 50 gallons) of value. Disadvantage — stains wools.

Prevention: Very seasonal, difficult to assess value of control measures. Dipping off-shears in 0.5 per cent zinc sulphate is a preventative. Research in progress to see if it can be prevented by treating faces of ewes prior to lambing.

Phalaris Staggers

Symptoms: Causes loss of muscular co-ordination. Affected sheep lose use of legs and walk on knees. Easily excited, tremble, make unusual movements such as nodding head. Symptoms are permanent. Very few recoveries.

Immediate action: Remove from phalaris. Dose with 'cobalt bullets' if put back on phalaris.

Prevention: Give 'cobalt bullets' before going on to phalaris. Staggers usually occur in autumn.

Photosensitisation

Symptoms: A form of sunburn. Affects light-coloured skin. Occurs with or without jaundice.

Plants responsible: St John's Wort, trefoil subterranean clover, millet, summer grass, lantana, rye grass, heliotrope, yellow vine, wild radish, lupins, kale.

Immediate action: Remove affected animals from sunlight. Allow grazing at night. Dust animals in contact with lampblack. Give 57 grams sodium thiosulphate in 284 ml water as a drench.

Prevention: To protect freshly shorn sheep, dip 'off-shears' and add 2 kg lampblack per 45 kg dipping fluid. Alternatively, dust face and back with lampblack. Use a hessian bag.

Pink Eye

Symptoms: Symptoms appear suddenly. Watery discharge from eye and a desire to get away from light. Eyelids intensely reddened. Opaque film covers the eye. May get ulcer in centre of eye.

Immediate action: Treat affected and non-affected sheep with a methylene blue-zinc sulphate spray.

Prevention: Caused by a virus, no method of prevention.

Pizzle Rot

Symptoms: Starts with swelling of prepuce of wethers. Scab forms on exterior, extends to interior. Semi-solid foul pus forms. External orifice becomes blocked. Condition lost.

Immediate action: Draft off affected sheep. Put balance on poor feed. Treat with Ropel. Syringe out affected sheep with bluestone, 227 grams to 4.55 litres (1 gallon).

Prevention: 'Ring' wethers and keep off rich clover pastures.

Plant Poisoning

Symptoms

Sudden Death	a.	Prussic acid poisoning	
	b.	Nitrite poisoning	
	c.	Other sudden poisonings	
Delayed Death	d.	Continuous staggers ⎫	Scouring, jaundice, lung congestion,
	e.	Staggers on driving ⎬	milk fever, photosensitisation,
	f.	Nervous fits ⎭	blindness, gangrene

Immediate action:

(a) Prussic acid poisoning. Give the following by intravenous or intraperitoneal injection: 10 cc 10 per cent sodium nitrate, 20 cc 10 per cent sodium thiosulphate. Also give 14 grams sodium thiosulphate in 142 ml water by mouth.

(b) Nitrate poisoning. Give 10 cc of 1 per cent methylene blue as an intravenous injection.

(c) Oxalata poisoning. Give 80 cc 25 per cent calcium borogluconate by subcutaneous injection.

(d) Other plant poisons. No treatment.

Prevention: Keep hungry stock away from areas where poison plants grow. Hungry stock eat plants they would not normally eat.

Pregnancy Toxaemia — see Twin Lamb Disease

Pulpy Kidney

Symptoms: Sudden death of big lambs three to ten weeks old. Adult sheep occasionally affected. These show more protracted symptoms including staggers, scouring and dullness.

Immediate action: Change paddock. Put on to poorest feed available. If losses do not stop, inoculate with 'pulpy kidney' serum.

Prevention: To protect young lambs up to marking, vaccinate with 'pulpy kidney' vaccine fourteen days prior to lambing. To protect them after marking, vaccinate lambs at marking time and repeat in three weeks.

Rye Grass Staggers

Symptoms: Symptoms vary from slight stiffness to inability to move without falling. Run with stiff leg action, head in air. Very few deaths.

Immediate action: Remove quickly from rye grass. Most sheep recover. Dangerous phase of grass usually passes in fortnight.

Prevention: Cause unknown. No prevention.

Scabby Mouth

Symptoms: Mainly affects lambs. Highly contagious, runs a course of four weeks. First week: lips swell and small pustules appear. Second week: pustules form into scabs, extend over nose, and under jaw. Third week: scabs dry and lift. Fourth week: scab falls off.

Immediate action: Vaccinate all sheep in contact with scabby mouth vaccine. Do not treat affected ones unless face badly swollen from secondary infection. Removal of scabs delays healing.

Prevention: Vaccinate lambs at marking time.

Sheep Measles

Symptoms: Small cysts occur in body muscle. Larval stage of tapeworm which occurs in foxes and dogs. Cysts approximately 6mm in diameter and found in heart, diaphragm, head muscles and tongue. Sometimes found through whole carcass.

Immediate action: Treat dogs with arecoline hydrobromide.

Prevention: Destroy foxes and treat dogs regularly with arecoline hydrobromide (every seven weeks). Boil meat if it contains cysts before feeding to dogs.

Shelly Hoof

Symptoms: Cause a half-moon shaped cavity in wall of foot. Packed with dirt. Causes lameness if moist and putrefying.

Immediate action: Trim the foot, open cavity and scrape out dirt. No other treatment necessary.

Prevention: Keep feet well trimmed.

Subterranean Clover Disease

Symptoms: Clover dangerous from germination to wilting stage. Causes birth of dead or very weak lambs. Many ewes unable to give birth to lambs. Lambs decompose. Leads to cystic condition of uterus. Affected ewes become sterile. Lambing percentages fall. Also causes bearing trouble and carriage of tail high. Wethers come in milk. Develop a swelling below tail due to a 'false bladder'. Rams not affected.

Immediate action: Remove affected sheep from clover. Keep close watch on ewes to remove dead lambs. Wethers recover without after-effects.

Prevention: Dispose of affected ewes as they will not breed. Do not allow ewes or wethers to graze sub-clover in green growing stage. Keep up top-dressing to encourage grass growth and prevent clover dominance.

Swelled Head

Symptoms: Affects young rams. Caused by black disease germ. Face and head swells, eyelids and nostrils close, lips swell. Usually die unless treated. Must differentiate from black leg and 'yellow big-head'.
Immediate action: Give 500 000 units penicillin and repeat daily for four days.
Prevention: Inoculate young rams at weaning with black disease vaccine.

Tetanus

Symptoms: Extremely fatal. Occurs four days or longer after shearing, dipping, mulesing, marking. Causes extensive stiffness of whole body.
Immediate action: Treatment of no value. 100 units of tetanus anti-toxin given immediately prior to shearing, etc. gives three weeks' protection.
Prevention: Vaccinate with tetanus toxoid at least three weeks before shearing, dipping, etc. Need vaccine and anti-toxin for lambs at marking. Do not hold sheep longer than necessary in sheep yards. Mules and lamb mark in temporary yards.

Twin Lamb Disease

Symptoms: Mainly affects ewes carrying twin lambs within six weeks of lambing. Affected ewes dull, listless, sleepy and blind. Stand on their own, die in one to three days.
Immediate action: Give 113 grams glycerine in equal amount of water as a drench. Effective if ewes in good condition on good feed. If ewes in low condition, glycerine is usually not effective.
Prevention: Pregnant ewes need sufficient feed in last weeks of pregnancy to improve in condition. Supplementary feeding essential if pastures fail.

Vitamin A Deficiency

Symptoms: Green feed is the natural source of vitamin A. As pasture dries off, vitamin A declines. Affects lambs dropped in drought, also weaners. Fertility of rams also affected. Affected sheep blind at dusk.
Immediate action: Give vitamin A as follows. Lambs: Dose at marking with 250 000 units; Weaners: Dose after four months without green feed with 500 000 units; Rams: Dose with one million units two months before mating.

Vitamin D Deficiency

Symptoms: Causes unthriftiness and rickets in weaners, particularly ram weaners. Occurs in winter months, especially if grazing crops such as oats.
Immediate action: Give one million units vitamin D as an injection or drench.
Prevention: Give one million units to fast growing weaners at beginning of winter if weather abnormally cloudy and weaners grazing on oats.

'Yellows'

Symptoms: Two causes — chronic copper poisoning from subterranean clover and heliotrope poisoning. Often combined. Symptoms come on suddenly. Acute jaundice. Death within twenty-four hours. Symptoms more prolonged as outbreak progresses.
Immediate action: When cause is *copper poisoning* remove flock into fresh paddock. Provide a lick of one bag salt, one bag gypsum and 454 g (1 lb) sodium molybdate or provide hay plus sodium molybdate and Glauber's salts (see local veterinarian for quantities).

When cause is *heliotrope poisoning* losses usually occur after heliotrope has disappeared. Nothing can be done.

Prevention: With *copper poisoning* do not use excessive quantities of bluestone for any purpose where bluestone applied to pasture. Prevent clover dominance by top-dressing late in autumn. Check for molybdenum deficiency and use superphosphate plus molybdenum if necessary. Provide lick already prescribed.

Heliotrope poisoning can be prevented only by avoiding grazing on heliotrope.

Yellow Big Head

Type of photosensitisation.

Symptoms: Mainly caused by grazing on summer grass. Also caused by yellow vine, millet and phosphorus poisoning. Head swells, eyes and nostrils close. Losses may be heavy. Post-mortem shows jaundiced condition of carcass.

Immediate action: Keep affected animals in shade.

Prevention: Must avoid plants responsible, such as summer grass — *Panicum effusum* — in young stages of growth. Native millet and crop millet are suspected of causing the disease.

34

POISONOUS PLANTS

The diagnosis of poisoning in sheep by plants is very difficult, obscure and often impossible. Deaths from plant poisoning can be sudden or delayed for a long period, the effect of the plant poisoning not being noticeable in some cases for weeks or months after the particular plant has been eaten.

Some plants are poisonous in some localities and yet quite harmless in others. Some are toxic during all stages of growth, and others only develop toxic principles at certain stages of growth. Some plants are poisonous in the early stages of growth and quite harmless when two or three weeks old. Others may be quite harmless until the flowering or seeding stage is reached, then they develop toxic principles. Others again, particularly some of the cyanogenic plants (prussic acid poisoning), may be perfectly harmless during all stages of normal growth, and only develop toxic principles when they receive a check in growth due to hot, dry weather or severe frosts.

The poison is not confined to any one part of the plant. It may occur in the leaf, stem, bark, seed, fruit, or in roots or bulb. Some sheep are more susceptible to poison than others. A number of sheep of the same age, size, condition and state of health may eat the same quantity of poison plant; some of these may die suddenly, others may become ill and linger for a day or so and then die, others may show signs of illness and recover, and yet others may show no ill effects at all.

Plants may cause injury to sheep in several different ways.

1. The actual toxic material contained in the plant may directly cause death.
2. The toxic principles of the plant may not be sufficient to cause death, but may produce serious injuries to the animal.
3. Injury may be caused by mechanical irritation arising from sharp points or seeds

of some plants, such injuries often being the means through which germs gain entrance to the body.

In general, the following points are important in helping diagnosis:
1. the previous history of the case;
2. the suddenness of the attack, rapid course and sudden death;
3. connection between the symptoms and the food eaten;
4. simultaneous sickness of a number of animals without noticeable contagion;
5. complication of gastric and nervous symptoms;
6. the characteristic symptoms of certain poisons and the change in the organs;
7. physical, chemical, and pathological proof of the presence of poison.

As a general rule, sheep avoid poisonous plants, but certain conditions may cause sheep to overcome their natural avoidance of harmful plants, such as:
1. During drought, owing to lack of feed, sheep are forced to eat plants that under ordinary conditions they would not touch.
2. During a flush season plants are often intermixed with pasture so that sheep are unable to avoid consuming a certain amount of the poison plants when feeding.
3. When travelling, losses from poisoning often occur, as sheep, when on the roads, will often eat plants which they would not touch when under paddock conditions.
4. When turned on to pasture empty and hungry, as after a train journey or after being yarded for a considerable time, sheep, in their anxiety to fill themselves and stop the pangs of hunger, are liable to eat poison plants.
5. Sheep new to a district often become poisoned through eating plants with which they are not familiar. Again a particular species of plant that is not poisonous in the district in which the sheep are bred may be so in the district the sheep are moved to.
6. Mineral deficiency, by causing a craving for certain elements, is often responsible for sheep eating plants which under normal conditions they would not touch. Mineral deficiency is responsible for sheep acquiring a depraved appetite for bones and rotten carcasses, the eating of which is often followed by toxic paralysis.

Cyanogenic plants are particularly dangerous, as many of them may be eaten without harmful effect during the greater part of the year, and at other times may become so poisonous as to cause sudden death. The poisonous principle of cyanogenic plants is contained in a glucoside which, acted upon by a ferment in the plant, liberates hydrocyanic or prussic acid, one of the most powerful poisons. It is widely distributed throughout the plant world, and is one of the most rapidly acting of all drugs. It is usually so rapid in its action that remedies are of little value.

Symptoms of Hydrocyanic or Prussic Acid Poisoning
Owing to the rapid action of the poison, symptoms are not often observed in cases where a lethal dose has been taken. Should symptoms be manifest they are usually as follows: unsteady gait, staggering, falling, difficult breathing, dilation of the pupils and rolling of eyes, groaning, convulsions and finally coma and death. The first symptoms may be seen within a few minutes of eating the poison plant and death may follow in from a few minutes to an hour. The breath of animals in cases of hydrocyanic acid poisoning has a smell similar to that of almonds.

The Effects of Poisonous Plants
The effects produced by the ingestion of poisonous plants are extremely varied depending upon the poison contained in the plant. Some plants cause rapid death due to the toxic action of poison. Others cause acute gastro-enteritis due to the irritant

action of the particular poison, whilst others again affect the nervous system or cause locomotory disturbance (staggers and uncoordinated movement).

The New South Wales Poison Plants Sub-committee (Finnemore, Hindmarsh and Anderson) has grouped the main poison plants of New South Wales under the following eight headings. Most of the additional plants reported poisonous in other states and New Zealand will fit into these groups:

1. Plants Causing HCN Poisoning

The cyanogenic plants constitute a very large group, differing widely in botanical relationships. They include the following species: *Sorghum* spp., *Heterodendron oleaefolium* (rosewood), *Cynodon incompletus* (blue couch), *Eremophila maculata* (native fuchsia), *Chenopodium carinatum* (boggabri), *Euphorbia drummondii* (caustic weed, milk weed), *Acacia glaucescens* (coast myall, sally wattle), *Indigofera australis* (native indigo), *Eleusine indica* (crowfoot grass), *Lotus australis* var. (birdsfoot trefoil), *Eucalyptus cladocalyx* (sugar gum).

Another plant which has been responsible for sudden mortality simulating HCN poisoning is *Silybum marianum* (variegated thistle). It is now considered that the deaths are due to nitrite poisoning, since this plant contains much nitrate which is readily reduced to nitrite.

2. Plants Causing Gastro-Enteritis

Many plants can be included in this section and it may be noticed that frequently lesions occur in organs other than the digestive system in animals in which enteritis is the prominent symptom. Examples of the more common plants in this section are: *Trema aspera* (peach-leaved poison bush), *Diplarrhena moraea* (native lily), *Homeria collina* (cape tulip), *Caesia vittata* (blue grass lily), *Macrozamia heteromera* (burrawang palm) seeds, *Asclepias curassavica* (red-flowered cotton bush), *Castanospermum australe* (black bean) green seeds, *Solanum sturtianum* (sturt's nightshade), *Marsdenia rostrata* (poison vine), *Myoporum deserti* (dogwood, ellangowan poison bush), *Nerium oleander* (oleander), *Pimelea pauciflora* (scrub kurrajong), *Ricinus communis* (castor oil plant).

3. Plants Causing Locomotory Disturbance

These plants produce the condition referred to as staggers, shivers and rickets (this latter term should be reserved for the food — or vitamin — deficiency disease, rachitis). Some of these cause inflammatory and degenerative changes of internal organs, although the chief symptoms are uncoordinated movement and inability to travel: *Malva parviflora* (small-flowered mallow), *Lamium amplexicaule* (deadnettle, henbit), *Stachys arvensis* (stagger weed), *Cheilanthes tenuifolia* (rock fern), *Echinopogon* spp. (rough bearded grasses), *Macrozamia spiralis* (burrawang palm) fronds.

4. Plants Causing Photosensitisation

Of this group the most commonly known are: *Medicago denticulata* (burr medic), *Hypericum perforatum* (St John's Wort) and *Panicum effusum* (summer grass). Many other plants may cause photosensitisation in the immature, rapidly growing stage, although they are valuable grasses and herbage. These have not been mentioned specifically as they are not a common cause of ill health, except in a few isolated localities.

5. Plants Affecting the Nervous System

Swainsona spp. (Darling pea), *Alstonia constricta* (quinine bush), *Passiflora alba* (wild passionfruit), *Atalaya hemiglauca* (white wood). The latter is apparently harmless to stock in New South Wales, although affecting animals in other parts of Australia.

6. Plants Causing Rapid Death on Account of Alkaloid Content

Astropa belladonna (deadly nightshade), *Conium maculatum* (hemlock, carrot fern), *Duboisia myoporoides* (corkwood), *Nicotiana suaveolens* (native tobacco), *Solanum cinereum* (narrawa burr).

7. Plants Containing Nitrate

These plants may cause rapid death by the reduction of nitrate to nitrite after ingestion, and include such plants as *Silybum marianum* (variegated thistles) and *Salvia reflexa* (mint weed).

8. Plants Causing Pneumonia

The most important are *Berbesina encelioides* (crown beard), and *Wedelia asperrima* (yellow daisy).

The plants quoted under the various headings are only examples. There are many others which could be listed under these headings. From time to time plants come under suspicion as being poisonous and a great many have been tested out by the various state Departments of Agriculture, and the CSIRO.

Usually, to test a plant for toxicity, the method adopted is to feed it to one or more animals, and preferably to the species of animals for which field observations have suggested it to be poisonous. Chemical analysis is also used to determine the presence of poisons in sufficient amounts to cause death. This is readily done with well-known plant poisons but naturally cannot be employed when the toxic principle of the plant is not known.

Plants That Cause Felt Balls

- crimson clover (*Trifolium incarnatum*)
- cape weed (seeds) (*Cryptostemma calendulaceum*)
- brome grasses (*Bromus*)
- wild oat (awns) (*Avena fatua*)
- foxtail grass (*Chaetochloa*)

Treatment of Plant Poisoning

The opportunities and usefulness of treatment are limited by several factors:

1. Death may occur before any remedy is available. Poisoning may occur on the stock route or in the station, far away from any chemist.
2. In very few cases is the actual antidotal remedy available. In most cases it is a matter of treating the symptoms and when seen the animal is often 'too far gone' for any remedy to be likely to save it.
3. Many plants cause such extensive or permanent damage to tissues that no remedy can offer any hope of cure after symptoms appear.
4. The number of animals affected may be too great for individual treatment.
5. Lack of yards and races that are necessary for drenching.

GLOSSARY

Allelomorph or allele	Genes which occupy the same position on corresponding chromosomes. For instance, if a gene is found one-third of the way along a chromosome, then its allelomorph will be found one-third of the way along another member of the chromosome pair.
Atavism	Reversion to an earlier type. Reappearance of ancestral form.
Barbé	One term used to describe the mean length of fibres in a top. It is a weight-biased distribution.
Basil	The finished dressed leather obtained from sheep skin.
Carding	After wool is scoured and dried it is fed into a carding machine which opens up the wool into an even layer, removing as much burr and seed as possible and drawing the fibres parallel to each other to form a single continuous strand of fibres called a 'sliver'.
Chine	Backbone or spine.
Chromosomes	Thread-like bodies of nuclear material found in the nucleus of a cell which carry genes and which segregate during the production of sex cells. Each nucleus in the body cells of the sheep contains fifty-four chromosomes which are arranged in pairs, the two members of a pair being similar in shape.
Clean colour	The colour of wool after scouring. Clean colour is measured in terms of brightness and yellowness, both of which can affect dyeing potential. See Colorimeter.
Coefficient of variation (CV%)	A statistical term used to describe the amount of variation within a set of measurements; it is expressed as a percentage of the mean value. For example, for length it is used to denote either the variability to staple length in greasy wool, or the fibre length

variability in wool top. The higher the percentage, the more variable the length.

Colorimeter An instrument used to measure the clean colour of wool. In this method of testing, a scoured sample of wool is placed in the instrument and light is shone on it. The colorimeter then measures the light which is reflected from the sample, from which the values for brightness (expressed as either L^* or Y) and yellowness (expressed as either b^* or Y-Z) are derived. As L^* or Y increase, brightness improves. As b^* or Y-Z increase, yellowness increases.

Combing A process performed after carding and gilling (usually by rectilinear comb) to remove most of the short fibres (noil), neps and foreign matter, leaving the longer fibres lying parallel to the direction of the sliver. This product after two more gillings is called top.

Core sample The sample extracted from a bale of wool by inserting a sharpened hollow tube the length of the bale. The diameter of the core tube is normally 18 mm. Subsamples of the composite core sample are used for testing yield, mean fibre diameter and vegetable matter content. The core sample may also be used to measure clean colour.

Every bale in a sale lot which is sold by sample with objective measurement is core sampled.

Dermis The inner layer of the skin lying immediately beneath the epidermis and consisting of a loose network of collagen and elastic fibres which support the principal blood and nerve supply. Fat cells may often be found in the deepest zone of the dermis.

Development The capacity of the merino for growing what is apparently more skin than is required to cover the body smoothly, with production of folds or wrinkles. Varies in extent according to type of merino.

Epidermis The outer non-vascular layer of the skin, consisting of a horny layer of cuticle, an intermediate layer of active cells, and a foundation layer from which the hair roots generally start their development.

Fibre diameter The thickness of individual wool fibres. Wool is inherently variable in fibre diameter but the average fibre diameter of any sale lot is by far the most important characteristic in terms of processing value (hence price received at auction). Average fibre diameter is measured commercially by the Airflow machine and expressed in micrometres (or microns): one-millionth of a metre.

Gamete A reproductive cell of either sex. The sperm in the male and the egg in the female.

Gene The unit of inheritance which is transmitted to the next generation in the sperm cells. A gene interacts with its fellow genes, the rest of the cytoplasm and the environment to mould the character of an organism. The genes are arranged in a linear manner along the chromosomes.

Generation A period of one generation is the average age of the parents when their progeny is born.

Genotype The genotype of an animal is its total complement of genes or its total complement of any particular site on a chromosome.

Germplasm	The reproductive tissues set aside from the rest of the body or somatic cells. It is the material basis of heredity.
Gestation	Period of pregnancy, average in sheep 150 days (five months).
Gilling	In the worsted system, three gilling operations are normally carried out prior to combing and two gilling operations after combing. In the gilling machine slivers pass through pairs of rollers followed by moving combs and a second pair of faster-turning rollers, which stretch the sliver into thinner hanks, blend different quality wools and further align the fibres to make the final sliver suitably uniform for worsted spinning.
Grab sample	A sample of greasy wool taken at random positions, by mechanical means, from every bale in a sale lot. A minimum number of twenty samples must be taken from each sale lot. The composite grab sample is used for display purposes on the showfloor. The woolbuyers inspect the sample and, in conjunction with the objective measurements, place a value on the wool. If staple length and staple strength measurements are requested, the grab sample is subsampled for staple tufts by the mechanical tuft sampler (MTS), before being displayed on the showfloor.
Hauteur	Hauteur is a term used to describe the mean fibre length in the top. It is a length-biased distribution. Hauteur as determined by the Almeter is widely used commercially for the specification and trading of wool tops.
Heterozygous	When the two genes for a trait are different (e.g. Tt, Rr)
Homozygous	When the two pairs of genes for a trait are alike, the organism is said to be homozygous for that trait.
Hybrid	The offspring of two parents unlike one another in one or more heritable characters. A heterozygote.
Inbreeding	The mating of the closest relations.
Interbreeding	Crossbreeding, usually by the mating of halfbreds.
Kilotex (ktex)	The unit of measurement used to express linear density (or thickness), defined as grams per metre. The thickness of staples is measured (in kilotex) and used in the calculation of staple strength. For example, a staple of 100 mm in length with a clean weight of 0.1 g has a 'thickness' of 1 kilotex. Typical staples range from 1 to 5 ktex. A staple of pencil thickness is equivalent to approximately 1 ktex.
Line breeding	Breeding from the same blood or within the same family, but not from close relations.
Mean	Has the same meaning as 'average'. It is derived by dividing the sum of the individual values of a measurement by the number of individual tests.
Meiosis	In sexual reproduction, offspring are produced by the fusion of two gametes in fertilisation. The single-cell zygote then divides by mitosis to form the new individual. The two gametes contain only half the normal number of chromosomes. If they did not, the chromosome number would be doubled in the next generation. This type of cell division, which forms cells with half the number of chromosomes normally found in cells of the species, is called meiosis (from the Greek *meion*, meaning less). *Source: Core Biology,*

	Mudie and Brotherton.
Micrometre (μm)	A unit of measurement, used commercially to express the average fibre diameter. The unit micrometre is equal to one millionth of a metre.
Mitosis	Cell division by which each of the resultant daughter cells receives a full complement of chromosomes — one of each pair that existed in the original cell — before division.
Noil	The short fibre which is removed in the combing operation. It is a mixture of short and broken fibres, neps and small particles of vegetable fault. It is then used as one of the components in blends in the woollen system.
Oestrum	The period during which a ewe will accept a ram's service.
Out cross	The introduction of animals of the same species from a family outside the breeding line of the stud.
Ovum	Egg in female animal from which the young is developed (ova plural).
Phenotype	The phenotype of an animal is the composite of all its tangible features. The phenotype includes an animal's external appearance, measures of its productivity and its physiological characteristics.
Regain	Regain is the amount of moisture in the fibres, expressed as a percentage of the clean oven-dry weight. Standard regain is brought about when wool comes to equilibrium with air at 20°C and 65 per cent relative humidity.
	Processed wool is adjusted to a particular regain according to national and international agreements, for example, 18.25 per cent for dry combed tops and 16 or 17 per cent regain for scoured wool.
Resistance to compression	The force required to compress a standard mass of wool into a fixed volume. This is related to the handle and bulk of the wool, and is also positively related to the amount of crimp — highly crimped wool has greater resistance to compression than low crimp wool of the same diameter.
Romaine	The amount of noil (short fibre) produced during processing, expressed as percentage of the top and noil produced.
Sliver	A continuous band of carded, or carded and combed, wool in an untwisted condition.
Standard deviation	A measure of variability in a characteristic. For example, for fibre diameter, the standard deviation indicates how the diameters of individual fibres vary from the average. Standard deviation is the square root of the variance. For example, if the variance is 4, then the standard deviation is 2.
Staple length	The length of a staple from tip to base. For a sale lot, a minimum of fifty-five staples must be measured to conform to the Australian standard. The average staple length is then calculated and reported in millimetres.
Staple length variability	Refers to the variability in staple length within a sale lot. It is reported as the coefficient of variation which is expressed as a percentage (CV%).
Staple strength	The force or 'pull' (newton) required to break a staple of given thickness (kilotex). Staple strength is expressed in newtons/kilotex (N/ktex). It is a measure of tensile strength which is independent of mean fibre diameter and average thickness of staple. A mini-

mum of forty staples must be individually measured for strength to conform to the Australian standard. The average staple strength is then calculated (expressed as (N/ktex) and reported in the sale catalogue.

Usually about sixty staples are measured for strength but in some cases the number may be fewer due to individual staples being shorter than the gauge length of the instrument.

Tear Ratio of the amount of top produced to the amount of noil produced during processing.

Top A continuous untwisted strand of combed wool in which the fibres lie parallel, with short fibres having been combed out as noil. Top is raw material for worsted wool processing and is specified in terms of fibre diameter, fibre length distribution and mean, regain, etc.

Variance Expresses the distribution of values about the mean. It is used in calculations of standard deviation and coefficient of variation.

Vegetable matter base (VMB) Consists of burrs, grass seeds, thistles, hardheads, straw, chaff and small pieces of stick and bark. The vegetable matter base is established when a scoured sub-sample of wool is placed in a boiling solution of caustic soda (10 per cent). The wool is completely dissolved and the remaining vegetable matter dried and weighed. Vegetable matter base is expressed as a percentage of the weight of the greasy core sample.

The type of vegetable matter present is subjectively determined and expressed to the nearest 5 per cent of burr, seed/shive and hardheads.

Wool base The oven-dry weight of wool fibre free from all impurities (vegetable matter, moisture, dirt, grease, etc.). The wool base is expressed as a percentage of the weight of the greasy core sample. The wool base is not used in commercial trading because it is impossible to achieve during the processing but is used as the basis from which the commercial yields are calculated . (See Yield.)

Woollen yarn The wool used in the woollen system generally has a shorter mean fibre length (e.g. approximately 40 mm and less) than that used in the worsted system. The cloth produced is comparatively bulky, with many fibres extending from the surface. Woollen fabrics include tweeds, felts, flannels, blankets and knitwear.

Worsted yarn Made from top, worsted yarn undergoes more stages of processing (i.e. gilling, combing and drawing) than the woollen system. In worsted yarn the fibres are more parallel and more tightly spun, resulting in smoother, stronger yarn than the woollen system. The fabric produced is smooth, dense and retains its shape well. Products from the worsted system include suitings and knitwear.

Yield The amount of clean fibre, at a standard regain, that can be obtained from the greasy wool. There are four core test yields that are normally calculated from the wool base for commercial trading purposes.

1. International Wool Textile Organisation (IWTO) Schlumberger Dry Top and Noil Yield (SCH DRY)
2. IWTO Scoured Yield 17% Regain (SCD 17%)

3. Japanese Clean Scoured Yield (JCSY)
4. Australian Carbonising Yield 17% Regain (ACY)

Zygote The fertilised egg following the union of the spermatozoa with the ovum.

INDEX

Additional measurements 207–11
Anatomy and physiology of sheep 304–11
Anthrax 333
Australian and New Zealand breeds 45–55, 300
Australian carbonising yield 156
Australian clip preparation standards 108–16
Australian Wool Corporation type list 132–51
Automatic noble comb control 231

Bale markings 119
Barley grass 98
Bathurst burr 97
Biology of skin and wool fibre 69–78
Black disease 333
Black leg 333
Blood poisoning 334
Blowfly strike 313–5
Bluestone 93
Bogan flea 98
Border Leicester 25–6, 299
British breeds in Australia 23–43, 298–9
  long-woolled 23–4, 299
  mountain breeds 24–5
  short woolled 24, 299
Bulk classing

Carbonising 213–4, 230
Carding 221
Carpet wool breeds 53

Characteristics of wool in relation to classing 84–94
Character of wool 87
Charcoal 93
Cheesy gland 334
Chemical composition of the wool fibre 77–8
Cheviot 43–4
Classer's duties 117–21
Classer's specification 120–1
Classing
  British breed short-wool clips 113
  Carpet wool clips 114
  Long-wool clips 113
  Merino wool clips 110–3
  Superfine merino wool clips 112–3
Clip definition 108
Clip preparation 108
Code of practice 108
Colour of wool 89, 134, 212
Coloured fibres 91–2
Combing 222
Combing and carding types 131–2
Common burr medic 97
Composition of greasy merino wool 154
Computer selling 129
Copper poisoning 334
Core sampling 158–63
Corkscrew grass 98
Corkscrew or smaller crowsfoot 98
Corriedale 45–7

Cotted wool 92
Crossbreds 299
Crossbreeding 295–303
Crowsfoot, corkscrew or geranium 97–8
Crutching and wigging 269, 315

Defects in wool 90–4
Density of wool 89
Dipping of sheep 324–7
Dip stain 94
Discolorations and stains 92–3
Diseases of sheep 333–9
Dorset Down 40–1, 299
Dorset Horn 31–3
Drawing 224
Drought feeding 289–94
Drysdale 53–4
Dusty 93

Ear marking 259
Elasticity of wool 80–1
English Leicester 28–30, 299
Epididymitis 334
External parasites of sheep 312–323

Fat lamb areas in New South Wales 302
Fat lamb raising 295–303
Fellmongering 231
Felting 227–8
Felting capacity 79–80
Fern or vegetable stain 93
Ferility and mating of sheep 248–51
Fibre population 72–3
Fineness of wool 84–5
Flock 198
Fodder conservation 284–5
Food elements required by sheep 285
Foot abscess 334
Foot rot 335
Foot scald 335

Galvanised burr 98
Gill box 222
Grass seeds 263
Gromark 52–3

Hairiness 90
Hampshire Down 39–40
Heterotypic fibres 91
Hygroscopic power and heat of wetting 82–3
Hypocalcaemia 335

Internal parasites 328–32
Itch mite 321–3
International Wool Tetile Organisation (IWTO)
   Airflow method of measuring fibre diameter
      187–207
   Clean wool content 161, 163, 169–77
   Combined test certificate 161
   Core regulations 158–66
   Estimated commercial card sliver yield 165

Scoured yield 169–77
Scoured yield at regain percentage 165
Scoured yield certificate 161
Test house guarantees 161
Theoretical top and noil yield 163
Vegetable allowance (VA) table 165–6
Vegetable matter base 169–77
Wool base 169–77
Wool fibre diameter by projection microscope
   181–7

Japanese clean scoured yield 156

Knitting 227

Lamb marking 257–61, 315
Lambing 251–6
Length of staple 209
Length of wool 85–7, 134
Lincoln 30–1, 299
Lustre 80, 89

Marketing
   Fat lambs 300–3
   wool 122–30
Marl printing 236
Mating 250–1
Mean fibre diameter 133
Melange printing 231
Merino 298
   Australian 7–15
   Breeding 241–7
   Fine-woolled 16
   Medium woolled 16
   Mudgee type 2
   Peppin type 2
   Polled 21–2
   South Australian 21
   Strong woolled 17
   Vermont 5
Mildew 94
Mineral deficiencies 287–8
Mob 108
Moisture content 152
Mules operation 261, 316
Mycotic dermatitis and fleece rot 93, 335

Non-conductivity of wool 81
Non-Flammability of wool 82
Noogoora burr 97
New England crusher 98

Parasitism 264, 312–23, 328–32
Pasture improvement 279–84
Perendale 50–2
Philaris staggers 335
Photosensitisation 336
Picking-up lambs wool 100
Picking up, skirting and rolling 99–107
Pink eye or ophthalmia 263–4, 336
Pizzle rot 336

Plant poisoning 336
Poisonous plants 340–3
Poll Dorset 49–50
Polwarth 47–9
Pressing and branding 119
Property of wool and woollen goods 79–83
Pulpy kidney 337

Regain 152
Regain standards 163
Reserve price scheme 126
Rolling the fleece 106
Poly poly 98
Romney Marsh 26–8, 299
Rye grass staggers 337
Ryeland 35–6

Saffron thistle 98
Sale
  by description 129
  by private treaty 130
  by sample 124–6
  by separation 126
Sale catalogue 125
Sale with additional measurement 129
Sampling 180–1
Scabby mouth 337
Shearing 265–9, 270–3, 303, 315
Shearing sheds and yards 269, 273–8
Shed labour 107
Sheep areas of Australia 60–8
Sheep body louse 320
Sheep classing and merino breeding 237–47, 316
Sheep ked 319–20
Sheep measles 337
Shelly hoof 337
Shive 98
Shrink proofing 230–1
Shropshire 36–8
Si-Ro-Mark 228
Si-Ro-Moth'd 230
Sironised 230
Si-Ro-Set 230
Skirting 103–6, 108
Softness of wool 89
Soundness of wool 87
Southdown 33–5
South Suffolk 41–2
Spinning 224
Statistical method for wool measurement 178–81,

186–7
Stringy wool 92
Style 133
Subterranean clover disease 337
Sucking louse 320–1
Suffolk 38–9
Superwash 236
Supplementary feeding 285–6
Swelled head 338

Tar or paint stain 93
Tensile strength 134, 209–11
Test certificates (AWTA) 157
Tetanus 338
Tick stain 93
Tukidale 55
Twin lamb disease 338
Types and yield 131–57

Unevenness in fleece and staple 91
Urine stain 93

Vegetable matter in wool 95–8
Vitamin A and B deficiency 338

Washing yield 153
Weaning 262–4
Weaving 226–7
Webby wool 92
Weight book 119
Wild oat 98
Wool fibre structure 73–7
Wool follicle 69–72
Wool futures 130
Wool Marketing Act 129
Wool Scouring 214–20
Wool table construction 107
Wool testing 158–211
Wool textile manufacture 225–6, 234
Wool textile research 228–36
Wool valuation 153
Woollen yarn manufacture 225–6, 234
Worsted manufacture 221–5, 234

Yellow Big Head 339
Yellows 338
Yield 151–3
Yield and vegetable matter (VM) content 134
Yield measurement 155
Yolk or canary stain 93